The Mammals of Minnesota

The Mammals of Minnesota

Evan B. Hazard

Department of Biology, Bemidji State University

with illustrations by

Nan Kane

Published by the

University of Minnesota Press, Minneapolis

for the

James Ford Bell Museum of Natural History

Copyright © 1982 by the University of Minnesota
All rights reserved.

Published by the University of Minnesota Press
111 Third Avenue South, Suite 290, Minneapolis, MN 55401-2520
http://www.upress.umn.edu
Printed in the United States of America on acid-free paper
Seventh Printing 2008

Library of Congress Cataloging in Publication Data

Hazard, Evan B.
The mammals of Minnesota.
Bibliography: p.
Includes index.
1. Mammals—Minnesota. I. Title.
QL719.M6H28 599.09776 82–2706
ISBN 978-0-8166-0949-9 AACR2
ISBN 978-0-8166-0952-9 (pbk.)

The University of Minnesota
is an equal-opportunity
educator and employer.

For
Elaine

Acknowledgments

Authors of regional handbooks on groups of organisms are inevitably compilers. They may rely partly on their own work, but mostly they utilize resources provided by others. I cite published information throughout this book and am pleased to acknowledge other resources here.

The following curators in Minnesota and North Dakota graciously allowed me to examine their collections: Elmer Birney, James Ford Bell Museum of Natural History, University of Minnesota, Minneapolis; Ron Nellermoe, Concordia College; Robert Bellig, Gustavus Adolphus College; Merrill Frydendall, Mankato State University; Oscar Johnson, Moorhead State University; J. Frank Cassel, North Dakota State University; Charles Bruton, St. Cloud State University; Norman Ford, St. John's University; Lee Halgren and John Laundré, Southwest State University; Hollie Collins, University of Minnesota, Duluth; David Hoppe, University of Minnesota, Morris; Robert Seabloom, University of North Dakota; Ben Thoma, Willmar Community College; and James Opsahl, Winona State University. Elmer Birney also provided records of Minnesota specimens from the University of Kansas, the University of Michigan, the U.S. National Musuem, and the Field Museum of Natural History. Russell E. Mumford supplied records from Purdue University, as did John O. Whitaker, Jr., from Indiana State University, and Wayne J. Mollhoff from his own collection. Robert Bright of the Bell Museum provided data on archaeological bison remains.

Elmer Birney reviewed much of the manuscript in its earlier stages, and he and Harrison B. Tordoff, Director of the Bell Museum, provided facilities as well as advice and encouragement throughout the project. Among the students at the Bell Museum (most of them now widely dispersed) who were a great help were Donna Day Baird, Lawrence Heaney, Katie Hirsch, Nan Kane, Richard Lampe, Gerda Nordquist, and Robert Timm.

Carrol Henderson, Minnesota Department of Natural Resources, provided many distribution records. Other MDNR biologists who supplied information were Bill Berg, Patrick Karns, David Kuehn, and Jon Parker. L. David Mech, North Central Forest Experiment Station, John Mathisen, Chippewa National Forest, David Bosanko, University of Minnesota Biological Field Station at Itasca State Park, and Lee Pfannmuller, Nature Conservancy, also provided information and suggestions.

Diane Morris, North Country Curator-Naturalist, helped build the collection at Bemidji State University and supplied many distribution records. Over the years, several undergraduate assistants at Bemidji State have been an indispensable help in curating the collection, among them Mary Ellen Simon, James Howe, Kathryn Keating, Frances King, Christine Weir-Koetter, Gary Young, Mark Van Every, and Angela Doroff. The State University System aided me in the development of a file of statewide information on mammal distribution through a one-quarter sabbatical and a faculty research grant. The data compiled are available at Bemidji State and are continually used by scientists in Minnesota, adjacent states, and Canada.

Two reviewers who provided particularly useful comments were J. Knox Jones, Jr., Texas Tech University, and James N. Layne, Archbold Biological Station, Lake Placid, Florida. In addition to doing the scratchboard illustrations, Nan Kane gave me encouragement and advice in my first venture in scientific illustration. The people at the University of Minnesota Press have been most helpful; Vicki Haire edited the final copy thoroughly and sympathetically.

My wife, Elaine, read the manuscript critically at several stages, but that, of course, is the least of it. Without her support, this book would have been impossible.

E. B. H.

Department of Biology
Bemidji State University

Table of Contents

List of Figures

List of Maps

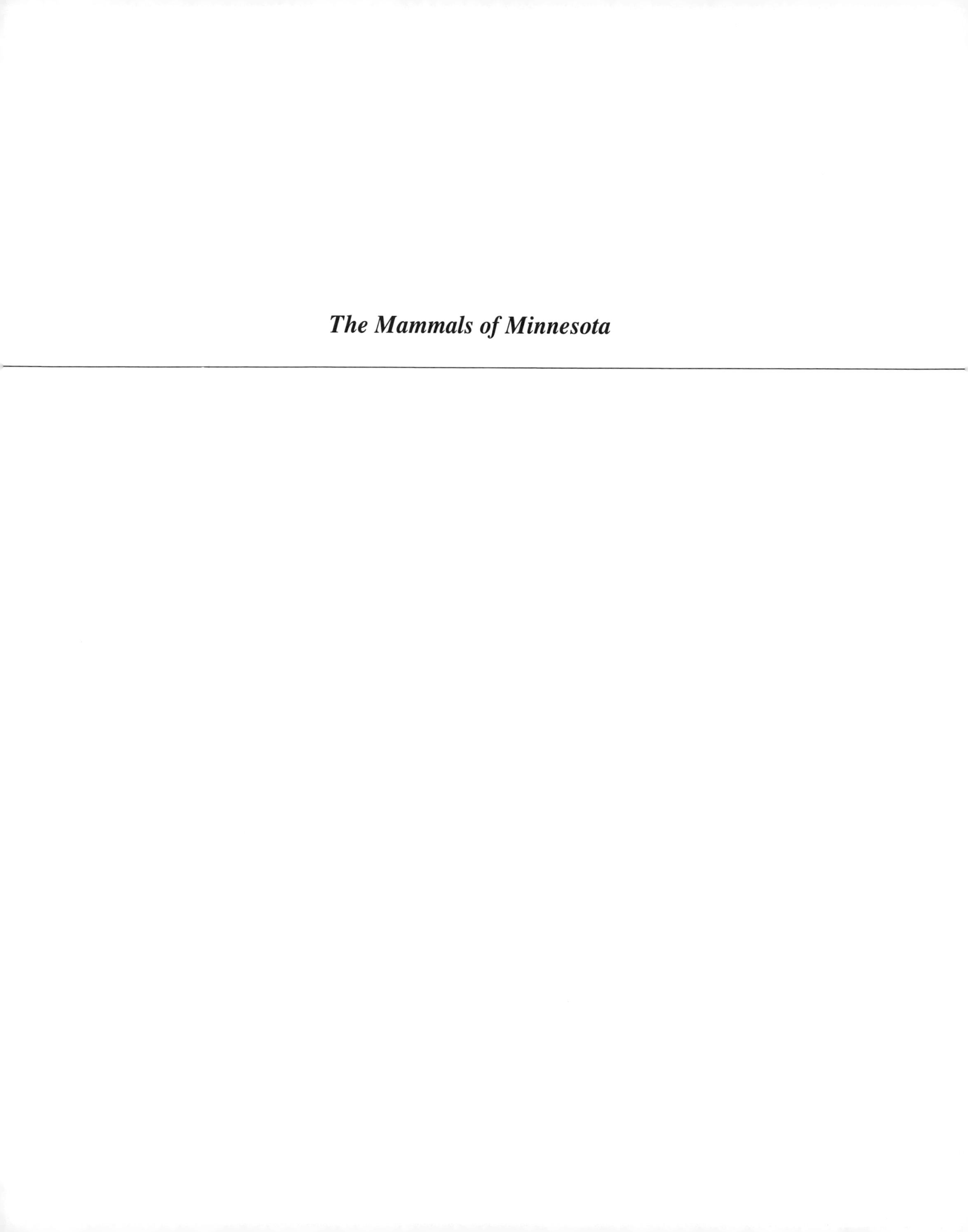

The Mammals of Minnesota

Introduction

The Scope of This Book

This book is about the wild mammals that inhabit Minnesota today or that have done so in the recent past. Its primary aims are to describe these mammals briefly, to aid in their identification, to present their known distribution in some detail, and to provide an introduction to their natural history. It also provides information on the interaction of these species with people, and lists references on the mammals of this state as well as some more general works. I hope that *The Mammals of Minnesota* will be useful to the general reader and to students in biology, zoology, mammalogy, wildlife management, and ecology, as well as to professional wildlife biologists, mammalogists, and ecologists. Since general readers and beginning students may be unfamiliar with many terms in common use among professional biologists, I include a glossary, beginning on page 201.

There have been several earlier works with titles similar or identical to this one (Ames 1873; Herrick 1892; C. E. Johnson 1916; M. S. Johnson 1930; Surber 1932; Swanson, Surber, and Roberts 1945; Gunderson and Beer 1953), all now out of print. These have been useful in the development of mammalogy in Minnesota. Longley and Wechsler (1977), a 28-page overview of the mammals of the state, is currently available from the Minnesota Department of Natural Resources (MDNR). Henderson (1979) reviews the nongame mammals of the state, and Henderson and Reitter (1979a-1979j) are 10 brief guides to the occurrence and status of nongame mammals by regions. Henderson (1980) is a description of the MDNR Nongame Wildlife Program.

The present work is distinctive in that the species accounts are more detailed than in the earlier works, distribution is plotted to township rather than simply to county, and the survey of the literature is extensive. The literature survey is particularly important because no single volume, no matter how detailed, can present all that is known of the mammals of a state, nor can it substitute for the experience of delving into the primary literature.

The Class Mammalia

Mammals are vertebrates that suckle their young and that have hair at some stage in their lives. Mammals, like birds, are endothermic, or "warm-blooded"; they maintain a relatively constant body temperature by regulating internal heat production and the loss of that heat to the environment (Whitrow 1971). Some do this more efficiently than others, and some switch to a lower temperature or to a fluctuating one close to the ambient (external) temperature during hibernation. Most mammals, including all species in Minnesota, bear their young alive.

Mammals have three sound-conducting bones in each middle ear (the malleus, incus, and stapes), and only one bone in each side of the lower jaw, the dentary, which articulates with the squamosal bone of the skull. Other vertebrates have but one middle ear bone (the stapes), and the articulare, one of several lower jaw bones, articulates with the quadrate at the back of the skull. At a relatively early stage in their growth, mammalian embryos have an articulare and quadrate, which later develop into the malleus and incus.

With some exceptions, mammals are diphyodont, having two sets of teeth, a milk (or deciduous) dentition and an adult (or permanent) dentition. The teeth are generally heterodont. The milk dentition primitively comprises incisors, canines, and premolars, and the adult dentition comprises incisors, canines, premolars, and molars, the molars having no milk predecessors. In many mammals, the dentition is highly modified, and one or more of the categories of teeth may be lacking. A few mammals are homodont, all of the teeth being much like one another, and a few have no teeth, but neither condition occurs in any species living in Minnesota.

Mammals also have a bony shelf in the roof of the mouth, the hard (or secondary) palate. Crocodiles and alligators are the only nonmammals having a secondary palate, but the anatomical and developmental details are different. The phalangeal formula of mammals is typically 2–3–3–3–3 (two bones in the pollex, or thumb, and in the hallux, or big toe, three in each of the other digits), in contrast to the reptilian formula: 2–3–4–5–4 (or 3). The only variation on this formula in Minnesota mammals is loss or reduction of some digits. Mammalian limbs typically are positioned directly below the body, rather than sprawling to the side as is common in living reptiles.

These characters and others neatly separate extant species of mammals from other extant vertebrates (Anderson and Jones 1967; Gunderson 1976; Vaughan 1978). However, those characters involving soft parts which rarely fossilize are of little help in separating early mammals from their reptilian ancestors and contemporaries. Those that do fossilize readily (dentition, palate, ear-jaw complex, phalangeal formula, erect limbs) change gradually in the geologic record of the Therapsida, the order of mammal-like reptiles that gave rise to the mammals. Further, they do not all simultaneously reach the condition we regard as mammalian (Hopson 1970; Hopson and Crompton 1969; Romer 1966, 1969). Therefore, the separation of the two classes must be arbitrary. The development of a dentary-squamosal jaw articulation is generally considered to be what distinguishes mammals from mammal-like reptiles. This may be of little help with some early Mesozoic specimens that are missing the rear of the skull. Pronounced heterodonty, a complete secondary palate, and a mammalian phalangeal formula occur in some fossils classified as reptiles on the basis of the ear-jaw complex, and there is evidence that some of the same animals were endothermic and suckled their young (Bakker 1971, 1975a,b; Crompton and Parker 1978; Hopson 1973; Ricqlès 1974). If so, it might be better to classify them as mammals, but that would not change the animals, just the classification. In any case, late Triassic fossils indicate a gradual change, not an abrupt one, in several characters used to distinguish mammals from reptiles.

The first fossils designated as mammals occurred in the late Triassic Period, about 200,000,000 years ago (see table 1). During the Jurassic, the number of species increased, and small mammals were relatively abundant during the Cretaceous, from 141- to 65,000,000 years ago. Primitive members of several extant orders are known from the Upper Cretaceous (Dawson 1967). To avoid a common misunderstanding, I should note that these were not simply ancient representatives of present-day orders. Some of these fossils are more like each other than like Recent (Holocene) species of the orders to which we assign them. They are classified as members of separate orders because of the subsequent evolution of their descendants. On the basis of their own morphology, some Cretaceous mammals assigned to different orders could equally well be placed in a single family (Van Valen and Sloan

1977). Unfortunately, our hierarchical system of classification has no provision for indicating the existence of such transitional groups.

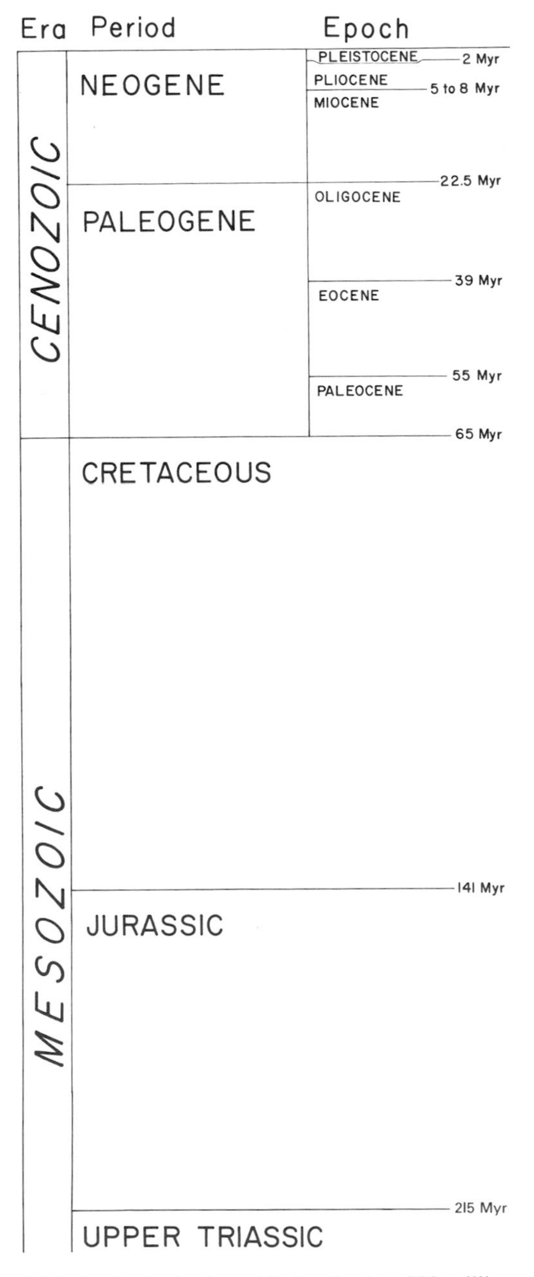

Table 1. Geologic timetable for the last 200 million years (Myr).

These Cretaceous mammals were small. It was not until the beginning of the Cenozoic Era, after the last dinosaurs had gone extinct, that some groups of mammals attained large size and that mammals, with birds, became the dominant land vertebrates. Those interested in reading further about early mammals and their ancestry should see the references cited above and also: Allin 1975; Case 1978; Crompton and Jenkins 1973; DeMar and Barghusen 1972; Greaves 1980; Hopson 1966, 1969; Kermack 1967; Lillegraven et al. 1979; Manley 1972; McKenna 1969; McNab 1978; Orians 1969; Pond 1977; Satinoff 1978; Szalay 1977.

Mammals constitute one of seven extant classes of vertebrates. The class Mammalia, comprising more than 4,000 extant species, is in turn divided into subclasses, infraclasses, and orders, the arrangement varying from one authority to another. The classification below combines elements of: Hall and Kelson 1959; Jones et al. 1975; Kirsch 1977b; Miller and Kellogg 1955; and Simpson 1945. It omits extinct groups. Starred (*) orders have wild representatives in Minnesota.

Class Mammalia .... mammals
  Subclass Prototheria ... egg-laying mammals
      Order Monotremata .. platypus, echidnas
  Subclass Theria .... live-bearing mammals
    Infraclass Metatheria .... marsupials
      *Order Polyprotodonta ....opossums
      Order Paucituberculata .....rat opossums
      Order Diprotodonta .... phalangers, wombats, kangaroos, etc.
    Infraclass Eutheria .... placental mammals
      *Order Insectivora .... insectivores
      Order Dermoptera ...."flying lemurs"
      *Order Chiroptera ....bats
      Order Primates ....lemurs, monkeys, apes, men
      Order Edentata .... sloths, anteaters, armadillos
      Order Pholidota ....scaly anteaters

*Order Lagomorpha .... rabbits, hares, pikas
*Order Rodentia .... rodents
Order Mysticeti ...... whalebone whales
Order Odontoceti .... toothed whales
*Order Carnivora .... carnivores
Order Tubulidentata .... aardvark
Order Proboscidea .... elephants
Order Hyracoidea .... hyraxes
Order Sirenia .... manatees
Order Perissodactyla .... odd-toed ungulates
*Order Artiodactyla .... even-toed ungulates

Habitats of Mammals

Mammals have probably occupied the area that is now Minnesota for most of their history. However, geological events in Minnesota were such that no terrestrial or freshwater sediments of Mesozoic or Cenozoic age survive except for deposits of the last few tens of thousands of years. During the Late Cretaceous and the Cenozoic, there were frequently land connections between North America and Eurasia, and Africa was sometimes connected to Eurasia. Australia, Antarctica, and South America were isolated from the northern continents, and eventually from one another, during most of the same period. South America became reconnected to North America only about 3,000,000 years ago. (See Wilson, 1976, for a variety of articles on continental drift and plate tectonics, and their relationship to the history of organisms.) As a consequence of its geologic history, North America shares most of its mammalian orders, many families, some genera, and even several species with Eurasia (Kurtén and Rausch 1959; Rausch 1953). Its mammal fauna is less like that of Africa and South America, and is distinctly different from that of Australia. The natural vegetation types (e.g., coniferous forest, prairie) that mammals inhabit are determined largely by climate, topography, and soils. The topography and soils of Minnesota, and the climate until several thousand years ago, have been strongly influenced by the Pleistocene glaciations (Bryson et al. 1970). Most of the state is covered by a variety of glacial deposits and formations left by various substages of the last of four great Pleistocene glaciations, the Wisconsin. The northern parts of St. Louis, Lake, and Cook counties have some glacial drift, but are characterized largely by outcrops of Pre-Cambrian rock that were scoured by the ice (Grout et al. 1959; Schwartz and Thiel 1963; Sharp 1953; H. E. Wright 1972). Glacial action and glacial deposits have been major factors in determining elevation and slope of the land, drainage patterns, and the locations of Minnesota's many lakes. Weathering of the various glacial deposits has produced most of the soils of the state, and is still a major factor in the maturation of these young soils in habitats unaltered by man (McMiller 1947). For discussions of Pleistocene conditions and faunas, and of subsequent changes, see: Graham 1976; Haag 1962; C. W. Hibbard 1970; C. W. Hibbard et al. 1965; Hoffman and Jones 1970; Lundelius 1974: and Stauffer 1945.

The last ice sheets probably left Minnesota about 10,000 years ago. The area immediately south of the glaciers supported a tundra vegetation, and farther south a boreal spruce forest grew (Wright 1968). As the glaciers receded, the spruce forest covered the state. As the climate grew warmer and drier, this forest was succeeded in part by other forests, principally pine in the north and hardwoods in southern Minnesota. Tallgrass prairie advanced into Minnesota from the west and south and, by the height of the warming trend some 6,000 years ago, extended about 120 km farther northeast than the present prairie-forest edge (H. E. Wright 1972). Forests subsequently moved west again; the forests along the present prairie-forest ecotone often grow on deep, rich prairie soils.

Map 1 shows the probable distribution of vegetation just before people of European descent settled the area that is now Minnesota. The

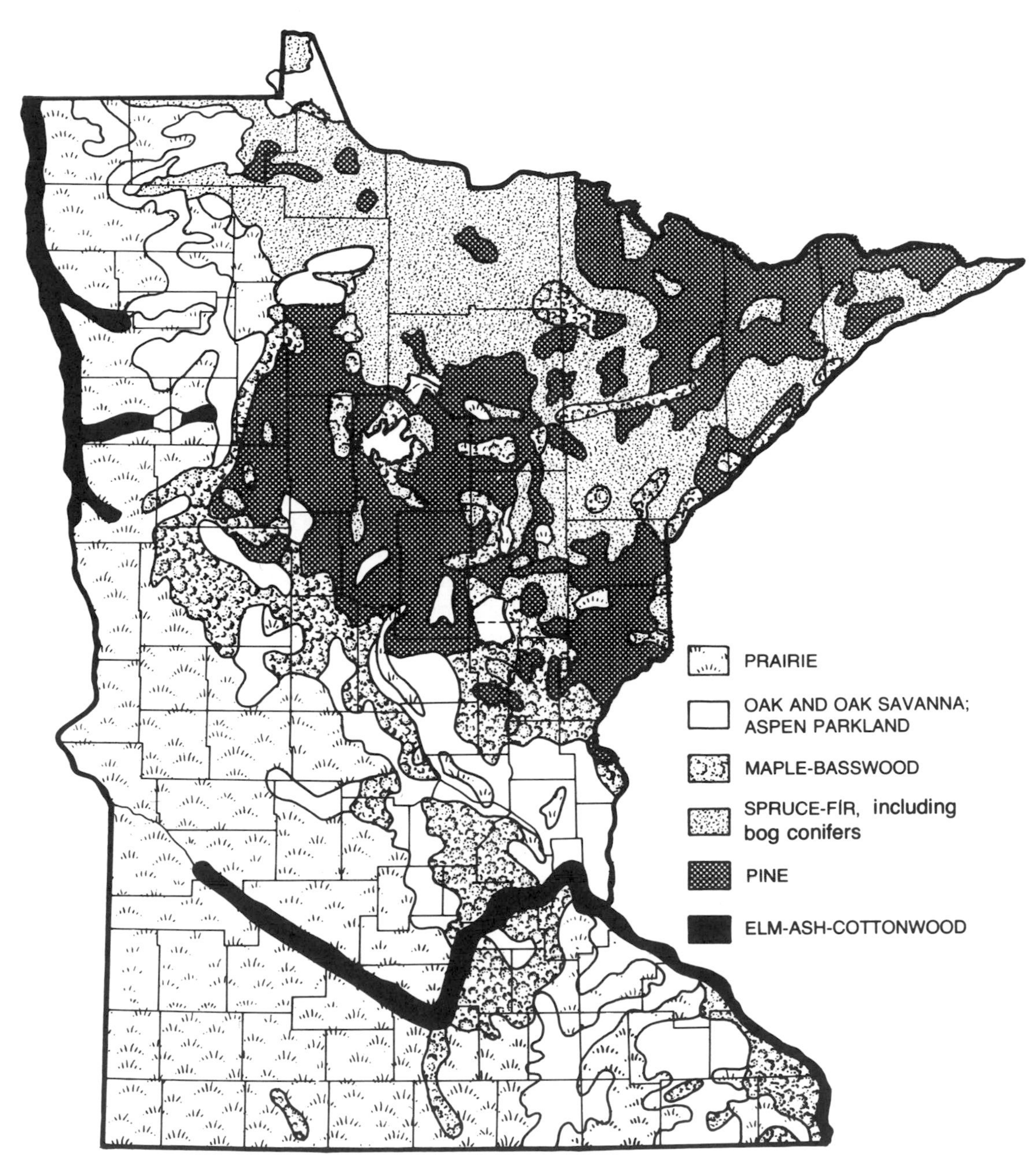

Map 1. Presettlement vegetation types in Minnesota. (Adapted by Patricia Burwell from an unpublished map drawn by F. J. Marschner in 1930 for the United States Department of Agriculture.)

original map from which it is adapted (Marschner 1930) divides the vegetation types somewhat further into 16 subtypes. The printed version of the original is approximately 122 cm wide by 137 cm high, and resolves detail down to approximately 1 km. For example, throughout the prairie counties, there are numerous hardwood groves and oak-brush openings associated with drainage patterns, and most of them are too small to be recorded even on the original map. These provide suitable habitat for species such as various shrews, eastern chipmunks, fox squirrels, white-footed mice, and red-backed voles which would otherwise be restricted to the areas mapped as forest.

Ecologists refer to the major vegetation types of the world as biomes, and generally agree that three of these meet in Minnesota: the boreal forest (needle-leaved evergreen forest), the broad-leaved deciduous forest, and the prairie. These biomes generally do not begin and end abruptly, except where associated with an abrupt change in topography; rather, they grade into one another in ecotones. Examples in Minnesota include oak savanna, aspen parkland, and mixed coniferous-deciduous forests. Drainage patterns, topography, and the composition of the parent material (clay, sand, gravel, bedrock, etc.) from which the soil develops will result in local subdivisions of the habitat. Poor or blocked drainage gives rise to a variety of wetlands, an important group of habitats in Minnesota (Cowardin and Johnson 1973; Marshall and Miquelle 1978).

Fire is also an important natural ecologic factor (Green and Janssen 1975; Heinselman 1973; Krefting and Ahlgren 1974; Moyle and Moyle 1977; Stewart 1951; Swain 1973; Wright 1974). Trees such as aspen and pine, especially jackpine, are adapted to conditions following fires, and deer and various small mammals do well in the successional vegetation that follows forest fire. The tallgrass prairie that covered much of western Minnesota would have been largely replaced by forest were it not for periodic fires. Probably some of these were deliberately set by earlier inhabitants of the area, whose ancestors invaded North America during the waning centuries of the Wisconsin glaciation. It may thus be inaccurate to refer to map 1 as a map of natural vegetation, if by natural we mean unmodified by man.

Settlement in the last two centuries, and the recent processes misleadingly called "development," have drastically modified the habitats of mammals. Most of the former prairie is under cultivation. Small untilled areas within the prairie region, protected from fire, have reverted to brush and woodland. People have drained many prairie wetlands, channeled streams, and dammed rivers (J. S. Choate 1972; Possardt and Dodge 1978). They have also cut most of the virgin forest, primarily for lumber but partly to clear it for farmland. Fire, often drastic, followed the logging. Much of the cleared northern land, including abandoned farmland, has gone through various stages of succession, with some of the intermediate stages favoring invaders from the south, such as white-tailed deer. Even-aged planted stands of commercially desirable trees, for example red (Norway) pine and jackpine, have replaced other former forests.

Two of the most drastically modified habitats are annually tilled fields, typically grown to a single crop each year, and cities. These provide suitable habitat for fewer species than did the habitats they replaced, but some mammals do well in each: for example, prairie deer mice and plains pocket gophers in tilled land, and gray squirrels, some bats, and introduced house mice, rats, cats, and dogs in cities. People generally create less diverse plant communities than the ones that existed prior to settlement, and thereby decrease the diversity of mammals in various habitats. In some areas, however, by interspersing open and wooded areas in a formerly continuous habitat, man has increased the species diversity and allowed some mammals, such as Franklin's ground squirrel, to expand their geographic ranges (Robins 1971).

Scientific Collections of Mammals

Collections of organisms of any sort—algae, grasses, beetles, or mammals—serve many purposes. The most obvious is identification; no matter what aspect of the natural history of mammals interests them, students must know one kind of mammal from another. Taxonomists classify organisms and study phylogenetic relationships primarily on the basis of specimens in collections. Collections enable us to investigate many aspects of genetic variation in mammals:

variation within and between populations, and variation related to age, season, and sex. Collections permit identification of the remains of mammals that predators use as food. And, particularly at a time when we have altered and continue to alter habitats, collections allow us to follow expansions and contractions in the geographic ranges of mammals. Thus, it is obvious that the date and locality on a label are as important as the specimen the label is attached to.

The uses listed above are research uses. Collections also aid in teaching, primarily in college mammalogy and vertebrate zoology courses. The American Society of Mammalogists' Advisory Committee for Systematic Resources in Mammalogy (1974) recommends that, in order to protect research specimens, research and teaching collections be kept separate. Because of limited facilities and time, a small institution may be unable to maintain two separate collections and may, in fact, carry on relatively little research utilizing the collection. Such a teaching collection should, however, be maintained and curated as though it were a research collection, with the same care given to preservation, to complete and accurate labeling, and to the keeping of field notes and of a specimen file or files. Learning to maintain a research collection is a legitimate teaching objective, and it can be achieved in small institutions as well as large ones.

Any biology or zoology department in a college or university, even if it does not offer a course in mammalogy, should have at least a small, carefully maintained reference collection of study skins, skulls, and perhaps skeletons of local mammals, as well as the skulls of common domestic mammals. This will permit students and faculty to identify unknown specimens, both for their own biological studies and as a service to the public. High school biology teachers should also consider possible uses of a reference collection in their teaching. Students generally find science more interesting if it somehow relates to their personal experience, and a high school collection of known specimens may give a new dimension to the outdoor world with which at least some students are familiar.

It is common, in college or university mammalogy courses, to require that each student prepare a collection of a certain number of specimens, generally of study skins plus skulls. This is often a reasonable requirement, although it has been questioned (Carlock 1974). Any student planning to do professional work in mammalogy should be able to collect and prepare specimens. Others may have a strong interest in the subject but find the killing of mammals personally distasteful. In my own course, such students satisfy an alternative requirement in lieu of preparing a collection. Although most small mammals are common, both students and instructors will want to insure that collecting does not endanger populations of any locally rare species. This may be a particular problem in urban areas. It should go without saying that the collector will respect game laws, public refuges, and private property.

I think it is undesirable to require high school students to submit collections of mammals in their biology courses, although collecting may be appropriate in a superior student's individual research project. Too often, high school biology teachers uncritically transfer their own college experiences to the high school classroom. There are too many high school students, especially in "built up" areas, to permit collecting of wild mammals as a class requirement. Furthermore, unless collecting is carefully integrated into the syllabus by a teacher who has a thorough understanding of the scientific role of collections, student collection may become wasteful busywork and encourage a lack of respect for wildlife. Whereas a reference collection of mammals in a biology classroom is often desirable, high school student collections, except in special circumstances, are not.

Anderson (1965), DeBlase and Martin (1974), and Hall (1962) give detailed instructions on collecting specimens, taking measurements, and recording data in field notes and on labels. They also indicate the necessary equipment and procedures for preparing study skins. There are several variations on the basic techniques. One is to use a stick of appropriate size (such as an ice-cream stick) to provide a rigid support within the cotton body. Some preparators use an applicator stick as a support for study skins, particularly of bats, allowing it to extend below and past the tip of the tail. A standard technique for bird skins, this allows one to examine a specimen without handling the skin itself.

Minnesota, and most of the United States west of the Appalachians, is divided into townships, squares about six miles on a side, each designated with a Township number north or south of, and a Range number east or west of, standard meridians. Each township in turn comprises 36 numbered sections of one square mile each. In populated areas, roads often follow section lines. My own preference is to use this system to indicate location on labels, in field notes, and on specimen cards. For example, for a shrew caught near Spider Lake (just east of Itasca State Park), the location would be: Minnesota, Hubbard Co., T. 143 N, R. 35 W, sec. 28. It might also be useful to add: Spider L., approx. 13 km (8 mi) WSW Lake George, 1 km W. co. rd. 3. (Lake George is the nearest post office *in the same county*.) Detailed county maps may be bought at county courthouses, and compilations of maps of all 87 counties are available commercially (Miles and Yeager 1979). Upham (1920) is a useful source for locations of obscure Minnesota place-names.

Additional References: Barthelemy 1971; Chelberg 1972; Choate and Genoways 1975; Genoways et al. 1976; Williams et al. 1977.

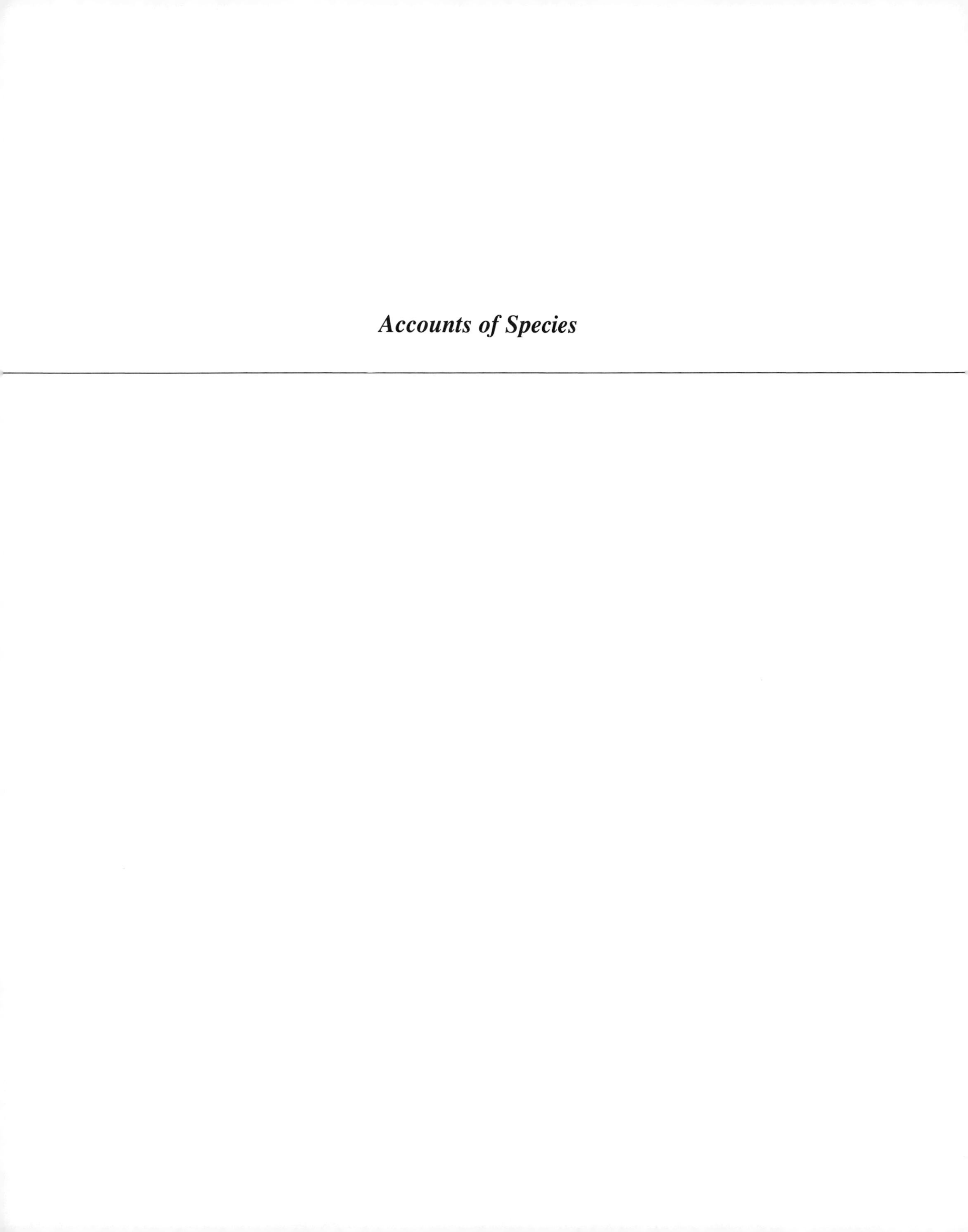

Accounts of Species

Accounts of Species

This section comprises accounts of the wild species of mammals that occur, or have occurred, in Minnesota in Recent times, along with introductions to the orders, families, and genera to which these species are assigned. Each species account includes measurements, a description of appearance, comments on range and habitat, a distribution map, sections on natural history and interaction with people, and a list of references. Many of the accounts include a drawing of the mammal in its natural surroundings and/or a skull drawing. Skulls labeled BSCVC are in the vertebrate collections of Bemidji State University; those labeled MMNH are in the collection of the James Ford Bell Museum of Natural History. The sequence of accounts follows Jones et al. 1975.

Where possible, the measurements given are based on at least 10 adult Minnesota specimens, and include the arithmetic mean, or average (unless fewer than six or seven measurements were available), and the range of variation. For some species, very few data are available. The standard measurements are total length, from the tip of the snout to the bony tip of the tail (not including hairs at the tail tip); tail length; hind foot, measured from the heel to the tip of the claw of the longest toe; ear from notch, measured from the notch at the base of the ear opening to the highest point of the fleshy part of the ear (again, excluding hair tufts, which may vary seasonally); and weight. Measurements in all species accounts are in metric units and are given in millimeters and grams unless otherwise indicated. Table 2 lists the metric units used, their abbreviations, their conversions to larger and smaller metric units, and their English equivalents.

Two additional measurements given for bats are the length of the tragus, a fleshy projection that arises from the base of the ear opening, and the length of the forearm. Unfortunately, these useful measurements for bats are often lacking, especially for older specimens. Ear from notch and weight of other mammals are also seldom recorded on older specimens. Large mammals, such as deer, which are generally uncommon in

Table 2. Metric units.

Unit (Abbrev.)	Metric Equivalent	English Equivalent
	Length or Distance	
meter (m)	—	3.28 ft.
decimeter (dm)	0.1 m	3.94 in.
centimeter (cm)	0.01 m	0.394 in.
millimeter (mm)	0.001 m	0.0394 in.
hectameter (hm)	100 m	109.4 yds.
kilometer (km)	1,000 m	0.621 mi.
	Area	
hectare (ha)	10,000 m^2 *or* 1 hm^2	2.471 a.
square kilometer (km^2)	1,000,000 m^2	0.386 sq. mi.
	Volume	
cubic centimeter (cm^3) *or* cc	0.000,001 m^3	0.155 cu. in.
liter (l)	1,000 cm^3 *or* 0.001 m^3	1.057 qts.
	Mass	
gram (g)	1 cm^3 of water*	0.0352 oz.
kilogram (kg)	1,000 g	2.205 lbs.

*At 20° C and standard sea level pressure.

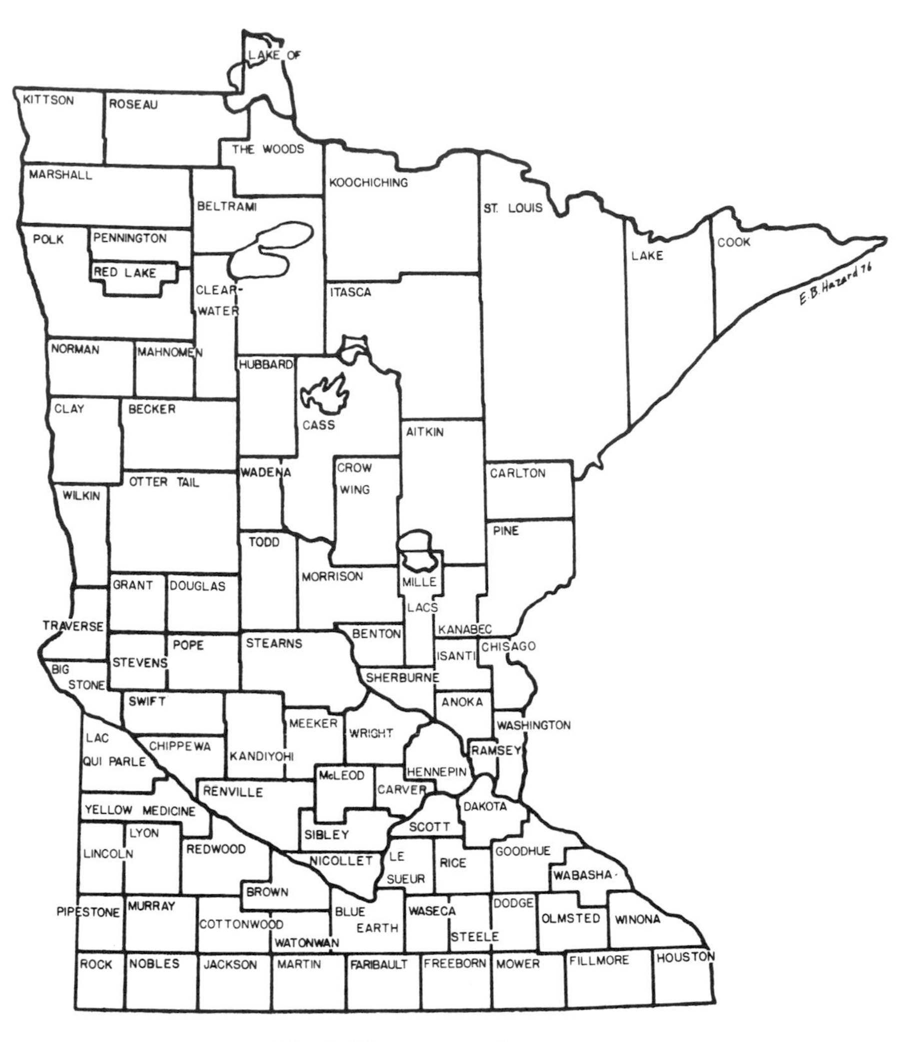

Map 2. Minnesota counties.

collections, are often accompanied by no measurements at all.

Most mammals, including all Minnesota species, have characteristic and relatively constant numbers of teeth. The primitive number for marsupials, on each side of the jaws, is five upper and four lower incisors, one upper and one lower canine, three upper and three lower premolars, and four upper and four lower molars, for a total of 50 teeth. The opossum has this primitive tooth formula, which is commonly written 5/4, 1/1, 3/3, 4/4 × 2 = 50. The corresponding primitive formula for placentals is generally given as 3/3, 1/1, 4/4, 3/3 × 2 = 44 (but see Schwartz 1978 for an alternative view). The star-nosed mole and domestic pig have 44 teeth (as do occasional horses); all other Minnesota species have fewer teeth. Each species account includes a tooth formula, unless it is constant for a genus or family, in which case it appears in the respective generic or family account.

The majority of species have fewer than the primitive maximum number of teeth. In the evolution of mammals, which teeth have been lost? There is normally only one canine per jaw, so no question arises if the canine is missing. Most mammalogists agree that, unless special circumstances indicate otherwise, incisors are lost from the back, premolars from the front, and molars from the back. To refer to a particular tooth, we use uppercase letters I, C, P, and M for upper teeth and lowercase letters i, c, p, and m for

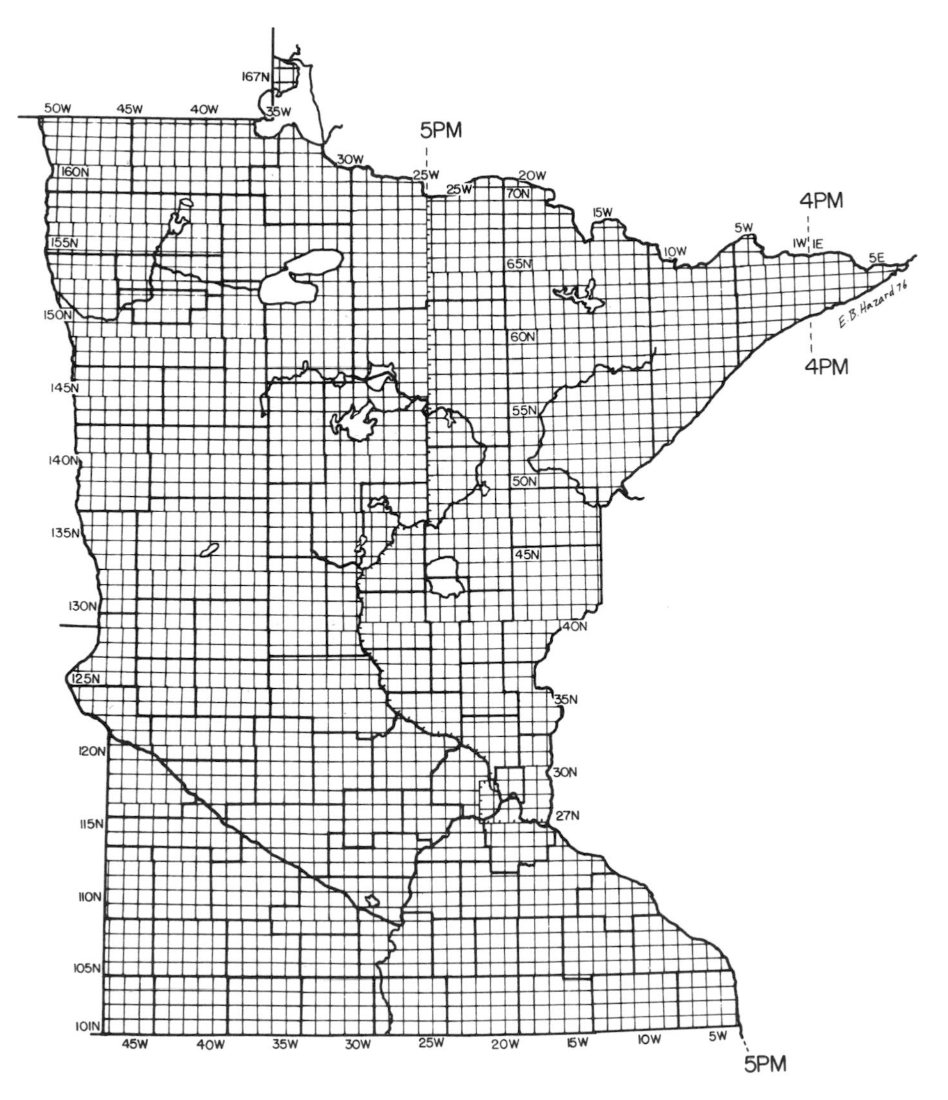

Map 3. Minnesota townships. (Map produced by the Department of Biology, Bemidji State University; based on an official Minnesota highway map with township overprint provided by the Department of Highways.)

lower teeth, each followed by a number (except for C or c) to designate the particular tooth. The abbreviated tooth formula for a typical human adult is 2/2, 1/1, 2/2, 3/3 × 2 = 32. People have lost I3, i3, P1, p1, P2, and p2. (Actually, these teeth were lost in ancestral Old World primates roughly 40,000,000 years ago.) A person who never grew wisdom teeth is also missing M3 and m3. I will use this system to refer to characteristics of specific teeth.

Map 2 shows Minnesota's 87 counties. It has been customary to record species distribution by counties, but this is misleading since county sizes can vary up to ratios of 39:1 (St. Louis:Ramsey). Becker County is larger than the State of Rhode Island, Beltrami is larger than Delaware, and St. Louis is larger than Connecticut. Ideally, townships are all the same size. Actually, they are only approximately equal because of adjustments made for the earth's curvature, some peculiar anomalies in surveying, areas cut off along three of Minnesota's borders, and discrepancies along the Fifth Principal Meridian. Nonetheless, the townships (map 3) provide a standard unit for recording distribution on a scale suitable for a statewide survey. A standard township, incidentally, is 93.24 km^2, close enough to 100 km^2 so that map 3 gives you an idea of how a 10 × 10 km grid would look.

The individual distribution maps (maps 4 through 76) are too small to show the township lines, but the locality symbols are centered on the appropriate township, if it is known. Stewart (1975) used a similar system to record localities for breeding birds in North Dakota, marking every township of record for each species. I have done this for rarer species, but for common ones (e.g., *Blarina brevicauda*), I have only indicated selected localities in order to avoid obscuring other features of the maps. The meanings of the locality symbols on the maps are:

● = specimens in collections, township known;

○ = other valid records, township known;

▲ = specimens in collections, known only to county;

△ = other valid records, known only to county;

◇ = skeletal remains of species no longer present.

As you can see on map 3, county lines often do not follow township lines. Further, Minnesota has two township grids that do not mesh, one east of the Fifth Principal Meridian (5PM) and one west of it. This meridian is indicated by tick marks on map 3. When a species is known from a township bisected by a county line, I use a partial circle to indicate the county from which it is known. When one is known from a township truncated by the Fifth Principal Meridian, I modify the circle to indicate which side it is from.

"Other valid records" comprise sight records by competent observers of easily identified species (e.g., opossum, but not ermine) under good seeing conditions; road kills (Case 1978); records of specimens handled but not preserved in professional collections; and Department of Natural Resources game, furbearer, and nongame records. I have been conservative, omitting doubtful records. This is true even for specimens (e.g., I will not record a locality for *Microsorex hoyi* if the specimen is a skin only).

The insets with the distribution maps give a rough approximation of the ranges of the species in North America. They are based primarily on: Banfield 1974; Bowles 1975; Burt 1957; Burt and Grossenheider 1976; Godin 1977; Hall and Kelson 1959; Hamilton and Whitaker 1979; Jackson 1961; Peterson 1966; and Youngman 1975. Bowles (1975) and Jackson (1961) map specific localities for specimens from Iowa and Wisconsin, respectively. I have mapped selected records from these works, as well as a few other records from bordering counties, in order to give a better understanding of the distribution of some species.

Perusal of the maps will indicate two general deficiencies in distributional data. First, many presumably widespread species—most of the bats, for example—are poorly represented. Second, many areas are not well collected. This became apparent when I prepared a rough composite of all the records of specimens that I had in 1975. Most of the records were from within 20 or 30 km of Moorhead, Bemidji, Collegeville, Mankato, Duluth, Minneapolis and St. Paul, Itasca State Park, etc.—the sites of collecting institutions. Some counties had only two or three records of mammal specimens; most townships had none. To my knowledge, there is only one thorough report on mammal distribution in any region of Minnesota, Timm's (1975) study of Cook County. This accounts for the concentration of symbols in that county on many of the maps. We need more such careful studies, by professional mammalogists or students under professional supervision. Counties or portions thereof might be suitable areas, but so might various physiographic or ecologic units: for example, the Wadena Drumlin Fields, the Anoka Sand Plain, or the Coteau des Prairies (Green and Janssen 1975; Schwartz and Thiel 1963). I hope the obvious gaps in the distribution maps in this book will stimulate further research. I would also like to know of any valid records that are not mapped.

ORDER POLYPROTODONTA
Opossums and Allies

Most authors have previously placed all members of the infraclass Metatheria in the single order Marsupialia. Others, however, have concluded that enough variation occurs within the group to justify dividing it into three or four orders. The common term "marsupial" will doubtless continue to be used, just as the term "placental" is employed for all orders in the infraclass Eutheria. The classification here follows Kirsch (1977b).

Female marsupials bear their young after a very short gestation period, and most have a pouch in which the nipples are located and within which the young nurse and continue their growth and development. One family occurs in North America.

Additional References: Birney et al. 1980; Collins 1973; Hunsaker 1977; Hunsaker and Shupe 1977; Kirsch 1968, 1977a, c; Lillegraven 1974, 1975; Low 1978; Parker 1977; Rowlands 1966; Sharman and Berger 1970; Stonehouse and Gilmore 1977; Tyndale-Biscoe 1973; Van Deusen and Jones 1967; Van Valen 197lb.

Family Didelphidae
New World Opossums

Opossums are small to medium-sized marsupials with five toes on each foot and with long pointed snouts. Their big toes have no claws and are opposable, and their tails are generally prehensile. The opossums that roamed North America in the Upper Cretaceous were not much different from those of today, and are assigned to the same family (Russell 1928). The Didelphidae, to my knowledge, is the only family of Cretaceous mammals with representatives alive today. Of the dozen genera that occur in Latin America, only one occurs in the United States (Van Deusen and Jones 1967).

Genus *Didelphis*

Didelphis virginiana
Opossum, Virginia Opossum

Measurements: Total length 759 (694–803); body length 465 (428–543); tail 295 (259–352); hind foot 67 (61–80); ear from notch 50 (44–54); weight 2.2–4 kg. (Opossum tails and ears are often shortened from being frozen.)

Description: An opossum is about as big as a large house cat or domestic rabbit. The underfur is white, and the long guard hairs have black tips, giving an overall gray appearance, with much individual variation. Opossums have long narrow muzzles and wide gaping mouths. The tooth formula (5/4, 1/1, 3/3, 4/4 × 2 = 50) is unique among mammals in Minnesota, and the total number of teeth exceeds that of any other Minnesota species. Opossums are stout, with relatively short legs. Opposable big toes and long, naked, prehensile tails aid them in getting about in the trees. Opossums have short, hairless ears. With ears and tail ill-protected from the cold, it is not surprising that the tips of these

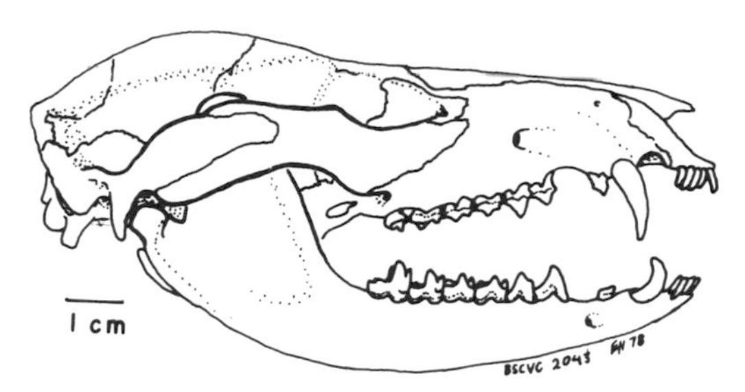

Figure 1. Skull of *Didelphis virginiana*, Virginia opossum, BSCVC 2043.

members are often frozen off in Minnesota specimens. Female opossums have a pouch in which the young are carried.

Until the early 1950s the Minnesota species was named *D. virginiana*. It then became generally regarded as a subspecies of *D. marsupialis*. It has since been shown to overlap *D. marsupialis* in Middle America and is again properly referred to as *D. virginiana* (Gardner 1973).

Range and Habitat: Although opossums move about readily on the ground, they are generally found in wooded or brushy areas. Opossums reach their northern limits in central and west-central Minnesota, and are not found in the northern coniferous forest. They have been expanding their range northward and westward in the last several decades, but are still not common north of Minneapolis-St. Paul (de Vos 1964; Gunderson 1961; Hazard 1963; Hibbard 1965, 1970; Long and Copes 1968). Opossums are more characteristic of various woody, brushy, and cultivated habitats in states to the east and south. They have also been introduced and have become established in the Pacific states (Burt and Grossenheider 1976).

Natural History: Opossums are omnivores. The teeth of *Didelphis* are not highly specialized and, in fact, closely resemble the tritubercular pattern typical of early mammals. Opossums have small, nipping incisors, sharp, piercing canines, and premolars and molars that can not only tear meat but also crush insect exoskeletons and chew soft plant material. They eat a wide variety of food: insects, other invertebrates, frogs, small birds and mammals, bird eggs, carrion, berries and other fresh or rotten fruits, and vegetables (Sandidge 1953). They sometimes raid henhouses. Opossums are nocturnal and thus likely to go unnoticed. They seek shelter in abandoned burrows, under stumps or fallen logs, in tree holes, and in deserted buildings.

Except for the mother-young relationship, opossums are generally not sociable. Gestation lasts 12 to 13 days. At birth the many tiny young climb unassisted into the pouch, where the successful ones attach to the nipples. Any young in excess of 13, the usual number of nipples, perish. The young remain in the pouch about two months. After emerging, they may be carried about by clinging to the mother's fur. They do not, popular legend and illustrations to the contrary, hang by their tails from the mother's tail, as she arches it over her back. The young are

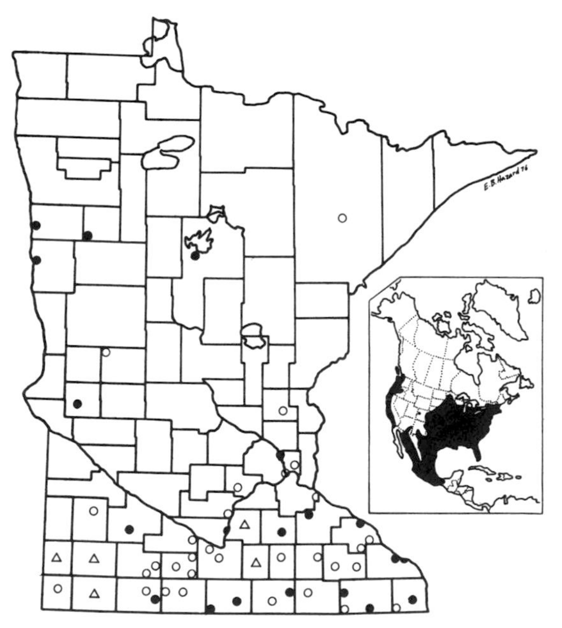

Map 4. Distribution of *Didelphis virginiana*. ● = township specimens. ○ = other township records. △ = other county records. (Map produced by the Department of Biology, Bemidji State University.)

Figure 2. *Didelphis virginiana*, Virginia opossum.

weaned when about 95 days old (Petajan and Morrison 1962).

Opossums are not as quick nor as efficient as other predators, nor as well protected from the cold as most furbearers. Their brains are relatively small. Jackson (1961) could fill the braincase of an opossum skull with only 26 navy beans, wheareas the braincase of a raccoon of a similar size held 139 navy beans. But opossums have a high reproductive rate for a medium-sized mammal, they eat almost anything, they seem to be capable of defending themselves against moderate-sized predators, and they can apparently survive substantial injuries. Of 95 specimens from eastern Kansas examined by Black (1935), 39 had broken bones that had healed. One specimen had survived breaks in nine ribs, a scapula, and the lower jaw.

Relationship to People: In Minnesota, opossums are seldom common enough to be economically significant. Their fur is usable but has little commercial value. Their occasional forays into the poultry yard are damaging, but hardly a threat to agriculture. In the South, they may do more economic damage, but they are probably also important disseminators of the seeds of a favorite food, the persimmon (Worth 1975). Opossums are not exciting game animals, but they are hunted for food, especially in the Southeast. I prefer roast opposum to another omnivore that is more widely respected as game in Minnesota, the raccoon. Every species, of course, is valuable because of the unique insights it affords into the nature and intricacy of life. The opossum, perhaps more than any other living mammal, gives us a look at the remote past, at a time when the land was dominated by giant reptiles, when mammals, though common, were mostly small, not very specialized, and not very bright.

Additional References: Barr 1963; Chesness et al. 1968; Fitch and Sandidge 1953; Fitch and Shirer 1970; Francq 1969, 1970; Hamilton 1958; Hartman 1952; Jenkins 1971; Layne 1951; McManus 1967, 1969, 1974b; Minkoff et al. 1979; Severinghaus 1975; Shirer and Fitch 1970; Verts 1963a; Werner and Vick 1977.

ORDER INSECTIVORA
Insectivores

The Insectivora is the third largest order of mammals. Most of the species are small, but some are as large as a rabbit. Insectivores are hard to characterize as a group, partly because some families of doubtful affinities are often classified in the order, and partly because the group retains many primitive characters that have changed little since the Cretaceous, some of which characters are also possessed by ancient species that are generally assigned to other orders (Findley 1967; Van Valen and Sloan 1977; Vaughan 1978). The ordinal name implies a diet of insects, and, indeed, insects and other invertebrates are common food items for members of the order. However, the group is not restricted to an insect diet, and insects are important foods for many other mammals.

Minnesota's insectivores are small mammals, weighing no more than 130 g, with sharply pointed snouts (except in the star-nosed mole), small or rudimentary eyes, and the pinna (the fleshy part of the ear) reduced or absent. All five digits are present on each foot. Two families of insectivores occur in Minnesota, the Soricidae and the Talpidae.

Family Soricidae
Shrews

Shrews are characterized by small size (in all Minnesota species), reduced but visible eyes, short ears often completely hidden in the fur, and the complete absence of a zygoma. Soricids are the most widespread insectivores, inhabiting most of Eurasia, Africa, North America, northernmost South America, and the western islands of the East Indies. They are absent from the Australian region. They range farther north than moles do, and have apparently moved through Beringia relatively recently (Burt 1958; Hoffman and Peterson 1967).

Additional References: Findley 1967; Pearson and Barr 1962; Rudd 1955; Vaughan 1978.

Genus *Sorex*
Long-tailed Shrews

Long-tailed shrews of the genus *Sorex* occur in both Eurasia and North America, and at least two species are common to both hemispheres. The tooth formula is 3/1, 1/1, 3/1, 3/3 × 2 = 32.

Sorex cinereus
Masked Shrew

Measurements: Total length 94 (83–104); tail 39 (36–41); hind foot 12 (11–12); ear from notch 6 (4–8); weight 4.1 (3.0–6.5).

Description: The masked shrew is one of Minnesota's smallest mammals, and the smallest widespread, common mammal in the area. It has the pointed nose characteristic of shrews, much more pointed than that of any mouse. The teeth, like those of American shrews generally, are brown at the tips. The tail is over one-third of the total length, and the soft body fur is dark brown above and dark gray below. The eyes and ears are inconspicuous. The masked shrew can be safely distinguished from most other local shrews by size, color, or tail length, but none of these traits will separate it reliably from the pygmy shrew, *Microsorex*. These two species can be distinguished only by comparing the teeth (see fig. 82, p. 186). It is therefore important to keep the skull of any specimen, even if it is damaged.

Range and Habitat: Sorex cinereus ranges over most of the northern United States and Canada, and there are a few isolated populations in Siberia (Hoffmann and Peterson 1967). In Minnesota, it occupies a variety of habitats. Iverson et al. (1967) found the masked shrew in both grassland and woodland in the prairie-forest transition zone in northwestern Minnesota. Others have collected specimens in open habitat as well as in deciduous and coniferous forests throughout the state. Spencer and Pettus (1966) reported that the masked shrew prefers moist habitats, but Tester and Marshall (1961) found it wherever there was adequate ground cover. Actually, the woodland form (*S. c. cinereus*) and the prairie form (*S. c. haydeni*) do not seem to interbreed in the Prairie Provinces, and perhaps should be considered separate species (van Zyll de Jong 1980). Masked shrews occur throughout Minnesota.

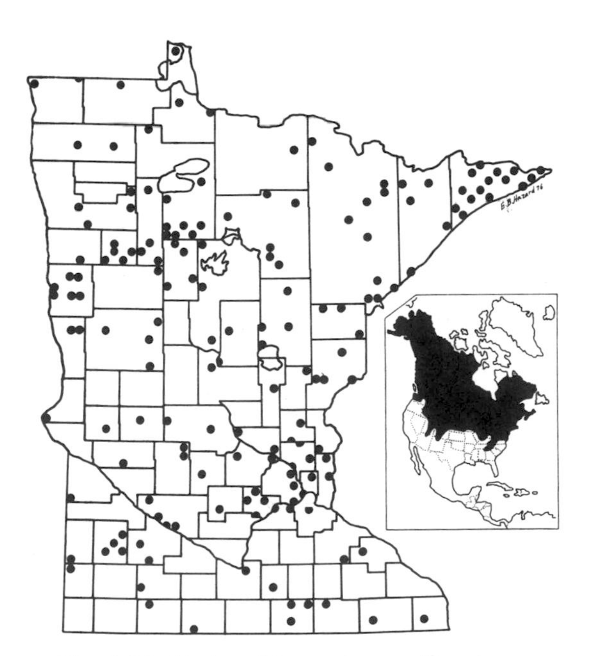

Map 5. Distribution of *Sorex cinereus*. ● = township specimens with intact unicuspids, selected locations. (Map produced by the Department of Biology, Bemidji State University.)

Natural History: Sorex cinereus, like other small shrews, is about as small as an animal can be and still maintain a constant body temperature well above that of its surroundings. Like hummingbirds, which are of comparable size, shrews have so little mass that they must feed frequently on high-energy food to produce enough heat. Small shrews must eat every few hours or they will starve, and typically they eat more than their body weight each day. Masked shrews eat adult and larval insects of many kinds, spiders, centipedes, sowbugs, snails, small mammals (mice and other shrews), and a small amount of plant material (Hamilton 1930; Whitaker and Mumford 1972a; Whitaker and Schmeltz 1973). Shrews are capable of killing mice that weigh much more than the shrews themselves, but some of the vertebrate flesh that they eat may be carrion. Shrews are active throughout the winter, finding their winter food in the leaf litter and upper soil levels under the snow. Feeding more or less continually, they are active day and night.

Although *Sorex cinereus* is one of our commoner mammals, its breeding habits are not known in detail. Gestation may be about 18 days. Litter size apparently ranges from four to 10, and a female may bear up to five litters in a year. The young probably leave the nest when 10 to 12 days old.

Shrews may be preyed on by a variety of animals, although their strong musky odor sometimes makes them unattractive to carnivorous mammals. In winter, I have tracked foxes that have killed shrews and then left them uneaten in the snow. However, shrews have also been found in the stomachs of predatory mammals. The odor does not seem to offend hawks and owls, and, in moist habitats, shrews are sometimes eaten by frogs (Marshall 1951a).

Relationship to People: People perhaps encounter *Sorex cinereus* (and other shrews) most often as something a cat drags in, and then they may mistake it for a mouse. The masked shrew takes its toll among insects of various sorts, and perhaps helps keep mouse populations in check. It does no harm to people or to domestic plants or animals, and it is not known to transmit rabies (Verts and Barr 1960).

Additional References: Brown 1967a; Buckner 1964; Buech, Siderits, et al. 1977; Doucet and Bider 1974; Forsyth 1976; Getz 1961c; Goodwin 1979; Graham 1976; Grant and Birney 1979; Kirkland and Griffin 1974; Manville 1949; Meierotto 1967; Pruitt 1954a; Vickery and Bider 1978; Wrigley et al. 1979.

Sorex palustris
Water Shrew

Measurements: Total length 149 (136–158); tail 67 (62–76); hind foot 19 (18–21); ear from notch 9 (7–11); weight 14.4 (11.4–19.1).

Description: The water shrew is much larger than any of our other long-tailed shrews, and its long tail quickly distinguishes it from the larger short-tailed shrew. Water shrews are distinctly bicolor—black above and silvery below. The fringes of stiff hairs on their hind feet, which increase the surface area of the feet and thus aid in swimming, are found in no other North American shrew.

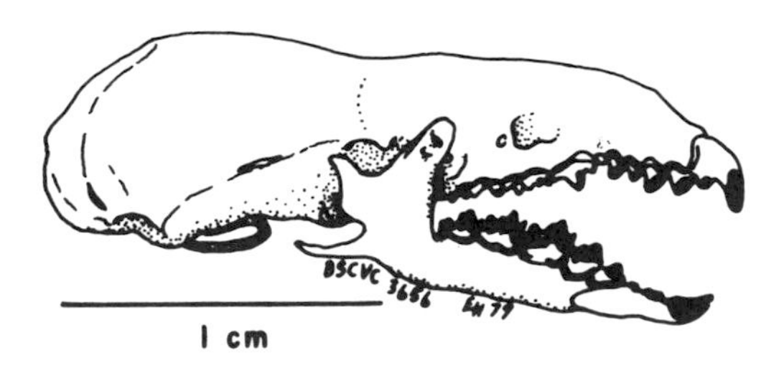

Figure 3. Skull of *Sorex palustris*, water shrew, BSCVC 3656.

Range and Habitat: Sorex palustris is a northern species, characteristic of the boreal and mixed forests of Canada and the northern United States, and of the western mountains. As their name implies, water shrews are generally found near water, especially running water (Conaway 1952; Spencer and Pettus 1966). In Minnesota, they inhabit the northcentral and northeastern counties, where they are often found at some distance from streams (Whitaker and Schmeltz 1973).

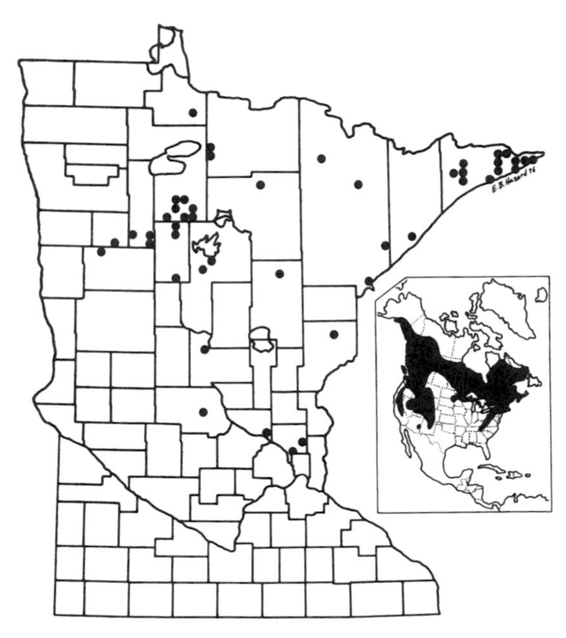

Map 6. Distribution of *Sorex palustris*. ● = township specimens. (Map produced by the Department of Biology, Bemidji State University.)

Natural History: Although water shrews live near water, the extent to which they depend on aquatic animal life for food is disputed. In some areas, they apparently feed primarily on aquatic animals, mostly invertebrates. Elsewhere, most of their diet seems to be terrestrial (Hamilton 1930). Slugs, earthworms, and spiders were the three most common foods in the stomachs of water shrews taken in St. Louis County in September by Whitaker and Schmeltz (1973).

Male water shrews apparently do not breed in their first summer. Females are reproductively active from late January to August. Litter size ranges from five to eight, with six being most common. In Montana, two or three litters may

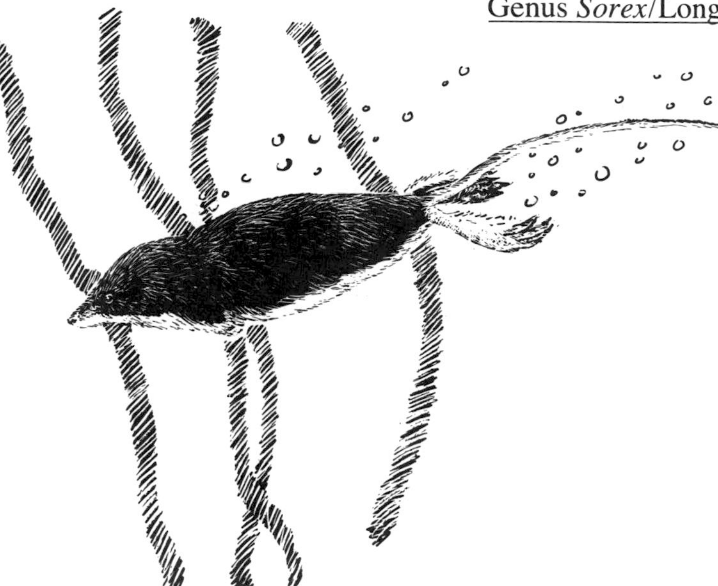

Figure 4. *Sorex palustris*, water shrew.

be produced per summer (Conaway 1952). Detailed studies of food habits and reproduction need to be done in the Upper Midwest.

Relationship to People: The range and habitat of *Sorex palustris* are restricted to areas largely unaffected by people, and the species probably is rare enough to be of little practical significance.

Additional References: Bailey 1929; Brown 1967a; Graham 1976; Kinsella 1967; Meierotto 1967; Wrigley et al. 1979.

Sorex arcticus
Arctic Shrew

Measurements: Total length 109 (106–112); tail 40 (36–45); hind foot 14 (12–15); ear from notch 8 (7.4–10); weight 9.1 (7.1–11.7).

Description: The arctic shrew is distinctly larger than the masked shrew and the pygmy shrew, and distinctly smaller than the water shrew. Like all of these, it has a long tail. Its most conspicuous characteristic, especially in adults in winter, is its tricolor pattern: blackish brown above, lighter brown on the sides, and pale brown below. This pattern is less well marked in subadults, which make up most of the late summer population. In summer, a small arctic shrew might be mistaken for a masked shrew, but skull size and dentition should serve to separate them (see fig. 82).

Range and Habitat: Bee and Hall (1956) and Hoffmann and Peterson (1967) regard *Sorex arcticus* as a Holarctic species which ranges from northeastern Eurasia through Alaska and across Canada and the upper midwestern states to the mouth of the Saint Lawrence. Jackson (1928) and Youngman (1975) consider the populations of Eurasia, Alaska, and the northern Yukon to be a distinct species, *Sorex tundrensis*, the usage followed here. Thus restricted, *S. arcticus* is a Nearctic, primarily boreal forest species which reaches its southern limits in the northern parts of the Upper Midwest. During the Pleistocene, this species ranged south to West Virginia (Guilday and Hamilton 1973). Within the coniferous

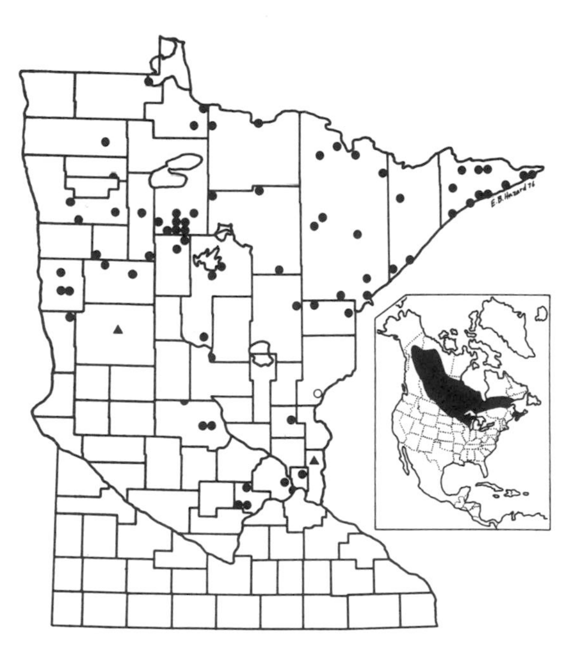

Map 7. Distribution of *Sorex arcticus*. ● = township specimens. ○ = other township record. ▲ = county specimens. (Map produced by the Department of Biology, Bemidji State University.)

Figure 5. *Sorex arcticus*, arctic shrew.

forest biome in Minnesota, it is often found in cedar, tamarack, and spruce swamps.

Natural History: Like water shrews, arctic shrews live in areas remote from man, and thus their habits are not well known. Buckner (1966) found arctic shrews less tolerant than masked shrews of wet conditions within Manitoba bogs. In fact, the two species seemed to be in competition, with *Sorex cinereus* being more abundant when a high water table apparently put *S. arcticus* at a disadvantage. Iverson et al. (1967), on the other hand, found masked shrews in both forest and grassland in the prairie-forest transition of northwest Minnesota, and recorded arctic shrews only from marshy areas near East Grand Forks.

Sorex arcticus probably breeds from March through September, and may have up to three litters of five to eight young per season. The females are sexually mature at four or five months of age. A female from an early spring litter can thus bear young during the same breeding season (Buckner 1966).

Relationship to People: Arctic shrews are uncommon in populated areas, and have little effect on people. They may be significant, along with other shrews, in controlling such pests as larch sawflies.

Additional References: Bailey 1929; Becker et al. 1976; Buckner 1964; Clough 1963; Graham 1976; Heaney and Birney 1975; Martell and Radvanyi 1977; Whitaker and Pascal 1971; Wrigley et al. 1979.

Genus *Microsorex*

Microsorex hoyi
Pygmy Shrew

Measurements: Total length 84 (75–91); tail 30 (27–34); hind foot 10 (8–11.5); ear from notch 4 (3–6); weight (3 specimens) 2.9–4.2.

Description: By average measurements, this is the smallest North American mammal, but its measurements broadly overlap those of the masked shrew. Because it is not safe to try to distinguish the two species without examining the teeth, I have rejected records based on skins alone. As indicated in the key, only three unicuspid teeth are easily visible when a *Microsorex* skull is viewed from the side. This is true of no other long-tailed shrew. The brown fur of the pygmy shrew, like that of the masked shrew, is lighter on the belly than on the back. The tooth formula is 3/1, 1/1, 3/1, 3/3 × 2 = 32. Diersing (1980) regards *Microsorex* as a subgenus of *Sorex*.

Range and Habitat: Microsorex hoyi ranges primarily through the boreal forests of Alaska and Canada, but it also occurs south of the Canadian border in the northern Rockies, in an isolated population in northern Colorado and southeastern Wyoming, and in the Upper Midwest. Long

(1972a) has named populations in the northeastern United States, the Appalachians, and Nova Scotia as a distinct species, *M. thompsoni*, but van Zyll de Jong (1976) disputes its status. In Minnesota, *M. hoyi* is known only from the northeastern half of the state. It may be taken in dry habitats, but is most often found in moist habitats not far from water (Long 1972b).

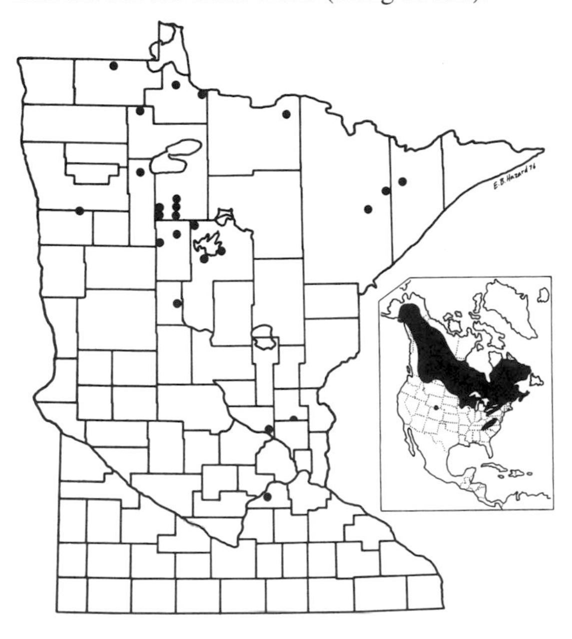

Map 8. Distribution of *Microsorex hoyi*. ● = township specimens with intact unicuspids. Iowa specimen reported in Bowles 1975. (Map produced by the Department of Biology, Bemidji State University.)

Natural History: The biology of the pygmy shrew is poorly known. Its food probably consists primarily of insects, other terrestrial arthropods, earthworms, and land snails. There are typically five or six young per litter, and there may be only a single litter per year (Long 1972b). *Microsorex hoyi* seems to be less abundant in most habitats than other shrews. This apparent scarcity may result partly from the species's avoidance of snap traps. When large tin cans—sunk in the ground and partly filled with water—are used instead of snap traps, the percentage of *Microsorex* in the catch is usually higher. Nonetheless, this shrew seldom appears to be abundant. Presumably, there is some aspect of the habitat that it exploits relatively efficiently, and this permits it to persist in competition with its more abundant cousins, but what this factor might be is unknown.

Relationship to People: Microsorex hoyi is of most interest to people as a species whose life history and ecologic role are largely unknown. It is probably of little economic importance.

Additional References: Bailey 1929; Becker et al. 1976; Brown 1967a; Buckner 1964, 1966; Graham 1976; Heaney and Birney 1975; Long 1974, 1976a; Prince 1941; Roscoe and Majka 1976; Spencer and Pettus 1966; Wrigley et al. 1979.

Genus *Blarina*

Blarina brevicauda
Short-tailed Shrew

Measurements: Total length 129 (124–134); tail 25 (22–27); hind foot 16 (16–17); ear from notch 7 (6–8); weight 27.3 (22.9–35.5).

Description: Because of its relatively large size and short tail, *Blarina* is the shrew most likely to be mistaken for a vole or a mouse. However, the dentition, pointed snout, tiny eyes, and absence of visible external ears clearly identify it as a shrew. Its size and short tail distinguish it from all other shrews of Minnesota. It is stocky, and its fur is generally slate-colored, but with consid-

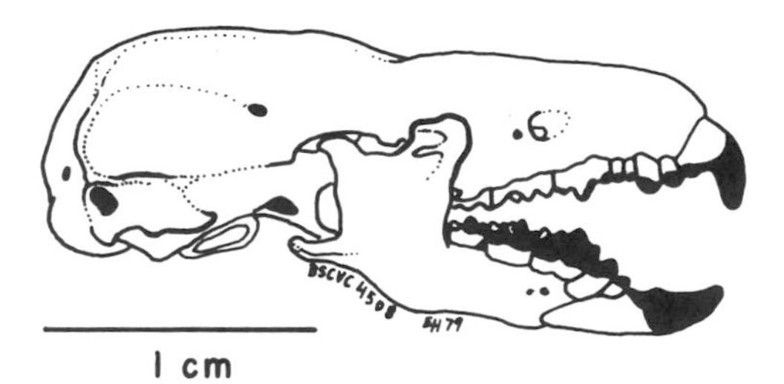

Figure 6. Skull of *Blarina brevicauda*, short-tailed shrew, BSCVC 4508.

erable variation, including albinism (Williams 1962). Younger animals are often darker, and newly molted animals are glossy. The skull, unlike that of other shrews, is sturdy and angular. The tooth formula is 3/1, 1/1, 3/1, 3/3, × 2 = 32.

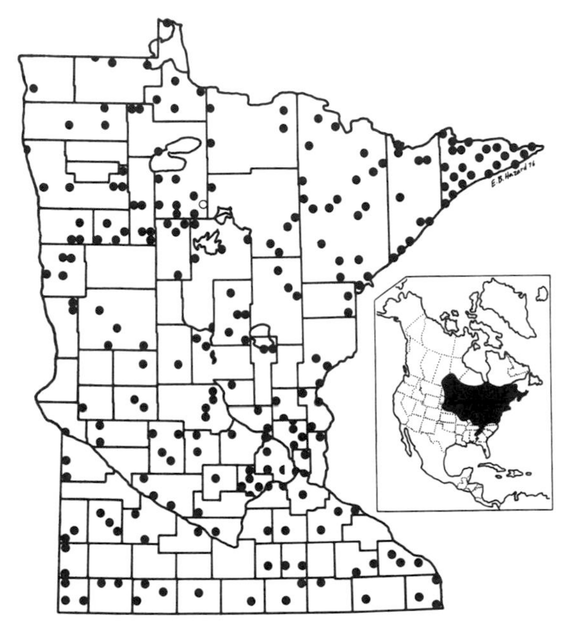

Map 9. Distribution of *Blarina brevicauda*. ● = township specimens, selected locations. ○ = other township record. (Map produced by the Department of Biology, Bemidji State University.)

Range and Habitat: Jones et al. (1975) recognize three species of *Blarina*. *B. brevicauda* occupies the north central and northeastern United States and adjacent southern Canada. This species occurs throughout Minnesota. *B. carolinensis* occupies the south central and southeastern states, and *B. telmalestes* is confined to coastal wetlands in Virginia and North Carolina. Burt (Burt and Grossenheider 1976) treats them as a single species. Short-tailed shrews occupy many habitats, both open and wooded. Although *B. brevicauda* perhaps favors moist habitats, it is probably less limited by lack of moisture than is *Sorex cinereus* (Getz 1961c). *Blarina* generally avoids open water.

Natural History: Blarina is one of our commoner small mammals, and may sometimes be the most abundant mammal in an area (Hamilton 1929). However, the short-tailed shrew is one of many mammals whose populations fluctuate markedly. *Blarina* may abound in an area one year and be seemingly absent the next. When short-tailed shrews are abundant, they are potentially significant predators both on invertebrates such as insects, earthworms, and snails, and on small vertebrates. They store snails underground, at least in winter (Shull 1907).

Like other shrews, *Blarina* is primarily a meat eater, but it differs from them in two ways. First, it is relatively large, which permits it to take larger invertebrate prey but also makes it a less efficient predator on the tiniest soil invertebrates. Large size enables *Blarina* to conserve heat more effectively than smaller shrews; it does not have to eat either so much in relation to its body weight each day, or so often. Captives have survived over 24 hours without food, which smaller shrews cannot generally do. The size of the short-tailed shrew also makes it a potential predator on mice and voles. In captivity, *Blarina* readily kills and eats adult meadow voles (*Microtus pennsylvanicus*). It may be a significant factor in regulating vole populations (Barbehenn 1958; Eadie 1944, 1948, 1952; Fulk 1972), although Hamilton and Whitaker (1979) doubt this.

Second, *Blarina* is venomous (Krosch 1973; Pearson 1942; Tomasi 1978). The submaxillary salivary glands produce a poison which, when it enters a wound inflicted by the teeth, will quickly kill a mouse or vole. The only other North American mammal known to produce such a poison is *Solenodon*, a large insectivore from the West Indies (Rabb 1959).

Captive *Blarina* will eat a variety of plant as well as animal foods. Their ability to vary their diets may account in part for their widespread occurrence and frequently high population densities (Martinsen 1969).

Blarina may mature more slowly than the smaller shrews. Buckner (1966) reports that females may not breed until their tenth month, but Dapson (1968) suggests that either sex may breed at under two months of age. Gestation periods of 17 to 21 days have been reported. Two to four litters of five to eight young each may be produced in a breeding season, which in Minnesota may last from May into August or perhaps later. The glands in the flanks, which produce the musky substance characteristic of shrews, are well developed in *Blarina*, especially in males during the breeding season. Although possibly a deterrent to predators, the substance is probably also a sex attractant. It produces a persistent, distinctive, and I think pleasant odor in the woods.

Relationship to People: The short-tailed shrew seldom does any direct injury to people or their property. Live specimens should be handled with care, since the poisonous bite can be painful and, perhaps only in sensitive individuals, can be at least temporarily incapacitating (Krosch 1973). *Blarina* is often abundant enough that it may be of some importance in limiting populations of herbivorous insects and rodents, thus being of benefit to agriculture and forestry. It is probably not a significant carrier of rabies (Verts and Barr 1960).

Additional References: Briese and Smith 1974; Buckner 1964; J. R. Choate 1972; Choate and Fleharty 1973; Christian 1950; DeMeules 1954; Ellis et al. 1978; Gaughran 1954; Genoways and Choate 1972; Graham and Semken 1976; Grant and Birney 1979; Hamilton 1930; Ingram 1942; Iverson et al. 1967; Jameson 1949; Kirkland and Griffin 1974; Meierotto 1967; Pearson 1944, 1946; Pruitt 1954b, 1959; Richardson 1973; Stehn et al. 1976; Tomasi 1979; Whitaker and Mumford 1972a; Wrigley et al. 1979.

Genus *Cryptotis*

Cryptotis parva
Least Shrew

Measurements: (one specimen from Homer, Minnesota): total length 80; tail 20; hind foot 11.5; weight 4.2–5.6 (Jackson 1961).

Description: The body of the least shrew is about the same size as that of the masked or the pygmy shrew, but the least shrew is a bit stouter and has a short tail. You can easily distinguish it from the

Figure 7. *Blarina brevicauda*, short-tailed shrew.

larger short-tailed shrew by its small size and its distinctly brown, rather than slate-colored, fur. The tooth formula is 3/1, 1/1, 2/1, 3/3 × 2 = 30. No other North American shrew has 30 teeth.

Range and Habitat: Cryptotis parva inhabits open grassy and shrubby areas in the eastern, southeastern, and middle United States, reaching Canada only on the north shore of Lake Erie. In Minnesota, specimens are known only from Winona County. Bowles (1975) records *C. parva* in Clay Co., Iowa, and suggests (his fig. 5, p. 33) that a population exists in adjacent southwestern Minnesota. The least shrew may be more common than trapping records indicate. In areas where few least shrews are trapped, their remains are sometimes abundant in owl pellets (Hamilton 1944). It would be interesting to examine owl pellets from southeastern Minnesota to increase our knowledge of the distribution of *C. parva* in that area. Only one species of *Cryptotis* occurs in the United States, but there are several others in Mexico and Central America. See map 11.

Natural History: Little is known of the biology of *Cryptotis parva* because it is seldom encountered even within its known range. Studies outside Minnesota (Chamberlain 1929; Hamilton 1944; Whitaker and Mumford 1972a) indicate that the diet is similar to that of other shrews, comprising invertebrates of various sorts. Individuals may eat three-fourths or more of their weight per day. Least shrews are active day and night, as are many shrews. They usually bear five or six young, which reach adult size in about one month.

Relationship to People: Too little is known of the biology of *Cryptotis* to be sure of its effect on people, but it is probably seldom of major, direct importance.

Additional References: Andrews 1974; Briese and Smith 1974; Broadbooks 1952; Choate and Fleharty 1973; Getz 1962b; Graham 1976; Guilday and Hamilton 1978; Martin 1967; Porter 1978; Verts and Barr 1960.

Family Talpidae
Moles

Moles are stout-bodied, short-legged, digging insectivores (Campbell 1939). They have short dense fur, no pinna, greatly reduced eyes, and powerful forepaws with broad, flattened palms. Both genera of moles in Minnesota are restricted to eastern North America. Moles of the Pacific states belong to different genera, and those of Eurasia to still other genera (including *Talpa*, for which the family was named). Moles cannot persist in extremely arid habitats nor in permafrost areas. Hence the moles became isolated from one another millions of years ago as dry and cold climates became more widespread, and these isolated populations have had time to differentiate into several genera (Burt 1958). The most extensive works in English on the natural history of moles (Godfrey and Crowcroft 1960; Mellanby 1971) concern *Talpa europaea*.

Additional References: Findley 1967; Jackson 1915; Stroganov 1945; Vaughan 1978; Ziegler 1971.

Genus *Scalopus*

Scalopus aquaticus
Eastern Mole, Prairie Mole

Measurements: Total length 171 (155–185); tail 26 (21–30); hind foot 23 (19–25); weight 75–120 (Jackson 1961).

Description: The eastern mole is the larger of the two moles native to Minnesota. Dark brown to grayish fur with a silvery sheen covers the body except for the naked, pointed muzzle, the paws, and the short, nearly hairless tail. The fur will lie either forward or backward, which helps the mole move in either direction in its burrow. Friction against the tunnel walls wears down the fur.

In a specimen trapped during its molt, the molt line is conspicuous because the new fur is longer than the old. The tooth formula (Ziegler 1971) is 3/2, 1/0, 3/3, 3/3 × 2 = 36. The large front feet are wider than they are long.

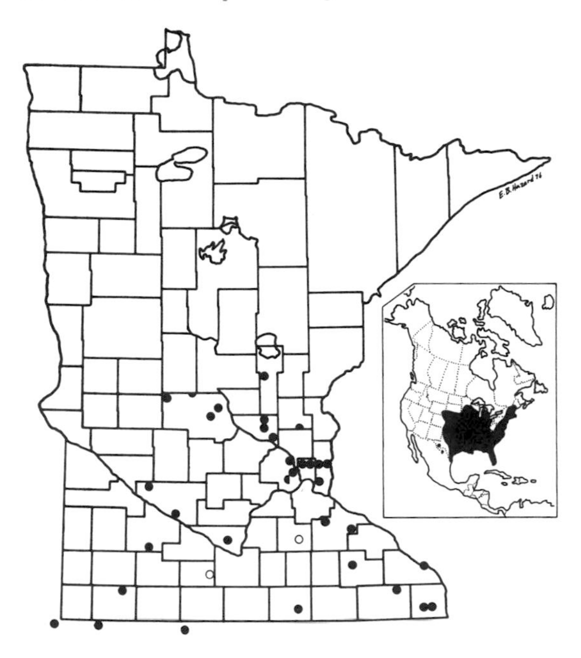

Map 10. Distribution of *Scalopus aquaticus*. ● = township specimens. ○ = other township records. Iowa specimens reported in Bowles 1975. (Map produced by the Department of Biology, Bemidji State University.)

Range and Habitat: Scalopus inhabits the eastern and central states, reaching its northern limit in southern New England, Lower Michigan, and central Minnesota. Eastern moles do best in relatively light, well-drained soils, and are more common in open areas than in woods. Brumwell (1951) found that moles tunneling in prairie will turn back if they encounter a patch of woods, but they may move into wooded ravines during droughts. Despite the misleading appellation *aquaticus*, they are not characteristic of water-logged habitats. The genus contains only one species.

Natural History: Eastern moles are strictly fossorial, living almost their entire lives underground. They dig temporary shallow tunnels, 2–3 cm below the surface, and a permanent system of tunnels 10–40 cm underground (Harvey 1976). Dispersal of this species to the north and west may in part be limited by the increasing depths to which the ground freezes in colder winters and in areas of thin insulating snow cover. Eastern moles dig shallow tunnels while foraging. They eat some underground plant parts, but their primary food is invertebrates that live in the soil. Earthworms are a favorite food, but insects (especially grubs) and other arthropods are also taken (Arlton 1936; Hisaw 1923a; Whitaker and Schmeltz 1974). Living underground protects moles from predators to some extent; nevertheless, weasels, snakes, owls (Hamerstrom and Hamerstrom 1951), and others do prey on them.

Eastern moles have but one litter of three to seven young per year, in the spring (Conaway 1959). Gestation lasts at least 28 days. The naked, helpless young are born in a nest chamber lined with dead grass or leaves. This chamber generally has several exits. The young mature rapidly and leave the nest when about one month old.

Relationship to People: Like most insectivores and other small mammals, *Scalopus* participates in the necessary cycling of materials and flow of energy in the ecosystem. It may be useful in preventing population explosions of some insect pests (Arlton 1936). Its tunnels, however, can be a nuisance in lawns, gardens, and golf greens. Moles are best removed from such areas by using mole traps rather than poisons.

Additional References: Bailey 1929; Brown 1972; Campbell 1939; Christian 1950; Davis 1942; Eadie 1954; Gaughran 1954; Guilday 1972; Gupta 1966c; Heaney and Birney 1975; Henning 1952, 1957; Hisaw 1923b; Jackson 1922b; Jones et al. 1978; Scheffer 1910; Stroganov 1945; Yates and Schmidly 1975, 1978.

Genus *Condylura*

Condylura cristata
Star-nosed Mole

Measurements: Total length 190 (182–199); tail 74 (70–82); hind foot 27 (25–29); weight 53.5 (47.1–60)

Description: The star-nosed mole is smaller and less robust than the eastern mole, although its total length is about the same because its tail is longer. Its fur is almost black and does not have a silvery sheen. The forepaws are not nearly as large as those of the eastern mole. The most distinctive characters of this mole are at either end. The nose terminates in a disk around which are set 22 pink, fleshy papillae. The long tail is

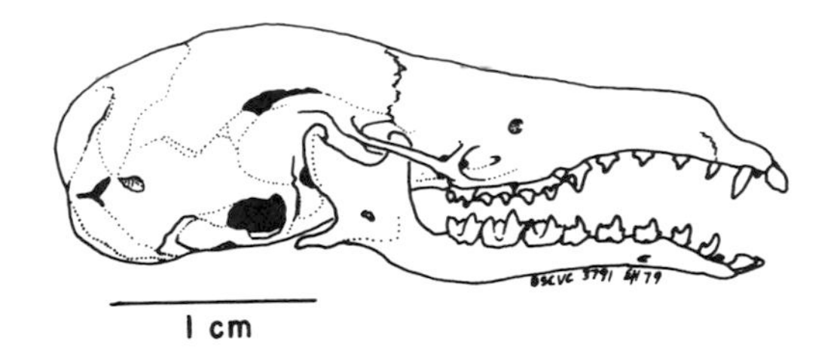

Figure 8. Skull of *Condylura cristata*, star-nosed mole, BSCVC 3791.

constricted at the base, somewhat swollen for about the first half of its length, and tapered toward the tip. The swelling is more pronounced in the winter. The tooth formula is 3/3, 1/1, 4/4, 3/3 × 2 = 44. The star-nosed mole is the only native mammal in Minnesota that apparently retains the primitive placental allotment of 44 teeth. Ziegler (1971) holds that the upper and lower first premolars of adult *Condylura* are actually persisting milk premolars.

Range and Habitat: The star-nosed mole inhabits southeastern Canada, the Great Lakes states, and the Atlantic coastal states from the Carolinas north. There are isolated populations in coastal Georgia. This species reaches its western limit in northern Minnesota and southeast Manitoba. It

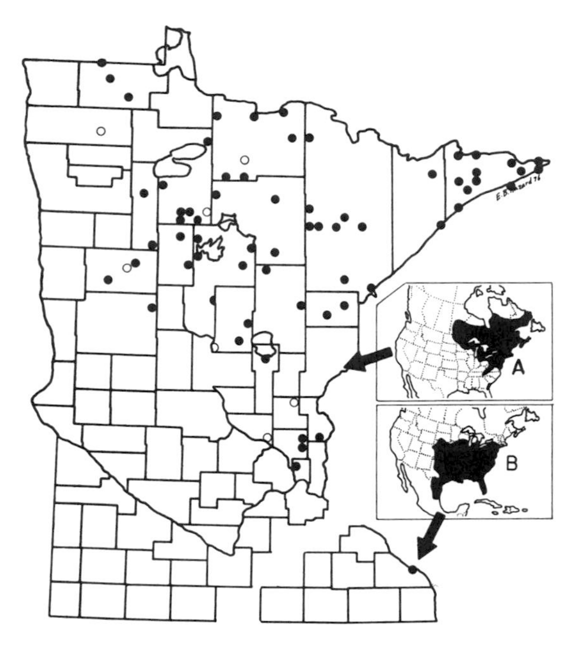

Map 11. A. Distribution of *Condylura cristata*. B. Distribution of *Cryptotis parva*. ● = township specimens. ○ = other township records. (Map produced by the Department of Biology, Bemidji State University.)

occurs in approximately the northeastern half of Minnesota where it is found primarily in wet habitats. Like the genus *Scalopus*, *Condylura* is monotypic.

Natural History: There have been relatively few studies of *Condylura*. The star-nosed mole is a less powerful digger than the eastern mole, but is an accomplished swimmer. It is active on the surface more often than the eastern mole. In winter, it may burrow under the snow or even move about on top of the snow. It burrows deeply in easily worked, often largely organic soil; its burrow often opens under water, and in winter *Condylura* may be seen swimming under the ice. Its food consists mainly of aquatic invertebrates (Hamilton 1931; Jackson 1961). It is active both day and night; like all insectivores in Minnesota, it is active through the winter.

The gestation period is believed to be about

Figure 9. *Condylura cristata*, star-nosed mole.

45 days. The nest is often built in a hummock or knoll, well above water level in a swamp. There is but one litter per year, of two to seven young, with an average of five. Young are born from March until August, but litters most often arrive in May. The young probably remain with the mother for a month or more. They reach adult weight in the fall, and are ready to breed the following spring (Eadie and Hamilton 1956; Hamilton 1931; Jackson 1961; Simpson 1923).

Relationship to People: Condylura generally lives in nonagricultural areas and is of little economic importance. It may damage lawns in low-lying areas. Even where it is common, most people do not know it exists. People usually encounter star-nosed moles as dead specimens found near swamps or on country roads. These may be left by predators, but often there is no mark on them. *Condylura* often occurs in dense local populations of 40 or more per hectare (Jackson 1961), but little is known of its social organization. The function and mode of operation of the 22 nasal papillae has not been elucidated, to my knowledge. When an adult that fell into my window well investigated a ball of chopped beef, it rapidly vibrated the entire array of papillae over the surface; the individual papillae did not appear to move independently. Obviously, *Condylura* deserves further study.

Additional References: Bailey 1929; Briese and Smith 1974; Hammerstrom and Hammerstrom 1951; Heaney and Birney 1975; Moyle 1975; Petersen and Yates 1980.

ORDER CHIROPTERA
Bats

Bats are the only mammals capable of powered flight. As in birds, the forelimb functions as a wing. The flight membrane of bats, however, consists not of stiff epidermal derivatives like the feathers of birds, but of a living layer of skin and associated tissues, the patagium. The patagium is supported by the arm, the greatly elongated second through fifth digits, the sides of the torso, the hind legs, and, in most bats, the tail (Norberg 1969, 1972). Although birds and bats are both diverse groups, bats are generally adapted for slower flying speeds than birds, and for greater mobility. Whereas most birds are diurnal, bats are primarily nocturnal and are most commonly adapted to catching flying insects on the wing at night, locating their prey primarily by echolocation (Simmons 1979a, b). Vaughan (1978) describes in some detail the adaptations of bats for flight, and the mechanism of echolocation.

Bats also differ from most birds in that they generally are endothermic only when they are active. Typical bats decrease their metabolic rate when they are at rest and allow their internal temperatures to drop almost as low as the temperature of their surroundings (Davis 1970; Vaughan 1978).

The order Chiroptera is the second largest order of mammals, numbering over 850 species. People in the northern United States and Canada are not likely to appreciate the diversity of bats, because relatively few species inhabit cool temperate and subartic regions (Koopman and Cockrum 1967). Bats are extremely successful in the tropics, however, and in some tropical areas outnumber all other groups of mammals put together.

The order consists of 18 families, one of which makes up the suborder Megachiroptera, the Old World fruit bats. The other 17 comprise the suborder Microchiroptera. Of these, only three families occur in the United States, and only one of the latter, the Vespertilionidae, occurs in Minnesota.

Many temperate zone bats migrate. Migration in bats is studied in much the same way as it is in birds, by banding individuals in one place and recovering them somewhere else. Except for bats that form large maternity and hibernation aggregations, however, the banding returns have not been adequate to permit the delineation of winter and summer ranges. The nocturnal habits of bats generally prevent the accumulation of visual records to supplement banding data.

Additional References: Barbour and Davis 1969; Carter 1970; Greenewalt 1962; Griffin 1970; Luckens and Davis 1964; Strickler 1978a, b; Wimsatt 1970, 1977.

Family Vespertilionidae
Vespertilionid Bats

This is the largest family of bats (nearly 300 species), and the most widespread. It is absent from the polar regions and some oceanic islands, but is otherwise cosmopolitan. Temperate zone vespertilionids, like many other microchiropterans, are typically heterotherms. Not only do they hibernate at reduced body temperatures during colder months, but during the daytime they seek cool shelter in which their body temperature, and consequently their fuel consumption, can decrease.

Most vespertilionids, including the seven Minnesota species, eat insects. They catch them primarily by echolocation, which they also use to avoid obstacles and generally to find their way in the dark. Although echolocation is the primary hunting and navigating mechanism in vespertilionids and some other microchiropterans, bats can see and vision is important, especially in homing in some species (Barbour et al. 1966; Davis and Barbour 1965).

Another feature common to Minnesota vespertilionids and many other bats is delayed fertilization (cf. delayed implantation, a characteristic of weasels and other members of the family Mustelidae, order Carnivora). Spermatogenesis occurs in male bats during the warmer months, and mating occurs in the fall. Live sperm are present in the uteri of females throughout hibernation, but ovulation and fertilization do not occur until spring at the end of hibernation. The young are born in early summer (Wimsatt 1945).

Bats are among the more important vectors of rabies (Constantine 1970). Although this problem is more serious in states farther south, the Section of Disease Prevention and Control, Minnesota Department of Health, has reported the occurrence of rabid bats in Minnesota (Hazard 1973). Rick Steece, of the Rabies Laboratory, Minnesota Department of Health, reports that, in Minnesota, rabies occurs in *Myotis lucifugus*, *Eptesicus fuscus*, and both species of *Lasiurus*, and that rabies also occurs, in other states, in the other three species of bats that inhabit Minnesota. He suggests that the low incidence reported in Minnesota may not reflect the actual frequency of rabies in natural bat populations, and recommends that "the general public should not attempt to handle or molest bats unless absolutely necessary" (Steece, personal communication 1977). The likelihood that bats may be rabid does not imply that there should be a concerted effort to eliminate bats, except in those rare circumstances where people are likely to come into physical contact with a bat. Bats are typically careful and skillful in avoiding contact with people, and they are unlikely to bite unless handled. Trappers, laboratory workers, and others likely to come into direct contact with bats, or with other wild mammals that may carry rabies, should contact the Minnesota Department of Health concerning the recommended pre-exposure regimen.

Additional References: Baker and Patton 1967; Bickham 1979; Dwyer 1971; Geluso et al. 1976; Griffin 1958; Jones et al. 1967; Kunz 1973; Mumford and Cope 1964; Mumford and Whitaker 1974; Pearson and Barr 1962; Stones and Wiebers 1965a; Vaughan 1978; Webster and Griffin 1962.

Genus *Myotis*
Mouse-eared Bats

Myotis lucifugus
Little Brown Myotis, Little Brown Bat

Measurements: Total length 91 (83–99); tail 38 (32–48); hind foot 10 (9–12); ear from notch 15 (14–16.5); tragus 4.5–6; forearm about 38; weight 9.8 (7.3–11.9).

Description: The little brown myotis is a glossy coppery brown above, with a dark spot on each shoulder. Its ventral surface is buffy gray. The ears are relatively short; when those of a fresh specimen are laid forward (not stretched), they extend to about the tip of the nose. *M. lucifugus* is most likely to be confused with *M. Keenii*, which has brassy fur, a longer and more sharply pointed tragus, and longer ears. The interorbital width of a little brown myotis skull generally is slightly more than 4 mm. The tooth formula of the genus is 2/3, 1/1, 3/3, 3/3 × 2 = 38.

Range and Habitat: M. lucifugus ranges south

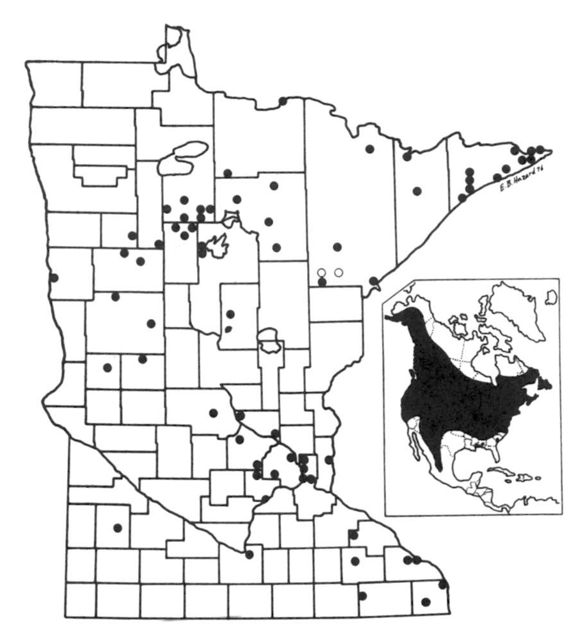

Map 12. Distribution of *Myotis lucifugus*. ● = township specimens. ○ = other township records. (Map produced by the Department of Biology, Bemidji State University.)

from tree line across Canada and Alaska into most of the contiguous United States. It is absent from Florida, the Gulf Coast, most of Texas, and parts of the extreme Southwest. Within its range, the little brown myotis obviously occupies diverse habitats. It generally prefers to forage near trees or near water.

Natural History: This is one of the better known American bats, yet many details of its life history are unknown. In winter, it hibernates in large groups in caves, mines, or other shelters where the temperature will not drop below freezing and where the humidity is high. Individuals may travel hundreds of kilometers to reach such sites, but the hibernacula used by populations from many areas are unknown (Davis and Hitchcock 1964, 1965; Evans 1934; McManus 1974c). In early summer, females form large maternity colonies, often in warm sites in buildings. Adult males may be solitary or colonial at this time (Davis et al. 1965; David and Hitchcock 1965; Krutzsch 1961; Wimsatt 1945). Individual little brown myotis fly in a set pattern repeatedly when they hunt (Hough 1957); it may be that most insectivorous bats do this. This species eats various insects, but prefers moths (Whitaker 1972a), mayflies in season (Buchler 1976), and beetles (Anthony and Kunz 1977).

Mating occurs in late summer or fall, but ovulation and fertilization do not occur until the next spring. The period of actual gestation is 50 to 60 days. In Minnesota, the young are born from late June to mid-July. Usually there is only one offspring, and there is only one mating cycle per year (Barbour and Davis 1969; Davis 1967; Kunz 1971). The youngster weighs about one-fourth as much as its mother at birth, can fly at about three weeks, and reaches adult weight at about one month. Twins occur rarely (Davis 1967, cited in Humphrey and Cope 1976).

Compared with other mammals of their size, bats have a low rate of reproduction. They also are unusually long-lived. One *M. lucifugus* is believed to have lived 24 years, and several are known to have lived 20 years or more (Griffin and Hitchcock 1965; Hall et al. 1957). Probably the low reproductive rate is correlated with relative security from predation in the maternity

Figure 10. *Myotis lucifugus*, little brown myotis.

colonies, and both the low reproductive rate and the long potential life-span are correlated with the relative immunity from predation enjoyed by flying vertebrates. Both birds and bats are long-lived in comparison with terrestrial mammals of comparable weights. In both, natural selection has probably favored potentially long-lived individuals because a high proportion of such individuals do live a long life, and thus are able to produce many offspring. In mice and other small mammals subject to heavy predation as adults, genes favoring great longevity would have little selective value (Cole 1954; Frazzetta 1975; Gadgil and Bossert 1970).

Relationship to People: Because they feed on insects and are often abundant near human dwellings, *M. lucifugus* should probably be considered beneficial. Most people have little use for them, however, perhaps because of unfounded fears and because they object to having bats in the attic or to the telltale stain of bat droppings left near the entry used by the bats. *M. lucifugus* has probably been the subject of more physiological studies than any other American bat.

Additional References: Belwood and Fenton 1976; Clark et al. 1978; Coutts et al. 1973; Davis 1964; Guthrie 1933a, b; Kunz 1971; Kunz et al. 1977; Martin and Fenton 1978; Rysgaard 1942; Schowalter et al. 1979; Stones and Wiebers 1965b, 1966; Webster 1968; Wimsatt 1944a, b, 1960; Wimsatt and Parks 1966.

Myotis keenii
Keen's Myotis

Measurements (seven specimens): Total length 89 (82–94); tail 40 (36–47); hind foot 10 (8.5–11); ear from notch 17 (16.5–19); tragus (one specimen) 11; forearm (three specimens) 36–38; weight (four specimens) 4.9–6.4.

Description: Keen's myotis is similar in size and color to the little brown myotis, but the dorsal fur is somewhat silky, rather than glossy, and has a brassy, rather than a coppery, hue. The ear is longer; when that of a fresh specimen is laid forward, it will extend some 4 mm past the nose. The tragus is distinctly pointed. The interorbital width is generally less than 4 mm.

Range and Habitat: M. keenii has a disjunct distribution. The larger of the two ranges covers the north central and northeastern United States, extends south through the Appalachians to the Gulf Coast, and north into south central and southeastern Canada. The smaller range is in the coastal Pacific Northwest. This bat probably occurs throughout Minnesota, though its distribution may be spotty. It favors forested habitats.

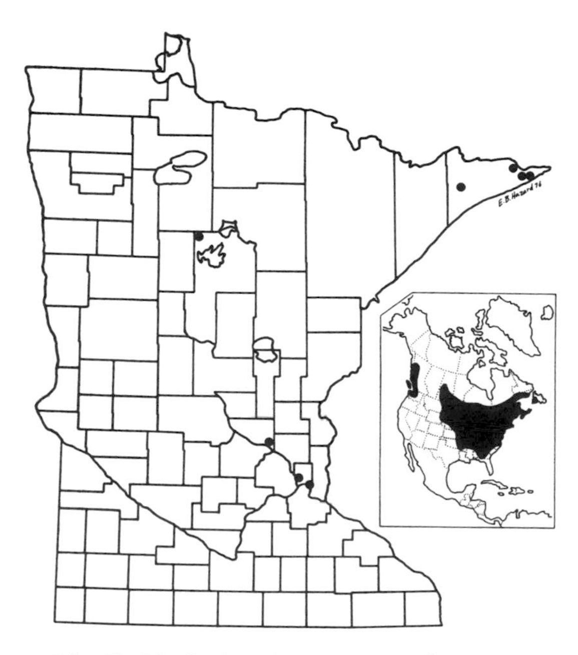

Map 13. Distribution of *Myotis keenii*. ● = township specimens. (Map produced by the Department of Biology, Bemidji State University.)

Natural History: Keen's myotis is less commonly collected through most of its range than the little brown myotis. In the winter it is often found hibernating in small groups in caves or mines occupied mainly by other bats. Some summer residents of Minnesota may migrate out of the state to hibernate, but hibernating individuals have been found as far north as St. Cloud (Goehring 1954). *M. keenii* forms small nursery colonies of up to 30 individuals (Mumford and Cope

1964), but some females may raise their young in isolation. A few females and their young are sometimes found in association with a large maternity colony of *M. lucifugus*. A female bears only one young. Longevity data are scarce, but *M. keenii* is known to live up to 18.5 years (Hall et al. 1957).

Relationship to People: We do not know enough about this species to evaluate its economic significance, if any.

Additional References: Barbour and Davis 1969; Findley 1972; Fitch and Shump 1979; Folk 1940; Griffin 1940; Hayward and Davis 1964; Hitchcock 1949, 1965; Kunz 1971, 1973; Rysgaard 1941, 1942; Stones and Branick 1969; Swanson and Evans 1936; Whitaker 1972a.

Genus *Lasionycteris*

Lasionycteris noctivagans
Silver-haired Bat

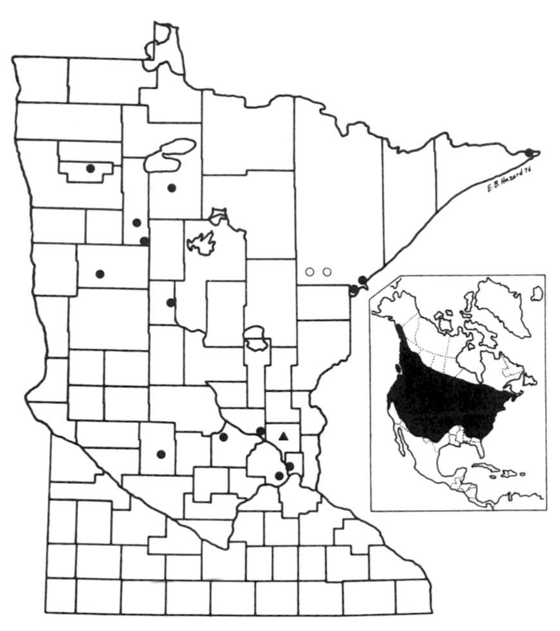

Map 14. Distribution of *Lasionycteris noctivagans*. ● = township specimens. ○ = other township records. ▲ = county specimen. (Map produced by the Department of Biology, Bemidji State University.)

Measurements (six specimens): Total length 93–105; tail 35–46; hind foot 10; ear from notch 13–18; tragus 5–6; forearm 42.5, 44; weight 9.9–17.2.

Description: The silver-haired bat is of moderate size, somewhat larger than either of the myotis in Minnesota but smaller than a big brown bat. Its fur is black (occasionally dark brown) with silvery tips. Its interfemoral membrane is lightly furred on the upper surface. The ears are round and hairless. The tooth formula is 2/3, 1/1, 2/3, 3/3 × 2 = 36.

Range and Habitat: Lasionycteris noctivagans ranges from southeast Alaska and the southern third of Canada into all but the southernmost parts of the contiguous states. There are also records from Mexico (Yates et al. 1976). Most females apparently migrate into the northernmost states and Canada to bear their young, whereas the males usually stay behind (Barbour and Davis 1969). Both sexes generally winter south of Minnesota, but there is one winter record from this state (Beer 1956). Within its range, the silver-haired bat is largely a forest inhabitant, preferring areas near small bodies of water.

Natural History: The silver-haired bat is primarily solitary, roosting singly or in small groups rather than in large colonies. Individuals roost in trees, under bark, in crannies in rock, and in more or less open outbuildings and woodpiles. There are a few records from caves (Beer 1956; Krutzsch 1966). *L. noctivagans* generally flies slowly and low to the ground. It emerges earlier in the evening than some of the larger bats. There are usually two young (Kunz 1971), which are probably born by late June in Minnesota.

The silver-haired bat, like many other species, has demonstrated homing ability. Banded individuals have returned to the original site of banding when released as much as 150 km away (Davis and Hardin 1967). Individuals may sometimes get off course, perhaps because of high winds. There are several reports of bats alighting on ocean-going ships. One male *Lasionycteris* was taken in August about 153 km SSE of the eastern tip of Long Island (Mackiewicz and Backus 1956).

Relationship to People: We know too little about the different ecologic niches of this and many other bats to evaluate their significance in the ecosystem and their direct or indirect effects on people. *Lasionycteris* seldom roosts in homes and thus is not a source of annoyance.

Additional References: Bailey 1929; Easterla and Watkins 1970; Gould 1955; Hayward and Davis 1964; Heaney and Birney 1975; Kunz 1973; Schowalter et al. 1978; Whitaker 1972a; Wimsatt 1945; Yates et al. 1976.

Genus *Pipistrellus*

Pipistrellus subflavus
Eastern Pipistrelle

Measurements: Total length 82 (79–88); tail 39 (35–42); hind foot 9 (8–10); ear from notch 13 (12–14); tragus 4.5 (3.5–6); forearm 35–36; weight 5.2 (4.2–7.2).

Description: The eastern pipistrelle is Minnesota's smallest bat. The overall color of its coat is usually yellowish-brown, but the individual hairs are distinctly three-banded: the base and tip are dark, the middle light. The ears are not particularly large, and the tragus is short and blunt. The basal third of the interfemoral membrane is furred. The tooth formula is 2/3, 1/1, 2/2, 3/3 × 2 = 34. No other Minnesota bat has 34 teeth.

Range and Habitat: Pipistrellus subflavus is found throughout most of the wooded parts of the eastern United States, Nova Scotia, New Brunswick, extreme southern Ontario and Quebec, and the east coast of Mexico. Its distribution is interrupted in the central Great Lakes region, and this bat seems to be absent from southern Florida. In Minnesota, it is restricted primarily to the deciduous woods of the southeast.

Natural History: The eastern pipistrelle winters in caves and comparable shelters; a storm sewer in St. Cloud is the northernmost known hibernaculum in Minnesota (Goehring 1954). It generally seeks a hibernation site with a temperature well above freezing and with a high relative humidity. Although in winter it may cluster in some numbers, in summer *P. subflavus* most often roosts singly. It may remain in hibernation well into May and typically bears two young sometime after mid-June. During the summer it frequents woodland edges near streams (Davis and Mumford 1962).

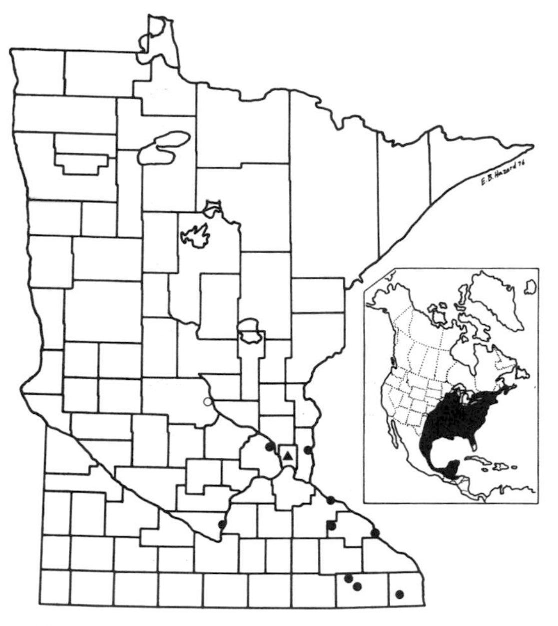

Map 15. Distribution of *Pipistrellus subflavus*. ● = township specimens. ○ = other township record. ▲ = county specimen. (Map produced by the Department of Biology, Bemidji State University.)

The eastern pipistrelle is generally observed to be one of the earlier evening foragers, as is its relative in the southwestern states, *P. hesperus* (Cochrum and Cross 1964). At times, pipistrelles and swallows feed on flying insects simultaneously (Van Gelder and Goodpaster 1952). Although the pipistrelle, like many bats, maintains a high and constant temperature only when active, it still takes a prodigious amount of food to sustain one. *P. subflavus* can consume about

3.3 g of insects in an hour's feeding, over half of an adult's body weight (Gould 1955).

Within the hibernaculum, which it shares with other bats, *P. subflavus* is apparently more sedentary than the other species, moving from site to site less frequently (Davis 1964). Long-term banding studies of hibernating colonies have revealed that males generally have a lower mortality than females and a longer maximum life-span—about 19 years as opposed to 10 (Davis 1966; Paradiso and Greenhall 1967; Walley and Jarvis 1971). The sex ratio among adults is about six males to four females, but there may be a greater preponderance of males in hibernacula because some females hibernate singly (Davis 1966).

Relationship to People: An abundant bat in many areas, *P. subflavus* may contribute significantly to insect population control. It does not roost in buildings in large numbers, so it is seldom considered a nuisance. In Minnesota the pipistrelle may be of little ecological importance.

Additional References: Barbour and Davis 1969; Davis and Wright 1967; Folk 1940; Greeley and Beer 1949; Hall and Dalquest 1950; Hayward and Davis 1964; Hitchcock 1949; Jones and Suttkus 1973; Kunz 1971; Myers 1978; Rysgaard 1942; Swanson and Evans 1936; Whitaker 1972a; Wimsatt 1945.

Genus *Eptesicus*

Eptesicus fuscus
Big Brown Bat

Measurements: Total length 115 (103–145); tail 42 (38–47); hind foot 12 (10–14); ear from notch 17 (14–19); tragus 4.5–9; forearm 44–50; weight 12.5–28.8.

Description: The big brown bat is Minnesota's second largest bat and the only common large bat in the state. The fur is uniformly brown, the wing and tail membranes are black, and the tragus is broad and blunt. The tooth formula is 2/3, 1/1, 1/2, 3/3 × 2 = 32.

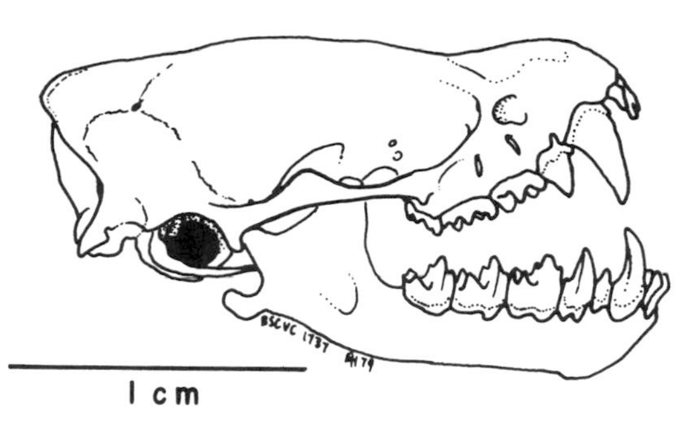

Figure 11. Skull of *Eptesicus fuscus*, big brown bat, BSCVC 1737.

Range and Habitat: Eptesicus fucus ranges from southern Canada through the 48 contiguous states into Mexico and Central America. It is also found in the West Indies. It probably occurs throughout Minnesota, though it is rarer in the north than in the south. It often takes shelter in human dwellings, and usually forages in open areas or areas with scattered trees. It is generally found within 0.8 km of water.

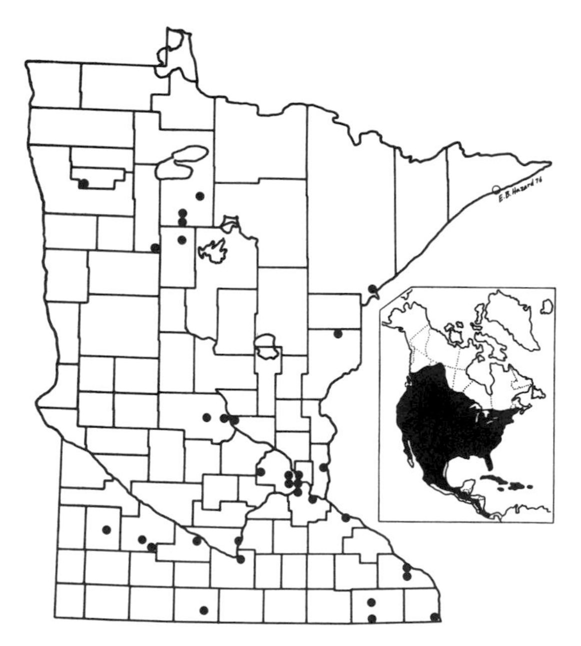

Map 16. Distribution of *Eptesicus fuscus*. ● = township specimens. ○ = other township record. (Map produced by the Department of Biology, Bemidji State University.)

Natural History: In the winter, the big brown bat hibernates in caves, mines, lofts, and attics, apparently seeking out dry places with near-freezing temperatures. It enters hibernation as late as November and emerges by late March or early April. Females establish maternity colonies in the spring in barns, attics, and other buildings. In many areas, *Eptesicus* is the most common bat. In Minnesota, the young are probably born in late June or early July. There are usually two young; in drier areas to the west, one is apparently the more common number (Brenner 1968; Kunz 1971, 1974; Mills et al. 1975; Phillips 1966). The young grow rapidly and are able to forage on their own in about a month. Hitchcock (1965) reports a longevity record of 19 years. *Eptesicus fuscus* eats relatively few moths (Black 1972), feeding primarily on beetles (Whitaker 1972a).

Winter colonies may be predominantly male or predominantly female; both situations have been reported (Goehring 1954, 1958; Mills et al. 1975). Maternity colonies contain few if any males; at this time, many males may roost singly. However, in a mine in Kansas used as a hibernaculum in winter, males apparently remain together all summer while the females leave for maternity colonies elsewhere (Phillips 1966). Occasionally, big brown bats will raise their young in the same structure that houses a nursery of little brown myotis. The latter species generally prefers warmer sites than do big brown bats, which tolerate suprisingly cold temperatures in winter (Rysgaard 1942). Individuals have survived temperatures well below 0° C (Goehring 1971a, 1972; Wetmore 1936).

Relationship to People: Eptesicus fuscus has probably benefited more than any other bat in Minnesota from the structures built by people. Its presence is seldom appreciated, partly for aesthetic reasons. Bats often defecate just as they enter their daytime roosts, so the area around the louver or entrance usually bears a conspicuous stain. However, many insects that *Eptesicus* eats are harmful, so the species should be considered beneficial overall. Like some other bats, it is harmed by residues of persistent pesticides and organochlorine compounds (Clark and Lamont 1976).

Additional References: Beer 1953a, 1955a; Beer and Richards 1956; Christian 1956; Cockrum and Cross 1964; Dapson et al. 1977; Davis et al. 1968; Evans 1934; Feng et al. 1978; Hamilton 1933a; Long and Severson 1969; Mumford 1958; Rysgaard 1941; Schowalter and Gunson 1979; Swanson and Evans 1936; Twente 1955.

Genus *Lasiurus* **Hairy-tailed Bats**

Lasiurus borealis **Red Bat**

Measurements: Total length 105 (94–116); tail 47 (42–52); hind foot 8–10; ear from notch 11–16; tragus 3–3.5; forearm (one) 40.6; weight 10.6–15.4.

Description: The red bat is medium-sized, generally smaller than a big brown bat. The ears are short and rounded. The interfemoral membrane is well furred above, and the tail is relatively long. In flight, the tail and interfemoral membrane extend straight back, giving the animal a characteristic broad V-shaped silhouette. The species is sexually dimorphic, with males being

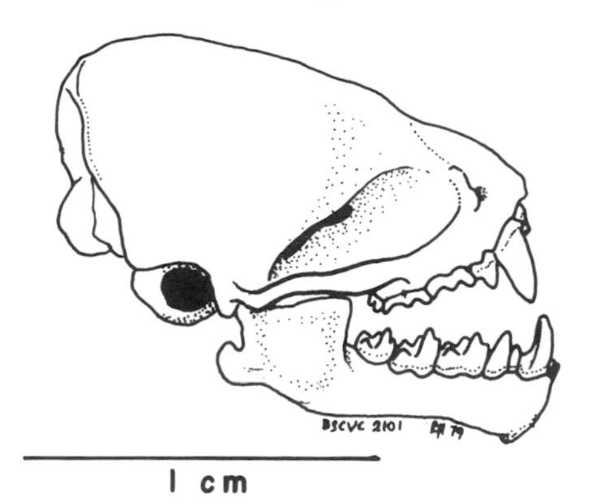

Figure 12. Skull of *Lasiurus borealis*, red bat, BSCVC 2101.

a bright brick-red and females a yellowish red. The tooth formula of the genus is 1/3, 1/1, 2/2, 3/3 × 2 = 32.

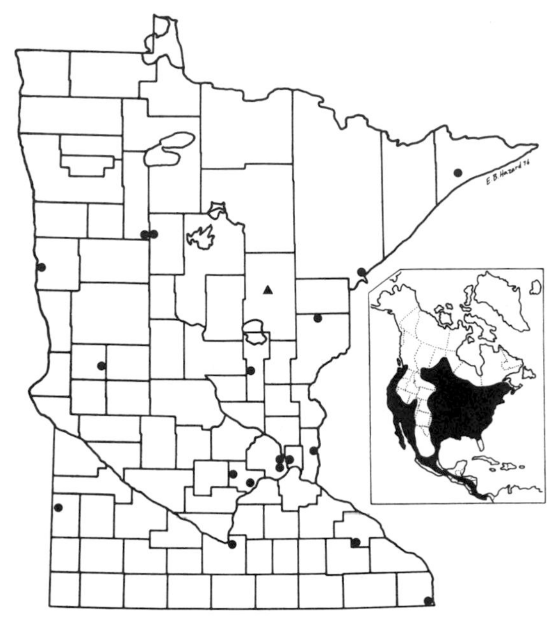

Map 17. Distribution of *Lasiurus borealis*. ● = township specimens. ▲ = county specimen. (Map produced by the Department of Biology, Bemidji State University.)

Range and Habitat: Lasiurus borealis ranges through most of the central and eastern United States and into south central and southeastern Canada. There is also a population in the southwestern United States. The eastern and western ranges join in Mexico and in southern British Columbia, and the range extends into Central America. *L. borealis* probably is migratory in much of the northern part of its range, but it is a tree bat, so banding data are meager. It favors open areas near trees—openings, woodland edges, fencerows, and tree-lined streets (Constantine 1966). In Minnesota, it probably occurs locally throughout the state in summer.

Natural History: The red bat has more young than most vespertilionids, with litters of two to four being the most common (Hamilton and Stalling 1972; Kunz 1971; McClure 1942; Quimby 1942b; Stains 1965). The bats mate in late summer or fall, but fertilization is delayed until spring. The actual gestation is about 80 or 90 days. Quimby (1942b) recorded a birth in Minnesota in early June. Females and their young generally roost singly, most often in trees. They prefer roosts with dense concealing foliage but with adequate flight space below (Constantine 1959). The young generally leave the mother when about a month old. Although some red bats migrate north to bear young, others remain in the central and southern states. The adult males at least sometimes migrate to different summer areas than do females.

Red bats eat a variety of insects, but seem to prefer moths (Ross 1961; Whitaker 1972a). Some predators also seem to have a liking for red bats. Opossums and sharp-shinned hawks take them, and blue jays apparently often prey on them, especially on the young (Allan 1947; Downing and Baldwin 1961; Elwell 1962; Hoffmeister and Downes 1964).

Relationship to People: In some areas, red bats are common enough to play a major role as insect predators. In Minnesota, they are apparently too rare to be of much significance.

Additional References: Constantine 1958; Davis and Lidicker 1956; Heaney and Birney 1975; Kunz 1973; LaVal and LaVal 1979; Mackiewicz and Backus 1956; Myers 1978; Quay and Miller 1955; Terres 1956.

Lasiurus cinereus
Hoary Bat

Measurements (seven specimens): Total length 138 (130–147); tail 57 (48–65); hind foot 11 (10–13); ear from notch 18 (17–19); tragus 4–8; forearm 52–55; weight 22.3–39.9.

Description: This bat is Minnesota's largest, and I think its most beautiful. The individual hairs in the pelage of most mammals are banded with a pattern of light, dark, or variously colored bands. (This is true even of many species that seem to have a rather uniform coat color.) The hair of a hoary bat usually has four such bands. Starting at the base it is dark brown, then yellowish, dark brown, and white. The white tips tend to lie in short broken rows, giving the bat a distinctive frosted appearance. The only other Minnesota bat faintly resembling it is the much smaller silver-haired bat.

As in the red bat, the ears are short and rounded, the tail long, and the interfemoral membrane fully furred above. Weights between 20 and 35 have been reported (Brisbin 1966), but they vary, depending on the time of year and the amount of fat stored prior to hibernation. A 39.9 g female taken in September near Bemidji, Beltrami Co. (BSCVC 3696), exceeded previously reported weights by nearly 5 g.

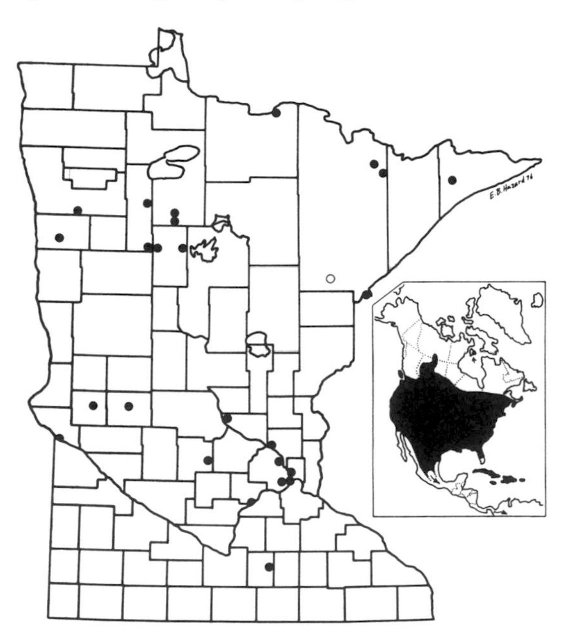

Map 18. Distribution of *Lasiurus cinereus*. ● = township specimens. ○ = other township record. (Map produced by the Department of Biology, Bemidji State University.)

Range and Habitat: Minnesota's largest bat is also its widest ranging. *Lasiurus cinereus* ranges north of tree line in parts of Canada, south throughout the contiguous states except for southern Florida, through Mexico and Central America, and into South America to Argentina and Chile. It is also native to Hawaii and the West Indies, and has been recorded in Iceland and Bermuda, but may not be resident there. *L. cinereus* is known to be strongly migratory, and waves of them move through certain areas, but records of banded migratory individuals are essentially lacking. Hoary bats are probably summer residents throughout Minnesota.

Natural History: Lasiurus cinereus begins to hunt later in the evening than do most bats, and so is seldom seen. It is a tree bat and generally solitary. Ranging as widely as it does, it will often actually be far from trees. It is known to roost in caves, rock crevices, and squirrel nests. Individuals that roost too deeply in caves apparently do not find their way out, and die there (Myers 1960). In Minnesota, as in other midwestern states, an adult hoary bat is likely to be a female (Constantine 1966; Provost and Kirkpatrick 1952). Most of the males that migrate north apparently concentrate in the Far West (Barbour and Davis 1969; Vaughan 1953). Females generally roost individually and bring forth their young (almost always two) between late May and early July. Development is apparently slower than in the red bat, and fall migration is later (Constantine 1966). The hoary bat is a strong flyer, known to attain a speed of over 21 km/hr (Hayward and Davis 1964).

The primary food of the hoary bat is insects. It hunts in open areas near trees where another large bat, the big brown, also hunts. There is evidence that the two have distinctive food preferences, the hoary concentrating on moths, the big brown concentrating on beetles and other insects (Black 1972). There is also evidence that the hoary feeds on smaller species of the family Vespertilionidae (Bishop 1947; Orr 1950), a predatory relationship paralleled in Minnesota by large and small species in the families Cricetidae, Canidae, Mustelidae, and probably others.

Figure 13. *Lasiurus cinereus*, hoary bat.

Relationship to People: People seldom knowingly come in contact with *L. cinereus*, although the species may not be as rare as it seems. As a predator on insects, it should be presumed to be economically beneficial.

Additional References: Beer 1954; Bogan 1972; Cockrum and Cross 1964; Constantine 1959; Findley and Jones 1964; Goehring 1955; Heaney and Birney 1975; Kunz 1971, 1973; McClure 1942; Munyer 1967; Whitaker 1972a.

ORDER LAGOMORPHA

Rabbits, Hares, and Pikas

The order Lagomorpha is relatively small, including only two extant families, the Ochotonidae (pikas) and Leporidae (rabbits and hares). Only the latter occur in Minnesota. At one time, biologists classified lagomorphs as a suborder of the order Rodentia. The consensus today is that most resemblances between the two groups, other than those typical of placental mammals in general, are the result of convergent evolution (McKenna 1969).

Many features, especially of the skull and skeleton, together distinguish the lagomorphs from other mammals. Only a few are given here. Lagomorphs have evergrowing incisors, separated from the cheek teeth in each jaw by a long space, or diastema. There are no canines. Lagomorphs share these characters with the rodents and with some members of other orders. The lagomorphs, however, have a second pair of small peglike upper incisors directly behind the first pair. The distance between the right and left upper cheek tooth rows is greater than the distance between the right and left lower cheek tooth rows. This permits occlusion of upper and lower teeth on only one side at a time. There is extensive fenestration of the facial region of the maxilla.

Additional References: Hall 1951a; Layne 1967.

Family Leporidae

Rabbits and Hares

The hares and rabbits constitute by far the larger of the two families in the order. They are characterized by relatively long hind limbs, furred soles, relatively long ears (always longer than wide), and short fluffy tails. In most species females average a bit larger than males. The supraorbital process is always prominent. The tooth formula of all leporids in Minnesota is 2/l, 0/0, 3/2, 3/3 × 2 = 28. The incisors are evergrowing and the front surface of the first upper incisors is grooved. The cheek teeth are rootless, high crowned, and lophodont. The premolars are molariform.

Newborn rabbits are altricial—born naked or nearly so, relatively helpless, and with the eyes shut. This condition is typical of many mammals: shrews, bats, moles, mice, dogs, and others. Young hares are precocial—born well furred, with the eyes open, and able to get about with a fair degree of coordination. This condition is less common, occurring in hares, whales (except for the fur of course), and hoofed mammals, among others.

Leporids are widespread, occurring naturally on all major land areas except southern South America, the Australian Region, New Zealand, and many islands of the East Indian Archipelago. Unfortunately, people have introduced various species outside their natural range. Two genera occur in Minnesota. One, *Sylvilagus*, occurs only in the New World. The other, *Lepus*, occurs in North America, Eurasia, and Africa.

A phenomenon common to many (perhaps all) leporids is reingestion (also referred to as refection, coprophagy, or caecotrophy) of soft fecal pellets (Hornicke and Batsch 1977). The rabbit droppings, or scats, encountered in the field are rather hard. However, rabbits also produce soft pellets, and these are eaten, usually at night, as they pass out of the anus. They are formed in the caecum where bacterial action has made available amino acids, vitamins, and monosaccharides which the rabbit assimilates when processing the material a second time. Reingestion has also been observed in some rodents and is probably a relatively common way for many mammals to benefit from the digestive and chemosynthetic abilities of their intestinal bacteria.

References: Hornicke and Batsch 1977; Layne 1967.

Genus *Sylvilagus*

Sylvilagus floridanus
Eastern Cottontail

Measurements: Total length 434 (405–478); tail 49 (33–65); hind foot 101 (92–109); ear from notch 60 (55–64); weight (eight specimens) 1,413.0 (1,247.4–1,701.0).

Description: The dorsal pelage of the cottontail is grayish brown; its grizzled appearance is due to black and gray tips on the hair. The sides are generally grayer than the back, and the rump is gray. The underside is whitish, and the nape of the neck is distinctly reddish. The underside of the short fluffy tail is white. The conspicuous white tuft of a tail, which has given the animal its name, is held erect when the animal runs. Summer and winter pelages are superficially much alike.

The skull may be distinguished from those of our other native lagomorphs by the fusion of the posterior ends of the supraorbital processes with the frontal bone, and by the separate, unfused, interparietal. The eyes of cottontails (and other rabbits) are set far to the sides of the skull. There is little overlap of the left and right visual fields, but the two bulging eyes together encompass almost a full 360° circle.

Range and Habitat: The eastern cottontail is native to southernmost Saskatchewan, Manitoba, Ontario, and Quebec, most of the central and eastern United States, much of Mexico, and parts of Central America. There is a small introduced population in Oregon and British Columbia. The eastern cottontail is absent from northernmost New England, and apparently also from northernmost Minnesota. It prefers partially open brushy areas, wooded swamps, woodland edges, and wooded fencerows, avoiding both prairie and closed forest. *Sylvilagus floridanus* is not common in intensively managed agricultural land but does well in residential areas where there is adequate cover.

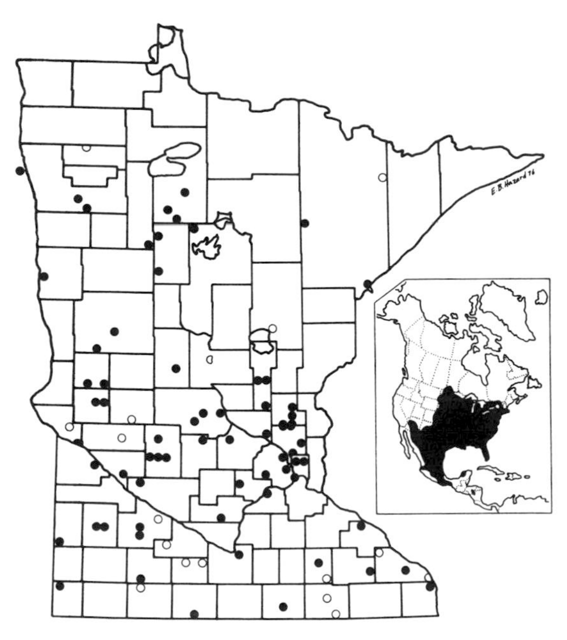

Map 19. Distribution of *Sylvilagus floridanus*. ● = township specimens. ○ = other township records, selected locations. (Map produced by the Department of Biology, Bemidji State University.)

Natural History: The eastern cottontail is essentially solitary and is largely crepuscular (Mech et al. 1966). It is also active at night. Adults associate only for breeding purposes. In Minnesota, females initially breed in late February or in March, and bring forth their first litter about a month later (Conaway et al. 1974). There may be seven or more litters per year in warmer states,

Figure 14. *Sylvilagus floridanus*, eastern cottontail.

but in Minnesota the upper limit is probably four. Four to eight or more young may make up a litter, four to six being common. In some areas, the female population is apparently in reproductive synchrony; litters are born to various females within periods of a few days at approximately two-week intervals. This synchrony is variously reported as lasting through the entire season or as breaking down after the first couple of litters (Conaway and Wight 1962; Conaway et al. 1963; Marsden and Conaway 1963; Pelton and Provost 1972). The naked, blind young are most often housed in a fur-lined nest in a "form," a slight depression in the ground, usually under shrubby cover (Ecke 1955). Their eyes open at about a week. The young are weaned in less than a month, and disperse soon after. Females are sexually mature and often breed at three months of age.

Sylvilagus floridanus is primarily a herbivore throughout the year, feeding on a variety of green vegetation in the summer and on twigs and young bark in winter. Rose (1973) calculated that, in the center of the species's range, in Ohio, an adult rabbit utilizes less than 1 percent of the total primary production of a hectare of old field succession vegetation. Individual cottontails range over areas from 1 to 4 ha; males generally range more widely than females (Trent and Rongstad 1974). Females tend to be territorial and to have mutually exclusive ranges, especially in late summer. Within the home range, the center of activity is generally near cover (Janes 1959).

Relationship to People: The cottontail is a popular game animal, especially in southern Minnesota. In the past, its fur had enough commercial value to make trapping profitable. Because of the cottontail's status as small game, sportsmen's groups have introduced rabbits from one part of the species's range into another (Chapman et al. 1977; Chapman and Morgan 1973). This probably has seldom improved hunting, but it has confused the study of natural variation in geographic subgroups of *S. floridanus*. The cottontail is not popular with gardeners and orchardists because it has a fondness for vegetable crops and for the bark of young fruit trees. It is the most serious vertebrate pest in at least one northern Minnesota garden.

Rabbits are prone to a variety of diseases, at least two of which, tularemia and Rocky Mountain spotted fever, are transmissible to man (Davis et al. 1970; Jacobson et al. 1978). Tularemia occurs in Minnesota (Loken 1980). It is unwise to bag a rabbit that is not lively, and it is safer to handle cottontails, as well as other game, with rubber gloves, especially if the skin of your hands is scratched or broken. Fortunately, tularemia is generally fatal to rabbits, and a live rabbit shot after cold weather has set in is unlikely to be infected. *S. floridanus* also harbors myxomatosis, a viral disease which usually does cottontails little harm unless they are weakened from other causes. Although native American leporids have evolved a high degree of resistance to myxomatosis, European species generally have not (except for a few populations in recent years). Introduced myxomatosis has been highly effective in reducing populations of *Oryctolagus cuniculus*, the Old World rabbit, in Great Britain and Australia. The various domestic breeds of rabbits are all derived from wild *Oryctolagus*, so anyone who keeps domestic rabbits as pets or for food should discourage contact between these highly susceptible animals and wild cottontails.

Additional References: Barkalow 1962; Burgdorfer et al. 1974; Casteel 1966; Chapman et al. 1980; Eberhardt et al. 1963; Edwards and Eberhardt 1967; Errington et al. 1940; Haugen 1942; Lord 1958, 1963; Marsden and Holler 1964; Rongstad 1966, 1969; Rose 1977; Schwartz 1942; Warren and Kirkpatrick 1978; Wight and Conaway 1961, 1962.

Genus *Lepus*
Hares

Lepus americanus
Snowshoe Hare, Varying Hare

Measurements: Total length 443 (370–454); tail 37 (24–55); hind foot 133 (121–141); ear from notch 71 (62–96); weight (seven specimens) 1,613.0 (1,275.8–2,115.0).

Description: Snowshoe hares average a bit heavier than cottontails and have markedly longer legs and larger feet. In summer, the dorsal pelage is a yellowish to grayish brown, grizzled but more uniform in appearance than that of cottontails. The underside is white. The nape is not colored differently from the rest of the back. The tip of the ear is black on its posterior surface. The tail is dark above and white below. In autumn, the dark outer fur of summer is replaced by longer, silky white hairs, which hide the gray underfur. Except for black ear tips and dark eye rings, *Lepus americanus* is totally white in winter over most of its range. The skull of an adult is smaller than that of a jackrabbit and about the same size as a cottontail skull. The supraorbitals are not fused posteriorly with the frontals but flare out posteriorly; the interparietal is fused to the parietals.

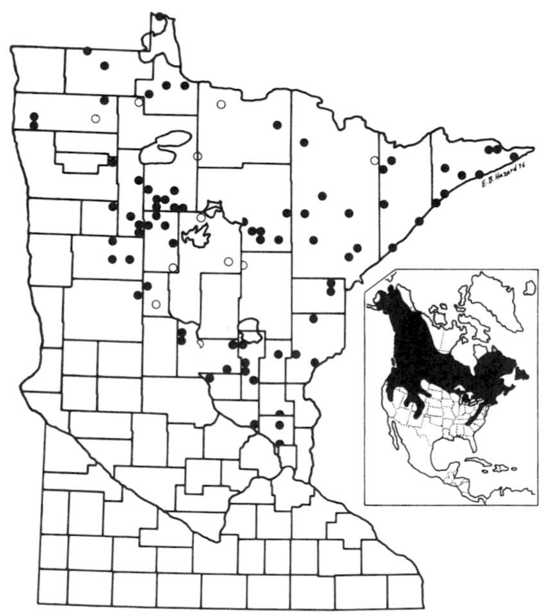

Map 20. Distribution of *Lepus americanus*. ● = township specimens. ○ = other township records, selected locations. (Map produced by the Department of Biology, Bemidji State University.)

Range and Habitat: The snowshoe hare ranges across Canada and Alaska, approximately to tree line. It also occurs in the northern contiguous states, and ranges south to northern California and New Mexico in the Sierras and Rockies, and to Tennessee in the Appalachians. It is found in northern and eastern Minnesota, and may range south to the Minnesota River. A skull in the Bemidji State University collection was found in Renville County, but, without further evidence, I do not consider this within the normal range of the species. The snowshoe hare generally lives in woods, thickets, and coniferous swamps and bogs. It is less likely to be seen in open country than the cottontail.

Natural History: Lepus americanus is generally solitary, though groups of adults have been seen together on occasion, especially when populations are high. The species is largely crepuscular and nocturnal, but may be abroad in the daytime on cloudy days. Snowshoes feed on a variety of forbs, grasses, and shrubs in the summer, and on the bark and twigs of deciduous woody plants and the foliage of various conifers in winter. They normally cannot reach food more than about a meter from the ground, but, as the snowpack accumulates, they can browse higher than 2 m above the ground. They eat many of the plants eaten by white-tailed deer, and the two species may be in competition for food, especially if the hare population is high (Telfer 1972a, 1974). Often, hares take more browse per unit area than do deer (Telfer 1972b).

Breeding begins in mid-March and continues into July or August. There are three or four litters per year, most often of three or four young each (Davis and Meyer 1973a; Dodds 1965; Keith et al. 1966; Keith and Windberg 1978). Gestation lasts 35 days. Young hares, unlike cottontails, are precocial, having a fur coat and open eyes at birth. Females with young are intolerant of other females. At least at some population densities, females tend to have mutually exclusive home ranges averaging under 3 ha. Male ranges overlap and average over 7 ha (Mech et al. 1966).

Population densities vary greatly; presumably territorial behavior intensifies when density is high. It has long been known that population densities of *Lepus americanus* fluctuate every six to 12 years, nine or 10 years being most common. Some authors report that cycles are independent from population to population, but others maintain that there is considerable synchrony of cycles over wide areas (Cox 1936; Davis and Meyer 1973a; Dodds 1965; Keith 1966; Keith et al. 1966; Keith and Windberg 1978). Changes in density are most pronounced in northern Canada, where high densities may reach 1300 hares per km^2, with lows of less than one per km^2. The extremes in Minnesota may be closer to those reported by Keith (1966) near Rochester, Alberta: 390 per km^2 and 22 per km^2, a ratio of only about 18.1. The decline from high to low density is usually more rapid than the subsequent increase.

The causes of these fluctuations in population density are only partly understood, but at least four factors are involved: a decrease in adult survival, a reduction in juvenile survival from a high of about 24 percent to a low of 3 percent, a 50 percent decrease in the number of young per litter, and a decrease in the number of females bearing four litters per summer rather than three. In part, these changes may result from hormonal or other physiologic responses to increased social encounters during high density periods. However, external factors are probably also involved (Keith and Windberg 1978). At peak densities in Alberta, hares apparently deplete their winter browse supply. This results in lower winter survival, especially among the young. The extent of the decline is aggravated as a result of heavy predation by relatively high predator populations that had previously thrived on dense hare populations. Subsequent decline of predator populations, accompanied by recovery of the vegetation, allows a gradual buildup of the hare population.

The availability of hares to predators such as lynx, bobcats, and barred, great gray, and great horned owls may greatly affect predator populations (McInvaille and Keith 1974; Nellis et al. 1972; Rusch et al. 1972). During a sevenfold increase in hare density in Alberta in a 100 km^2 area, the number of adult great horned owls increased from 10 to 18; hares increased in the diet of great horned owls from 23 percent to 50 percent and the percentage of owls nesting increased from 20 to 100 (Rusch et al. 1972). Apparently, snowshoes also fall prey to domestic cats on occasion (Doucet 1973).

Relationship to People: Lepus americanus was historically an important source of food and fur to Indians and settlers in the north woods. Today the fur is of only modest value, and few people in Minnesota rely on the hare for food. It is still a significant game animal, though for far fewer hunters than is the cottontail. It shares the cottontail's interest in garden crops and can be a problem for gardeners in rural northern Minnesota.

Additional References: Adamcik and Keith 1978; Adamcik et al. 1978, 1979; Aldous 1936, 1937; Bider 1961; Cary and Keith 1978; Davis and Meyer 1973b; Dolbeer and Clark 1975; Ferguson and Merriam 1978; Fox 1978; Green and Evans 1940; Heaney and Birney 1975; Holter et al. 1974; Keith et al. 1968; Keith and Surrendi 1971; Meslow and Keith 1968; Nellis and Keith 1968; Newson and de Vos 1964; Pease et al. 1979; Rongstad and Tester 1971; Rowan and Keith 1956; Rusch et al. 1978; Sherburne and Dimond 1969; Walski and Mautz 1977; Weinstein 1977; Windberg and Keith 1976a, b, 1978; Wolff 1978.

Lepus townsendii
White-Tailed Jackrabbit

Measurements (based on 104 specimens shot in December 1936 for the market, in Blue Earth Co., reported by Mohr 1943): Total length 625 (574–673); tail 91(64–108); hind foot 147(140–152)—I omit one 95 mm hind foot measurement as a probable misprint; ear from notch 104 (99–114); weight 3,357 (2,608–4,082).

Description: An adult white-tailed jackrabbit is too big to be mistaken for any other native rabbit. The minimum measurements all exceed the maxima for the snowshoe hare. In summer, jackrabbits are a brownish gray above and white or pale gray below. The tail is white below, and white or nearly white above. There is a conspicuous black patch on the back of the ear, on the outer half of the ear tip. In Minnesota, all jackrabbits are white in winter, except for the black patch on the ear.

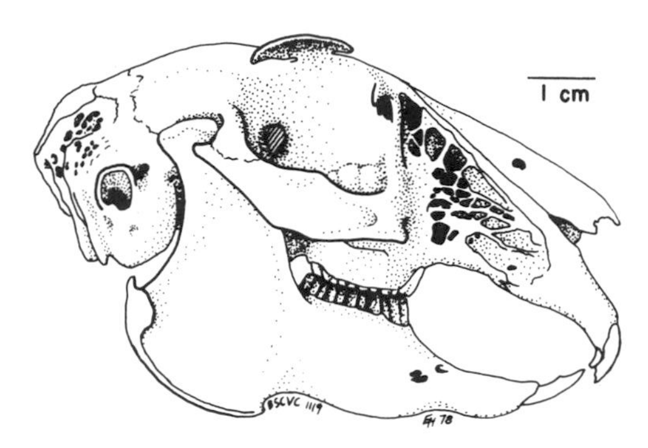

Figure 15. Skull of *Lepus townsendii*, white-tailed jackrabbit, BSCVC 1119.

Range and Habitat: Lepus townsendii is primarily a resident of the northern great plains and

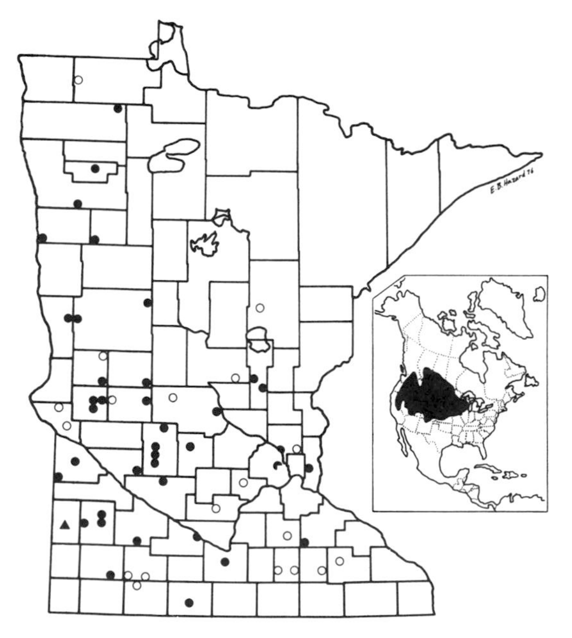

Map 21. Distribution of *Lepus townsendii*. ● = township specimens. ○ = other township records, selected locations. ▲ = county specimen. (Map produced by the Department of Biology, Bemidji State University.)

Figure 16. *Lepus townsendii*, white-tailed jackrabbit.

parts of the western mountain ranges, occurring from the southern Prairie Provinces to Utah and New Mexico, and from eastern Oregon and eastern California to Iowa, Wisconsin, and Minnesota. The species's widespread occurrence in Wisconsin is a result of natural immigration from the West plus many deliberate introductions (Jackson 1961). The white-tailed jackrabbit is absent from approximately the northeast quarter of Minnesota. It is a denizen of open country and rarely encountered in brushy areas.

Natural History: Jackrabbits are hares, if we accept the classical distinction between hares and rabbits (see p. 43). The female white-tailed jackrabbit prepares no real nest, giving birth to her litter, typically of three to five young (James and Seabloom 1969; Kline 1963), in a slight depression in the ground, probably a natural one. The gestation period is 30 days or more, and the reproductive cycles of females in a population are apparently synchronized, with four peak periods of birth in a season, at least in North Dakota (James and Seabloom 1969). The first litter in Minnesota probably arrives about the first week in April. I have not seen any accounts of females from early litters breeding in their first season. Although the young presumably are generally on their own in a few weeks, Sparks (1968) reports young to up three months old with milk in their stomachs.

Lepus townsendii feeds on a variety of grasses and on forbs such as clover and alfalfa in the summer, and on buds, twigs, and bark in winter. Jackrabbits are both crepuscular and nocturnal. In the daytime, they may sit in the shade of a shrub or simply lie low in tall grass, relying on speed rather than physical cover if flushed by a predator. Like most leporids, jackrabbits may crouch quietly until almost stepped on before bolting. Among the important predators of white-tailed jackrabbits are coyotes, bobcats, and great-horned owls (Gashwiler et al. 1960; Long and Kerfoot 1963). Farther west, golden eagles prey extensively on them (Engel and Vaughan 1966). The population density of jackrabbits varies cyclically and significantly influences the abundance of some predators.

Relationship to People: Jackrabbit fur is hardly worth enough to justify trapping, but jackrabbits provide sport and meat in parts of Minnesota. When abundant, they can cause damage to crops. However, they also help support predators which, if left to themselves, might do a more effective job of controlling potentially harmful rabbits and rodents than man does. By clearing the land and by deliberate introductions, people have extended the jackrabbit's range farther east. On the other hand, by intensive cultivation of the prairie and by urban "development," they have destroyed much of the jackrabbit's grassland habitat.

Additional References: Flinders and Hansen 1973; Grayson 1977; Hansen and Bear 1963; Leraas 1942; Mohr and Mohr 1936; Soper 1946; Watkins and Nowak 1973; Wrigley 1974.

ORDER RODENTIA
Rodents

Rodents make up the largest order of mammals, constituting about 40 percent of all living species. They are also the most widespread wild terrestrial mammals, occurring naturally on all the continents except Antarctica, and on many oceanic islands. They are the only terrestrial placental mammals that have successfully colonized Australia without the aid of people. Members of one family (rats and mice of the Old World family Muridae) also have followed people wherever they have gone, except to the moon and perhaps Antarctica.

Rodents are easily distinguished from all other North American mammals. The upper and lower jaws each bear a single pair of evergrowing incisors. There are no second or third incisors, or canines. The rostrum is always elongate, and there is a large diastema between the incisors and the cheek teeth. The maximum tooth formula is 1/1, 0/0, 2/1, 3/3 × 2 = 22, but many rodents have even fewer teeth. All North American rodents have at least three molars in each upper and lower jaw. The order has other distinguishing characteristics, including modifications of the jaw musculature associated with the specialized dentition and skull structure just described.

References: Anderson 1967; Dunmire 1955; Hart 1971; Lavocat 1974; Mascarello et al. 1974; Vaughan 1978; Wood 1955, 1959, 1965, 1974, 1975.

Family Sciuridae
Squirrels

Squirrels are neither unusually large nor unusually numerous. Nevertheless, the family Sciuridae includes our most familiar wild mammals, because many species not only tolerate people nearby but are also diurnal. Daytime activity puts squirrels in an unusual position relative to predators, one with both advantages and disadvantages. Diurnal squirrels are less likely to be caught by owls and by primarily nocturnal mammalian predators, but are more likely to be taken by hawks. However, most hawks migrate out of the north country in winter whereas most owls do not, so during winter in Minnesota, diurnal squirrels benefit from the absence of a significant group of predators.

Squirrels occupy a variety of habitats and ecologic niches, and their dentition is adapted to a relatively generalized diet. They always have one lower premolar and one or two upper premolars. The eyes are never reduced in size and are quite large in the two nocturnal flying squirrels.

The family takes its name from the genus *Sciurus*, the Latin word for squirrel. Mammalogists often refer to all members of the Sciuridae as squirrels, although "squirrel" is not part of the common name of many species, such as woodchucks and chipmunks. Some of the ground squirrels are commonly referred to as gophers, a name more properly restricted to pocket gophers. And pocket gophers are called "salamanders" in the Deep South, where "gopher" refers to a particular tortoise. (It is easy to see why scientists prefer universally accepted Latin names over names that vary not only between countries but between regions within a single country.)

References: Black 1963, 1972; Bryant 1945; Layne 1954b; McLaughlin 1967; Moore 1959b, 1961; Muchlinski and Shump 1979.

Genus *Tamias*

Tamias striatus
Eastern Chipmunk

Measurements: Total length 260 (250–272); tail 101 (93–106); hind foot 36 (33–38); ear from notch 18 (16–18); weight 114.7 (100.3–130.2).

Description: There are two chipmunks in Minnesota: the eastern chipmunk and the least chipmunk. Each has five dark brown stripes on the back and sides, separated by four light ones. In the eastern chipmunk, the larger of the two species, the body stripes do not extend to the base of the tail, and the rump is a solid brownish or reddish yellow. The stripes extend only indistinctly onto the face. The eastern chipmunk has only one upper premolar on each side, whereas the least chipmunk has two. The tooth formula of *Tamias* is 1/1, 0/0, 1/1, 3/3 × 2 = 20.

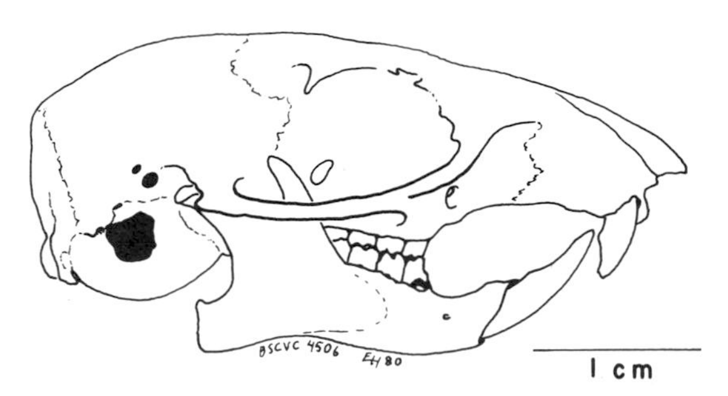

Figure 17. Skull of *Tamias striatus*, eastern chipmunk, BSCVC 4506.

Range and Habitat: Tamias striatus is the only member of its genus and occurs in the eastern half of the United States (except for the southern coastal plains) and in the southern parts of eastern Canada. An introduced population thrives in Newfoundland. Basically a woodland and brushland species, the eastern chipmunk is seldom found in wetlands or grasslands (Forbes 1966b, c; Iverson et al. 1967) but is prevalent in and around homes and outbuildings, wherever structures or landscaping provide shelter. It occurs in suitable habitat throughout Minnesota, except perhaps in the southwesternmost corner of the state (Ernst and Ernst 1972; Ernst and French 1977).

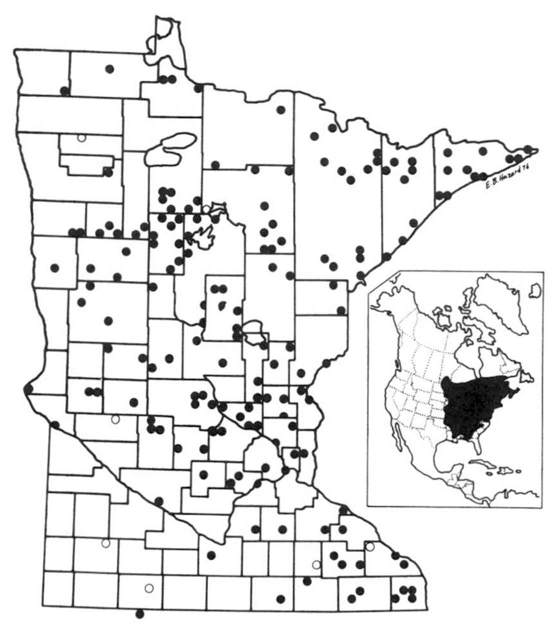

Map 22. Distribution of *Tamias striatus*. ● = township specimens, selected locations. ○ = other township records, selected locations. (Map produced by the Department of Biology, Bemidji State University.)

Natural History: Eastern chipmunks are common, conspicuous, and tolerant of human observers. Consequently, their habits are relatively well known. They are essentially woodland ground squirrels, and, like many other ground squirrels, they hibernate in the strict sense of the word. That is, during a period of the year, they go into a relatively deep sleep, and their body temperature decreases substantially. The period of hibernation varies with geography, age, and sex, and also from individual to individual (Brenner and Lyle 1974, 1975; Panuska 1959; Scott and Fisher 1972). In Minnesota, *Tamias* will usually be out of sight from some time in October to some time in March (Davis and Beer 1959). Males generally emerge from hibernation earlier in the spring than females. Chipmunks are relatively light hibernators, often emerging during warm spells in the winter.

Yahner and Svendsen (1978) report that, in southeast Ohio, there are two restricted and relatively constant mating periods each year, the first in mid-February to early March, the second in July. The animals emerge from hibernation in mid-February, mate, and then hibernate for a few weeks before resuming activity in the spring. The first litter is born in early April, the second in late August. Most females bear two litters. In Canada (Pidduck and Falls 1973; Smith and Smith 1975), the first breeding season is delayed if the spring climate is unfavorable; in such years there is usually no second breeding season. Chipmunks are the only hibernating sciurids that can breed twice a year (Yahner and Svendsen 1978). The breeding schedule of *Tamias* at various latitudes in Minnesota should be investigated.

As with most rodents and other small mammals, males take no part in raising the young. The young are born in a nest chamber in a relatively complex burrow which typically also has food storage chambers and two or more entrances. Simpler refuge burrows may also be dug within the home range of an adult chipmunk.

Eastern chipmunks are a delight to watch, and we tend to regard them as friendly creatures. Actually, *Tamias* is one of the more aggressive rodents (Wolfe 1966, 1969). An adult eastern chipmunk treats an area round its burrow as a territory, defending it against other adults of its species (Burt 1940; Dunford 1970; Forsyth and Smith 1973; Yahner 1978b). Chipmunks will fight vigorously for their territorial rights if necessary, and chases are common when populations are dense. However, chipmunks, like many animals, employ specific signals to announce their territorial boundaries and thus generally avoid combat. In the eastern chipmunk, the commonest signal is a relatively low bark, a single note repeated again and again. A typical summer day in a well-populated woods ends with a barking chorus which may last an hour or more; each animal sits near the entrance to its burrow both giving and receiving information about the social situation. This chorus may end quite abruptly, usually well before sunset, and then the chipmunks are gone.

Unlike tree squirrels, especially gray squir-

Figure 18. *Tamias striatus*, eastern chipmunk.

rels, which may occupy the same woods, eastern chipmunks seldom forage at dawn and sundown. They wait until the sun is well up and forage primarily in mid-morning, often with a decrease in average activity just after noon. In so doing, they may avoid interspecific conflict. When quarrels do occur, I have sometimes seen chipmunks aggressively chase the larger tree squirrels away (Hazard 1960).

Tamias, like other woodland squirrels, has a varied diet. The bulk of the food usually comprises various buds, seeds, and fruits in season. Typical foods are elm and maple samaras, hazelnuts, acorns, pine seeds, and black cherries. Eastern chipmunks also eat mushrooms of various sorts and are quite capable of obtaining animal food, both invertebrate and vertebrate. They are known to take small bullfrogs, red-bellied snakes, robins, juncos, house sparrows, starlings, and meadow voles. In Michigan, I saw an eastern chipmunk take a nestling cardinal despite vigorous harassment by the parents. Although chipmunks are certainly not as efficient predators as weasels (which are about the same size), they are several times more numerous and may well be equally significant predators on populations of various animals.

Chipmunks have internal cheek pouches which permit them to carry a large amount of food at once. It is common to see *Tamias* headed for its burrow with several acorns stuffed into each bulging cheek and another held between its incisors. Much of the relatively nonperishable food that chipmunks garner is stored in their burrows, largely for winter use. Although they are true hibernators, they do not put on as much prehibernation fat as do ground squirrels and woodchucks. They awaken frequently and feed on their underground larder intermittently through the winter. However, even nuts and other seeds eventually spoil, and *Tamias* must spend some time cleaning house. Many observers have noticed a lull in aboveground chipmunk activity in late summer; Wrazen (1980) suggests that this may be a period of burrow maintenance.

Relationship to People: Chipmunks are a source of much enjoyment, though people often have an unrealistically benign image of the personalities of these aggressive, often antisocial rodents. Eastern chipmunks are sometimes pests on farms, in gardens, and in food storage areas. They may, however, help keep harmful insects in check. Some people dislike chipmunks and other squirrels because they kill wild birds, but to me, this is an unrealistic and unreasonable view of nature. Predation is a universal and necessary part of nature, and, to enjoy wild things *as wild things*, one must accept the ways in which natural communities operate.

Additional References: Allen 1938; Blair 1942; Brenner 1973; De Coursey 1972; Dunford 1972; Elliot 1978; Forbes 1967; Kirkland and Griffin 1974; Kirkland and Kirkland 1979; Lang 1978; Nadler et al. 1977; Neumann 1967; Shackelford 1966; Smith and Smith 1972; Svendsen and Yahner 1978; Tryson and Snyder 1973; White 1953; Wrazen and Svendsen 1978; Yahner 1975, 1977, 1978a, c, d; Yerger 1955.

Genus *Eutamias*

Eutamias minimus
Least Chipmunk

Measurements: Total length 211 (202–220); tail 99 (94–107); hind foot 32 (30–33); ear from notch 16 (14–18); weight 44.8 (38.3–52.2).

Description: The least chipmunk is half the weight of the eastern chipmunk. It does not have a reddish rump, the stripes extend to the base of the tail, and the stripes on the face are more strongly marked than those of its larger relative. It has five upper cheek teeth (two premolars and three molars) rather than four. The tooth formula is 1/1, 0/0, 2/1, 3/3 × 2 = 22. Least chipmunks run with the tail held vertically; eastern chipmunks generally extend it straight back.

Range and Habitat: This is the most widespread North American chipmunk. It ranges from the upper Great Lakes region northwest to the Yukon and south into the mountains of the western

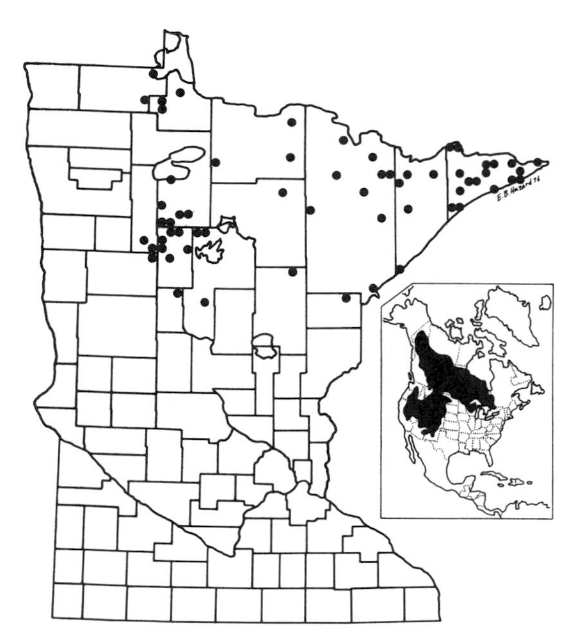

Map 23. Distribution of *Eutamias minimus*. ● = township specimens, selected locations. (Map produced by the Department of Biology, Bemidji State University.)

United States. There are many other species of the genus *Eutamias* in western North America, and others in Eurasia. (Some mammalogists consider *Eutamias* a subgenus of *Tamias*.)

Eutamias minimus is common in the boreal forest region of north central and northeastern Minnesota, often occurring in the same general areas as *Tamias*. The least chipmunk, however, is more partial to conifers, and particularly to open, dry, or disturbed areas within the coniferous forest biome (Forbes 1966c). It is less likely to be found under dense shrub cover or in mature closed forest than is the eastern chipmunk.

Natural History: The habits of least chipmunks, so far as we know, differ little from those of eastern chipmunks. Although least chipmunks are solitary, they are reported to be less aggressively territorial than eastern chipmunks (Jackson 1961). Least chipmunks are diurnal, omnivorous, and primarily terrestrial. Both species will climb trees on occasion, and both have been known to nest above ground (Broadbooks 1974; Genoways and Jones 1972). Broadbooks (1977) suggests that *Eutamias* females typically move their litters from a ground nest to a tree nest before the young disperse. *Eutamias minimus* feeds on a variety of plant and animal matter but is particularly fond of various seeds and nuts, and of insects and their eggs and larvae (Criddle 1943; Vaughan 1974). It doubtless feeds on some birds and their eggs. Like *Tamias*, it hibernates next to a substantial underground store of seeds which it has transported in its cheek pouches (Forbes 1966c).

The least chipmunk has a gestation period of 28-30 days. In Minnesota, the litter of four to seven young probably arrives most often from mid-May to early June (Forbes 1966a). Second litters may be less common than they are in eastern chipmunks; the chances of producing a second litter are greatest when the first litter is lost shortly after birth (Skryja 1974).

Relationship to People: In Minnesota, least chipmunks occur less commonly in association with people than do eastern chipmunks. They may occasionally damage garden crops or stored food but are likely to compensate for this by feeding on pests such as cutworms.

Additional References: Aldous 1941; Beuch, Siderits, et al. 1977; Ellis and Maxson 1979; Guilday and Hamilton 1978; Iverson et al. 1967; C. E. Johnson 1922; Manville 1949; Martinsen 1968; Meredith 1977; Nadler et al. 1977; Peterson 1966; Sheppard 1968, 1972a; Sutton and Nadler 1969; Wydeven and Wydeven 1976.

Genus *Marmota*
Marmots

Marmots are widespread in Eurasia and North America. Most species inhabit open country at high altitudes in mountainous regions. The woodchuck, as a denizen of wooded and brushy habitats at relatively low altitudes, is unique within the genus.

Marmota monax
Woodchuck, Ground Hog

Measurements: Total length 536 (501–594); tail 117 (95–140); hind foot 78 (70–85); ear from notch 27 (19–30); weight 1,985–4,167.

Description: The woodchuck is Minnesota's largest squirrel. It is stocky, with short ears, short legs, and a medium-length, somewhat bushy tail. There is much individual variation in pelage, but most woodchucks have a salt-and-pepper appearance on the top and sides, resulting from guard hairs with alternate light and dark bands overlying yellowish-tipped underfur. The top of the head is dark, and the underside and upper parts of the legs are reddish brown. The feet are black. Occasional individuals are totally black. The skull is distinguished by a flattened or depressed area between the orbits. In contrast to other squirrels, the enamel on the front of the upper incisors has little or no pigment. The tooth formula is 1/1, 0/0, 2/1, 3/3 × 2 = 22.

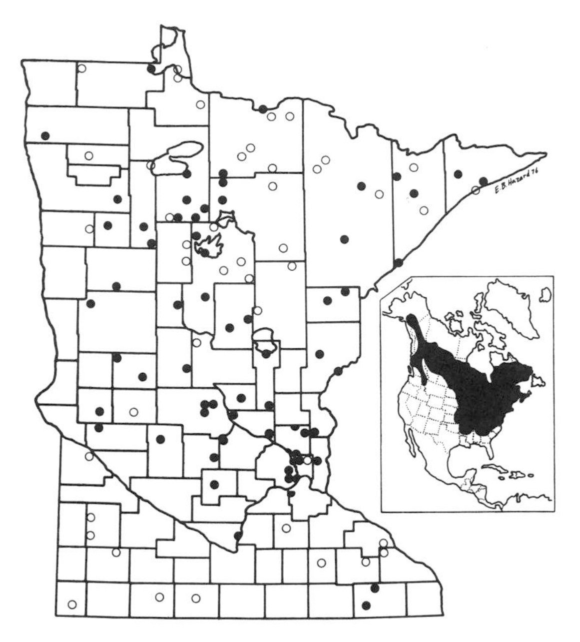

Map 24. Distribution of *Marmota monax*. ● = township specimens. ○ = other township records, selected locations. (Map produced by the Department of Biology, Bemidji State University.)

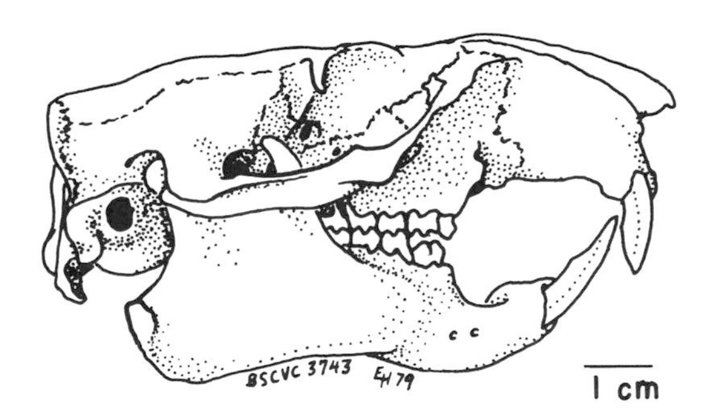

Figure 19. Skull of *Marmota monax*, woodchuck, BSCVC 3743.

Range and Habitat: Marmota monax ranges from the eastern seaboard of Canada and the United States (except for the Deep South) west and northwest within forested regions into the Yukon. It is primarily a mammal of woodlands, both coniferous and deciduous, though more characteristic of open woodlands, clearings, and edges than of deep unbroken forest. The woodchuck became more abundant in the deciduous and coniferous forest biomes after clearing and settling. Woodchucks do so well in some farming areas that they are often thought of as open country mammals, but they are seldom found far from woods or brush and are not characteristic of unbroken prairie.

Natural History: Marmota monax is relatively solitary; all other members of the genus are colonial (Barash 1974b; Bronson 1963, 1964; Merriam 1971). Each adult has an extensive burrow with one main entrance and one or more plunge holes; the latter are usually well hidden in brush (Grizzell 1955). Summer dens may be in the woods or in adjacent fields, and are most often on well-drained slopes (Doucet et al. 1974; Merriam 1966, 1971). In Minnesota most woodchucks probably enter hibernation in October, the older, fatter animals hibernating first. The hibernaculum is usually a woodland den with a single opening (Grizzell 1955). Woodchucks

Figure 20. *Marmota monax*, woodchuck.

emerge from hibernation in late February or in March; adult males leave their dens first. On emergence, woodchucks still have a substantial fat layer which serves as reserve food until green vegetation is abundant in May. Hibernation is regulated by an annual physiological cycle which involves changes in the concentration of sex hormones, other hormones, and the level of lipids and fibrinogens in the blood (Wenberg and Holland 1973a, b, c; Wenberg et al. 1973).

Woodchucks mate shortly after the females emerge from hibernation. Some yearlings are bred, but many females do not breed until their second annual emergence from hibernation (Barash 1974b; Christian et al. 1972; Snyder 1962). Mating occurs while food is still scarce, but green shoots are more abundant when the three to six young are born 31 to 33 days later (Grizzell 1955; Hoyt and Hoyt 1950; Snyder and Christian 1960). Because the young are born in the spring, they have ample time to grow and fatten before their own first hibernation. The young disperse when five or six weeks old (Barash 1974a; de Vos and Gillespie 1960).

Home ranges often overlap, and individuals apparently do not defend specific territories. Dominance relationships do exist among neighbors, however, for relatively subordinate animals generally avoid dominant ones. Aggressive encounters become less frequent as the summer progresses, during which time the basal metabolic rate gradually decreases and the woodchucks convert a progressively greater proportion of their food intake into subcutaneous fat (Bailey 1965a, b; Bronson 1964; Fall 1971).

The food of woodchucks consists almost

wholly of green vegetation (Fall 1971; Grizzell 1955; Hamilton 1934; Merriam and Merriam 1965). *M. monax* is reported to be diurnal, foraging in the early morning and late afternoon (but while the sun is up) in the warmer months, but primarily at midday in the early spring and in the fall (Bronson 1962; Hayes 1976). However, Hamilton (1934) reported that woodchucks are largely nocturnal when they first emerge from hibernation. Although their activities are primarily terrestrial, they occasionally climb trees (A. M. Johnson 1926; Hazard 1960), either to escape predators or to obtain a better look at their surroundings. *M. monax* will also voluntarily swim across small bodies of water (C. E. Johnson 1923b).

Relationship to People: M. monax can be a serious pest in the garden, and its burrows in the fields may interfere with agriculture and be dangerous to livestock. The fur has little value, but woodchucks are good eating. Woodchucks in Tennessee sometimes possess antibodies to tularemia and Rocky Mountain spotted fever (Burgdorfer et al. 1974); like other game, they can be vectors for human disease.

Additional References: Barash 1975; Chesness et al. 1968; Davis 1966, 1967a, b; Davis et al. 1964; Davis and Finnie 1975; Galbavy et al. 1972; Henderson and Gilbert 1978; Snyder et al. 1961; Weeks and Kirkpatrick 1978; Zalom 1977.

Genus *Spermophilus*
Ground Squirrels

Species of this genus are numerous and widespread in Eurasia and North America. For many years they were referred to as *Citellus* (see Hershkovitz 1949), and some European mammalogists still use that name. Typically, members of this genus hibernate; all three species in Minnesota do. The tooth formula for the genus is 1/1, 0/0, 2/1, 3/3 × 2 = 22.

Additional References: Bailey 1893; Davis 1976; Dunford 1977; Gerber and Birney 1968; Howell 1938; Nadler 1966; Nadler et al. 1971; Wobeser and Leighton 1979.

Spermophilus richardsonii
Richardson's Ground Squirrel

Measurements: Total length 300 (286–330); tail 73 (66–84); hind foot 45 (41–47); ear from notch 9–15; weight 249.3–397.7.

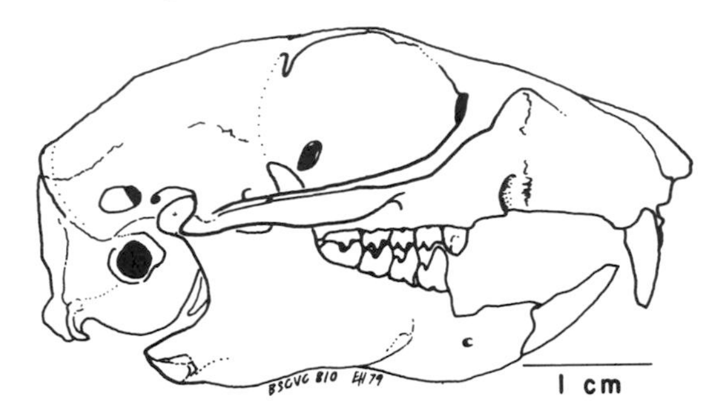

Figure 21. Skull of *Spermophilus richardsonii*, Richardson's ground squirrel, BSCVC 810.

Description: Compared with Minnesota's other ground squirrels, Richardson's is the plumpest and has the most uniform-appearing pelage. It somewhat resembles a small prairie dog. The pelage is a warm gray, becoming yellowish on the face, shoulders, and sides. The fur is distinctly less grizzled than that of Franklin's ground squirrel, the species you are most likely to mistake it for. Richardson's has a broad skull and a short tail, whereas the leaner Franklin's has a more elongate head and a longer, bushy tail. The two species are sympatric in western Minnesota, but the best place I know of to compare them at close range is along the footpaths in the Winnipeg Zoo. Male Richardson's ground squirrels are distinctly larger than females. On the prairies, this species is commonly called "flickertail" or simply "gopher."

Range and Habitat: S. richardsonii is a creature of the northern Great Plains, ranging roughly in an arc from western Minnesota north and west through the southern Prairie Provinces and the Dakotas and Montana, and then south and east along the eastern side of the Rockies into Colorado. In Minnesota, it is restricted to grasslands and croplands on the western edge of the state.

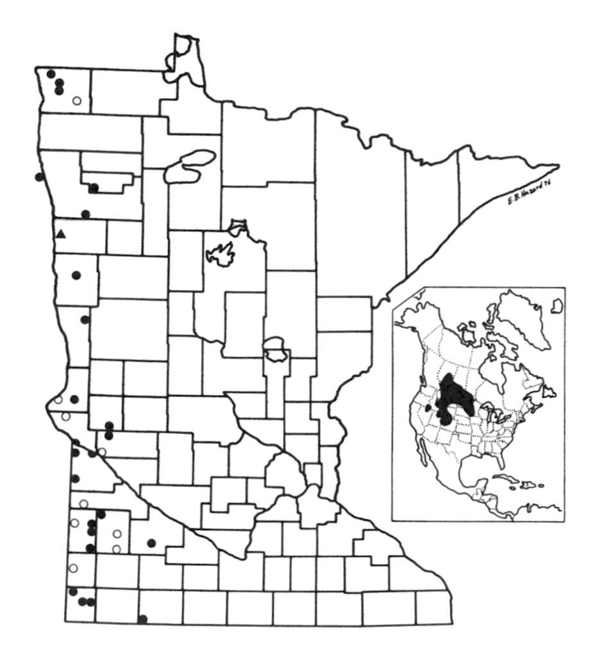

Map 25. Distribution of *Spermophilus richardsonii*. ● = township specimens. ○ = other township records. ▲ = county specimen. (Map produced by the Department of Biology, Bemidji State University.)

Natural History: Several workers have studied Richardson's ground squirrel intensively in the Prairie Provinces. Presumably, their findings apply in Minnesota, though it would be worthwhile to investigate the natural history of this species at the eastern limit of its range. Like most species of *Spermophilus*, *S. richardsonii* hibernates, relying on a heavy deposit of fat to provide the energy necessary for winter survival. The pattern of emergence and entry into hibernation is of interest, because it reduces competition for food between sex and age classes (D. R. Michener 1974; Yeaton 1972). Males emerge from hibernation in late March, and enter into hibernation again in early July. Females generally emerge in early April, at which time mating occurs. Gestation lasts 22 to 23 days (G. Michener 1980). Females reenter hibernation in late July, after their young have been out of the nest for a few weeks. From early August until mid-October, when the last of them has entered hibernation, the young comprise 100 percent of the active population. Each age group thus spends only four of the twelve months above ground, much of this time being devoted to fattening up for a prolonged stay underground. The major food of Richardson's ground squirrel is green vegetation, though it doubtless eats some insects. Within

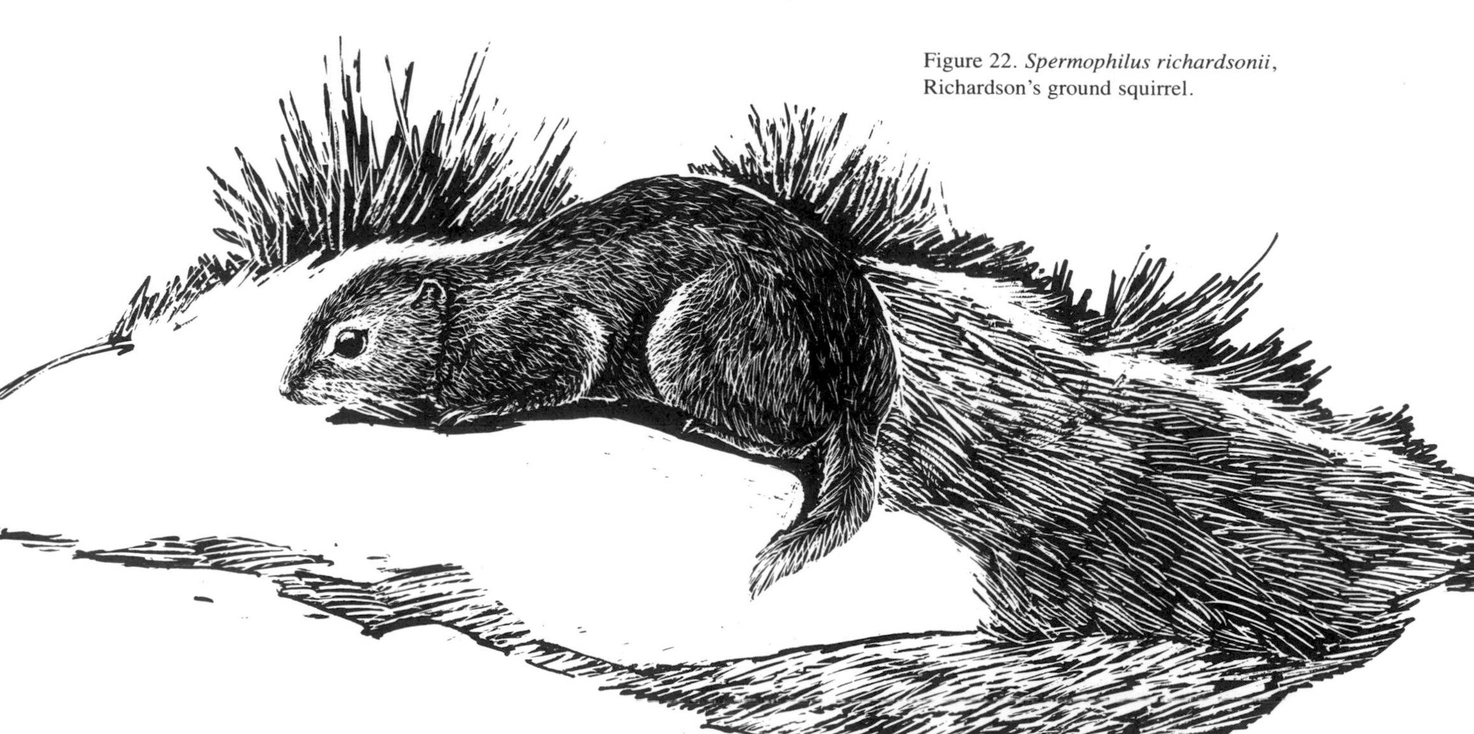

Figure 22. *Spermophilus richardsonii*, Richardson's ground squirrel.

each segment of the population, those that become heaviest have the best chance of overwinter survival. Nonetheless, males, which average heavier than females, have a lower rate of survival. The male to female ratio among the young is about 1:1, but among adults it is about 1:3.3 (D. Michener and G. Michener 1971; G. Michener 1979b; Sheppard 1972b).

Individual adult *S. richardsonii* insure themselves an adequate food supply by territorial defense of most of the home range. Each female territory is centered on a relatively moist depression in the prairie which supports a good stand of vegetation which serves as food for the lactating female and her litter of about six young (G. Michener 1973d; Sheppard 1972b). Females generally exhibit cohesive social behavior toward their own recently emerged young and are aggressive toward other young (G. Michener 1973a, b, 1974; G. Michener and Sheppard 1972). Yearling females typically establish their own territories near their natal site, which by itself might encourage in-breeding. Yearling males, however, move up to 10 times farther away from their natal site if their mother survives the winter (G. Michener and D. Michener 1973; Yeaton 1972).

Relationship to People: Richardson's ground squirrel is a serious agricultural pest in grain-growing areas. The species thrives on the products of human agriculture, growing fatter and attaining greater skeletal size in cropland than in untilled land (Sheppard 1972b). Its significance as a pest may be somewhat increased by human persecution of foxes, badgers, and buteo hawks which would normally prey on it.

Additional References: T. W. Clark 1970; Hansen 1962a; Hansen and Johnson 1976; Hansen and Ueckert 1970; Heaney and Birney 1975; Koeppl et al. 1978; McInvaille and Keith 1974; D. Michener 1972; G. Michener 1971, 1973c, 1977a, b, 1978; G. and D. Michener 1977; Nellis 1969; Quanstrom 1971; Sheppard and Swanson 1976; Sheppard and Yoshida 1971.

Spermophilus tridecemlineatus
Thirteen-lined Ground Squirrel

Measurements: Total length 282 (258–297); tail 93 (75–102); hind foot 37 (34–39); ear from notch 9 (8–10); weight 178.6 (151.0–204.8).

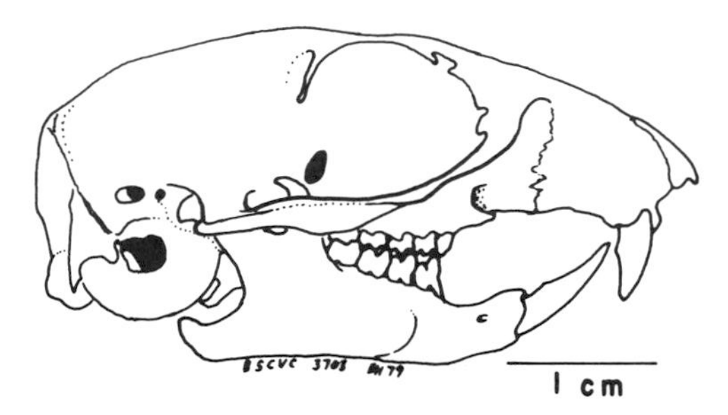

Figure 23. Skull of *Spermophilus tridecemlineatus*, thirteen-lined ground squirrel, BSCVC 3703.

Description: Thirteen stripes run from the nape of the neck to the base of the tail of this squirrel (often called the "striped gopher"). Seven narrow pale yellow stripes alternate with six broad dark brown stripes; each of the brown stripes has a row of pale yellow dots down the middle. The ears are short, the tail about a third of the total length, the body elongate, and the legs relatively short. Animals from eastern, partly wooded areas of the range are generally darker and more distinctly marked than those from farther west.

Range and Habitat: Spermophilus tridecemlineatus is a prairie mammal. Before Europeans cleared and cultivated the Midwest, it probably ranged from the grasslands of Texas, Illinois, and western Minnesota to the eastern edge of the Rockies from Alberta to Arizona. It now has expanded its range eastward into relatively open areas as far as western Pennsylvania, and into all but extreme northeastern Minnesota. Our alteration of the landscape has enabled the thirteen-lined ground squirrel to expand its geographic range despite the simultaneous alteration of previously suitable habitat into intensively farmed tracts. As some land in northern Minnesota un-

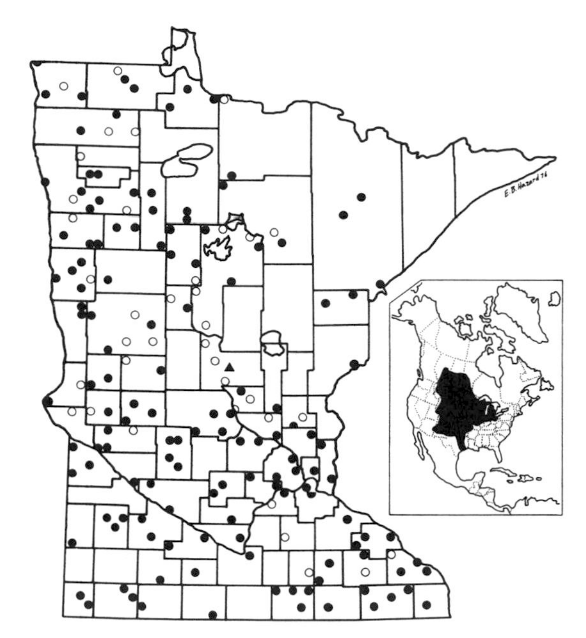

Map 26. Distribution of *Spermophilus tridecemlineatus*. ● = township specimens, selected locations. ○ = other township records, selected locations. ▲ = county specimen. (Map produced by the Department of Biology, Bemidji State University.)

suited for farming or grazing is allowed to return to its natural woody vegetation or is managed for forestry, this squirrel may disappear from part of its newly occupied range.

Natural History: In much of Minnesota, thirteen-lined ground squirrels are the most familiar wild mammals in such open areas as golf courses, lawns, highways rights-of-way, and pastures. They are strictly diurnal, and often relatively tame. Many other squirrels are active primarily early and late in the day, but *S. tridecemlineatus* is most active from midmorning to midafternoon (Hazard 1960). Like chipmunks and other ground squirrels, this species is a true hibernator. Individuals build up a heavy layer of fat in late summer (Lyman 1954) and store seeds in their burrows (Wistrand 1974). Adults generally enter hibernation by late August and emerge in late March and April, with males typically emerging earlier than females (Beer 1962; Rongstad 1965). Thirteen-lined ground squirrels hibernate for a longer season than do chipmunks. Mating may not occur until May (Rongstad 1965). The shorter active season permits only a single litter, commonly of six to ten young (Zimny 1965). Because the burrows are out in the open, it is easy to observe the young when they emerge at an age of four to six weeks. They often huddle together during the first days outside, sometimes closely accompanying the mother. Adults utter a short trill when alarmed, a sound which sends the young scurrying for shelter (Lishak 1977).

Spermophilus means "seed lover," and the major plant food of these squirrels is seeds, primarily of a great variety of herbaceous plants, but occasionally of trees and shrubs. They also eat green vegetation. However, like most squirrels, thirteen-lined ground squirrels do not eat only plants (Flake 1973). This species eats many insects, including such pests as grasshoppers, cutworms, and cabbage butterfly larvae, and will kill and eat mice and voles, ground-nesting birds, and lizards. When suitable insects are abundant, thirteen-lined ground squirrels generally eat more animal than plant food.

S. tridecemlineatus is more gregarious than some other squirrels, and colonies of the species seem to be common. However, these concentrations are not as highly organized socially as are those of truly colonial squirrels like the various prairie dogs and mountain-dwelling marmots (Wistrand 1974). At least in some situations, concentrations of ground squirrels are associated with "islands" of particularly favorable habitat, such as low moist swales in an otherwise dry field (Hazard 1960; McCarley 1966).

In an environment with limited cover, the thirteen-lined ground squirrel relies on camouflage and a low profile for concealment. Whereas a chipmunk moves from one refuge to another in rapid bounds, a ground squirrel scurries low to the ground on short legs and often seems to vanish when it stops abruptly, even out in the open.

Figure 24. *Spermophilus tridecemlineatus*, thirteen-lined ground squirrel.

This works well, especially if dry vegetation matches the yellow-brown pelage. However, it fails often enough for *S. tridecemlineatus* to be a major item in the diet of various soaring hawks. Many individuals are dug out by badgers (Leedy 1947), and some are taken by weasels, foxes, and bull snakes. Although the prairie offers adequate cover for this squirrel, it generally lacks vantage points. Having nothing to climb on for a look around, ground squirrels stand as tall as they can in the familiar "picket pin" posture, a pose not commonly seen in tree squirrels and chipmunks (Wistrand 1974).

Relationship to People: An omnivore such as this one will affect people in many ways. In some areas, thirteen-lined ground squirrels are agricultural pests, but they feed on weed seeds as well as grain, and their role in controlling insect populations may be significant.

Additional References: Bailey 1923; Brenner and Lyle 1974; Cothran et al. 1977; Evans 1951; Fisler 1969; Flake 1974; Foreman 1974; Grant and Birney 1979; Grant et al. 1977; Guilday and Hamilton 1978; Haigh 1979; Hohn and Marshall 1966; Kaufman and Fleharty 1974; Matocha 1977; Nadler and Hughes 1966; Rongstad 1968; Streubel and Fitzgerald 1978; Tester 1965; Whitaker 1972c.

Spermophilus franklinii
Franklin's Ground Squirrel

Measurements: Total length 374 (333–562); tail 127 (110–135); hind foot 51 (46–55); ear from notch 14 (12–16); weight 409.6 (292.3–576.3).

Description: Spermophilus franklinii is often called the "gray gopher." Its overall color on the back and sides is a speckled or barred gray, usually with an olive cast, especially on the rump. The head is a more solid gray, and the underside varies from yellowish white to varying shades of gray or olive-gray. The animal superficially resembles a gray squirrel but is smaller and has shorter ears, a shorter, less bushy tail, and a more pointed face.

Range and Habitat: Franklin's ground squirrel inhabits a northwest to southeast range within the northern tallgrass prairie, from central Alberta to the central Mississippi Valley. Within this range, it prefers brushy and partly wooded areas, as well as prairie edges, rather than open prairie. This preference, plus a tolerance for people, has made it a common mammal in campsites, state parks, and open dumps. Franklin's ground squirrel has moved north and east into the forest biomes as land has been cleared (Robins 1971).

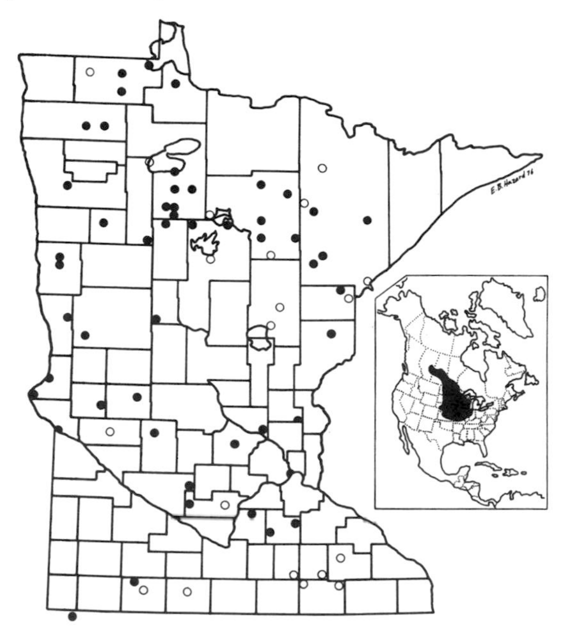

Map 27. Distribution of *Spermophilus franklinii*. ● = township specimens. ○ = other township records. Iowa specimen reported in Bowles 1975. (Map produced by the Department of Biology, Bemidji State University.)

Natural History: This is another semi-colonial squirrel, strictly diurnal, primarily terrestrial, and a true hibernator. A single annual litter of eight to ten young is usual. Hibernation extends from some time in August, September, or October until late March, April, or early May. Variation in emergence dates given in the literature seems, in part, to reflect annual variations in weather conditions. Males generally emerge earlier than females (Haberman and Fleharty 1971a; Iverson and Turner 1972a; Murie 1973).

Franklin's ground squirrel is omnivorous, varying its staple diet of seeds and other vegetation with whatever meat—vertebrate or invertebrate—it can find. It feeds on insects, frogs, various birds, birds' eggs, voles and mice, and even young rabbits on occasion. In some wildlife refuges, *S. franklinii* is a significant predator on the eggs and young of dabbling ducks (Sowls 1948). As with many squirrels, appropriately situated colonies or individuals live largely on the crumbs from our tables.

Relationship to People: Spermophilus franklinii may be a pest on the farm, in the garden, and in the henhouse, but it probably helps control some insect populations. In certain areas, such as wildfowl breeding grounds, it may be desirable to keep this species in check. As with most common mammals, it is neither desirable nor possible to eliminate the species, because the multiple effects of the necessary drastic control measures would likely create more problems than solutions. Except in intensively managed or cultivated areas, Franklin's ground squirrel should probably be allowed to play its dual role as prey and predator and, not insignificantly, as an enjoyable species to observe and study.

Additional References: Bailey 1893; Chesness et al. 1968; Errington 1937a; Haberman and Fleharty 1971b; Jackson 1961; A. M. Johnson 1922; Lyon 1932; Smith 1957; Turner et al. 1976; Wade 1930.

Genus *Sciurus*
Larger Tree Squirrels

The genus *Sciurus* occurs in North America, South America, and Eurasia, but the North American species are not native to other continents.

Sciurus carolinensis
Gray Squirrel

Measurements: Total length 479 (450–503); tail 211 (187–225); hind foot 65 (58–68); ear from notch 31 (22–39); weight 614 (540–713 [Thoma and Marshall 1960]); weight 625 (476–720 [Longley 1963]).

Description: The gray squirrel is primarily gray on the back and sides, but often with a brown cast, especially on the sides and hind feet, and sometimes on the face. The underside is whitish, often suffused with brown. The tail is long and bushy and, in juveniles, shows alternate dark and light bands in ventral view (Sharp 1958). Albinos are rare, but black individuals are common in some localities (Breckenridge 1947). The tooth formula is typically 1/1, 0/0, 2/1, 3/3 × 2 = 22; however, the small upper P3 is occasionally absent.

Range and Habitat: The gray squirrel is native to most of the woodlands of the eastern United Stated and extreme southern Canada. It has been successfully introduced into California, South Africa, and England (Barkalow and Shorten 1973; Elton 1958; Shorten 1954). Its numbers have decreased in much of the eastern United States, where it had been extremely abundant in the unbroken forest. However, it has extended its range westward in the natural river bottom forests of the prairie states. *Sciurus carolinensis* is most characteristic of mature hardwood and mixed hardwood-coniferous forests, especially those in which nut-bearing trees grow. In the northern coniferous forests and some mixed forests, the red squirrel is more common. In open woodlands, immature woodlots, and fencerows, the fox squirrel generally does better. However, the gray squirrel often coexists with one or both of these species, the relative abundance of each depending on local conditions. Gray squirrels are uncommon in the predominantly coniferous forests of northern and northeastern Minnesota (Timm 1975).

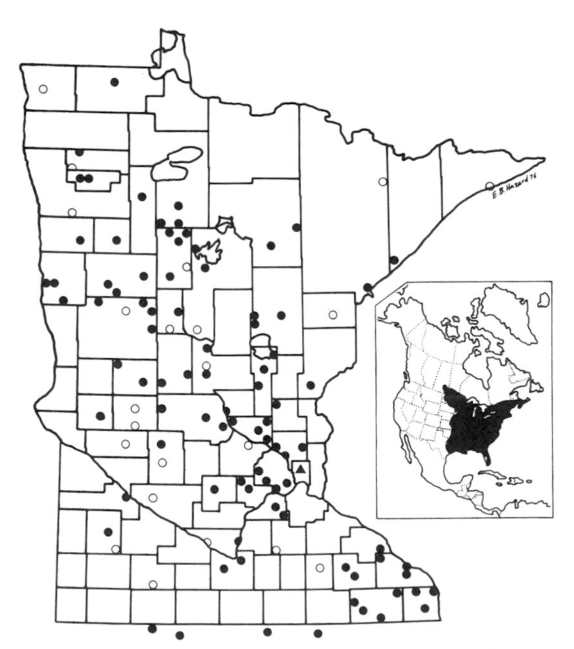

Map 28. Distribution of *Sciurus carolinensis*. ● = township specimens, selected locations. ○ = other township records, selected locations. Iowa specimens reported in Bowles 1975. (Map produced by the Department of Biology, Bemidji State University.)

Natural History: S. carolinensis is a popular game species and a conspicuous wild mammal in the cities, towns, and countryside of the densely populated eastern United States. Consequently, many game biologists and mammalogists have studied it. Although often abundant, it is a solitary, noncolonial species. Individuals do not usually defend extensive, well-defined territories (though some do), but they are generally intolerant of other nearby gray squirrels. Scolding and tail flicking are part of a complex of behavior patterns which gray squirrels use as threats and warnings, primarily among themselves, but also in interaction with other species, including us. These squirrels chase each other a great deal, yet this is at least partly ritual—they all seem to run at the same speed, and seldom is anyone caught. Actual fights between gray squirrels are rare, although they may be important in establishing a social hierarchy (Hazard 1960; Pack et al. 1967). Fights with other species are even more unusual, despite persistent tales to the contrary.

Gray squirrels are diurnal but have a greater preference for early morning and late afternoon hours than do fox squirrels and eastern chipmunks, both frequent residents of the same

Figure 25. *Sciurus carolinensis*, gray squirrel.

woods as the grays (Hazard 1960). Gray squirrels are generally out foraging before sunup, especially in fall and winter after a long night's fast, and have pretty much ceased foraging by 9:00 or 10:00 A.M., typically retiring to a tree hole or leaf nest. They become active again in mid- or late afternoon and often are still up after sundown, except in summer when the days are long. Their behavior in city parks, where people provide peanuts throughout the day, is much modified.

None of our tree squirrels hibernates. Well-fed squirrels stay in their nests or dens for several days if the weather is unusually cold or stormy, but there is no documented reduction in body temperature. If the mast crop has failed and the squirrels are lean, they have no fat reserve and must forage no matter how bad the weather. Squirrel populations plummet during such winters (Nixon and McClain 1969; Nixon et al. 1975). Mast—primarily acorns in Minnesota—is the staple of the diet, although gray squirrels eat a variety of plant and animal foods in season. Their diet includes spring buds and flowers of various trees and shrubs, elm and pine seeds, various berries, and acorns and other nuts. Gray squirrels bury mature nuts singly, not in large caches, in the fall. They apparently find many of these by smell later, in winter or spring. Some of those that they do not find eventually grow into mature trees. The unintentional tree planting of gray and fox squirrels is probably as beneficial to mast-bearing trees as the trees are to the squirrels (Smith and Follmer 1972). Gray squirrels also eat mushrooms, insects, and birds, especially nestlings. Various songbirds recognize tree squirrels as enemies and harass them. I have watched wood peewees drive gray squirrels out of trees so persistently that it was a wonder that the squirrels were ever able to return to their dens.

The ideal refuge and nursery for gray squirrels is a den hole in a live, hollow tree, and the best holes generally develop in white oaks. Squirrels nest in tree holes when they can, but if none are available or there are too few to go around, they build leaf nests. These may provide as much warmth as tree holes, but high winds may destroy them, and great horned owls sometimes attack such nests. So do those "sportsmen" who are more accurately called "slob hunters." The owl generally gets its prey if the nest is occupied. The slob hunter wastes much shot, because most nests are unoccupied, and he or she wastes squirrels too, because a squirrel shot in a nest often just dies there.

Sciurus carolinensis starts its breeding cycle early, the first mating chases generally occurring in late January (Longley 1963). Gestation takes 44 or 45 days, so the first litter, typically of two to four young, usually comes in March. There may be a second litter in July or August. The young often stay with the mother until just before the next litter is born; thus, a group of several squirrels may utilize one den or nest all winter. A female who is soon to have young will drive out the earlier litter. Once dispersed, the squirrels that survive establish new home ranges and remain there for one or more seasons. In the fall, however, many gray squirrels often abandon their home ranges and set up shop somewhere else. There is no well-established explanation for this "fall shuffle," and it is a commonplace and seemingly random event, with individual squirrels moving in all directions. In the nineteenth century, naturalists observed occasional mass emigrations of *S. carolinensis* in which hundreds of squirrels traveled through the countryside in one direction (Seton 1929). Perhaps this phenomenon is associated with dense populations in uncut mature forests; few such mass movements have been reported since the turn of the twentieth century (Jackson 1921; Fryxell 1926).

Relationship to People: Gray squirrels used to be taken for their fur but it is no longer of value. Today we value them as game animals and for their aesthetic appeal. They also play a role in ensuring the reproduction of the trees whose nuts they bury. Gray squirrels can be pests in the garden and cornfield, and in the attic if they keep house there. In this country they do little damage to trees, but in England introduced gray squirrels for some reason feed extensively on bark, causing a substantial economic loss (Shorten 1954). They have also been a partial factor in the decrease in numbers and range of the smaller native British squirrel, *Sciuris vulgaris*.

Additional References: Ackerman and Weigl 1970; J. M. Allen 1954; Baker 1944; Barkalow and Soots 1975; Barnett 1977; Brown and Yeager 1945; Connolly 1979; Cowles et al. 1977; Flyger 1960; Fogl and Mosby 1978; Hibbard 1954, 1955, 1956a; Ludwick et al. 1969; Moore 1959a, 1968; Mosby 1969; Nixon and McClain 1975; Nixon et al. 1968, 1975; Packard 1956; Shorten 1957; Taylor 1966, 1977; Thompson 1977, 1978.

Sciurus niger
Fox Squirrel

Measurements: Total length 550 (510–621); tail 246 (220–265); hind foot 73 (68–76); ear from notch 28 (22–31); weight 755 (633–930 [Thoma and Marshall 1960]); weight—southern Minn. average 827, central Minn. average 863 (Longley 1963).

Description: In Minnesota, the most distinctive characteristic of *Sciurus niger* is its orange color. The feet, underside, tops of the ears, and tops of the tail are dull orange. On the back and sides, the same color is mixed with gray. The fox squirrel is one-third again as heavy as the gray and averages a bit stockier. Black individuals do occur in Minnesota, but they are nowhere abundant. The tooth formula is 1/1, 0/0, 1/1, 3/3 × 2 = 20.

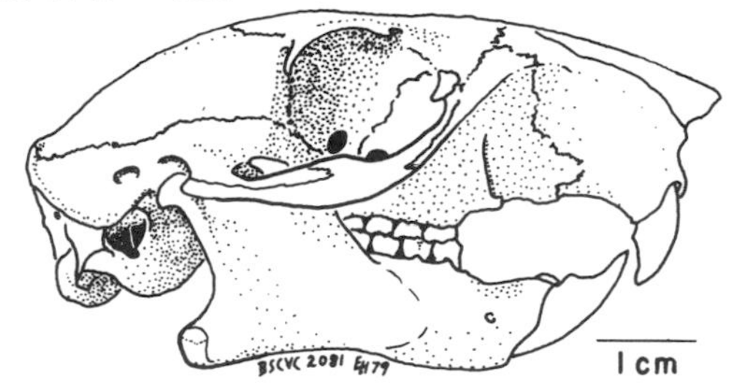

Figure 26. Skull of *Sciurus niger*, fox squirrel, BSCVC 2081.

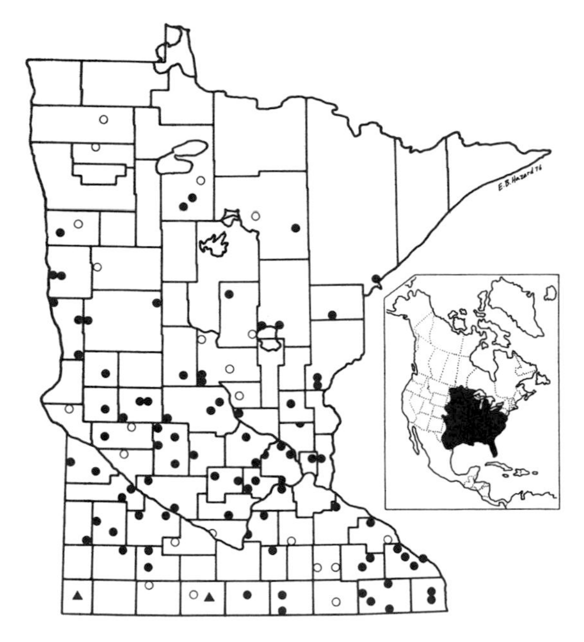

Map 29. Distribution of *Sciurus niger*. ● = township specimens, selected locations. ○ = other township records, selected locations. ▲ = county specimens. (Map produced by the Department of Biology, Bemidji State University.)

Range and Habitat: Fox squirrels range farther west than do gray squirrels, but generally not as far north. *S. niger* occurs sporadically in northern Minnesota and North Dakota, and even southern Manitoba (Wrigley et al. 1973), but it is rare in the northern third of Minnesota and apparently absent from the extreme northeastern counties. Fox squirrel populations in the eastern United States have increased as the continuous forest has been replaced by scattered groves, fencerows, and open woodlots, habitats resembling the parklands and oak openings where this species originally prospered in the Midwest. They have also spread from the natural wooded bottomlands of the prairie states into the groves and shelterbelts planted in the former prairie. *S. niger* shows pronounced geographic variation. Subspecies in the southern and eastern states differ markedly from our local fox squirrel in pelage and occupy distinctive ecologic niches (Moore 1957).

Natural History: Fox squirrels are the commonest tree squirrels in much of their range and important game animals. Their biology is generally similar to the gray squirrel's and it is not always clear why one species may be more successful than the other in a particular place. Fox squirrels, at least in the Midwest, seem better adapted than grays to more open and drier conditions; to younger, smaller, more isolated woodlots; and to using corn as a food source (Allen 1943; Brown and Yeager 1945). Grays do better in older, more extensive forests and are either more suited to mixed coniferous-deciduous forest or to cold weather or both. Both species do well in towns.

Both squirrels often live in the same woods, and one thing about their relation is clear: direct conflict between gray and fox squirrels has little to do with their relative success in a given habitat (Hazard 1960). Conflicts between the two are rare, but, when they occur, the more aggressive fox squirrel usually drives the gray off. Despite this, in suitable habitat the gray squirrel persists as the more abundant species. In such an area, the gray may monopolize all the tree holes, the relatively few fox squirrels utilizing leaf nests only. Within much of their common range, the relative densities and continued presence of both species may depend partly on land management. When we open up the forest, decrease the size of woodlots, and plant corn, fox squirrels prosper. When we abandon farmland and allow woods to mature, grays begin to regain their former dominance.

The breeding pattern of fox squirrels resembles that of grays. Mating begins in January, gestation lasts about 45 days, there are generally two to four young, and a female bears one or two litters per year. Spectacular mating chases involving several males and one or more females occur in both species when populations are dense. Much mating probably occurs in brief single-pair encounters, however, and thus goes unnoticed.

Fox squirrels depend primarily on mast for food (Short 1976; Smith and Folmer 1972), but eat a variety of buds, fruits (including cockleburr, which is toxic to many mammals [Gates and Gates 1975]), and other vegetation, and will take meat—invertebrate or vertebrate—when they can. They are perhaps more partial to corn than gray squirrels are. Like grays, they scatter-hoard, burying acorns and other nuts singly, and finding them later by smell (Smith and Follmer 1972; Stapanian and Smith 1978). They are not such early risers as grays. In a woods occupied by both species, fox squirrels begin to forage an hour to two after the first grays are up, and continue foraging into late morning. Fox squirrels generally resume foraging earlier in the afternoon than grays—they may even be active at midday—and retire well before sundown (Hazard 1960; Hicks 1949). If two fox squirrels come too near each other while foraging, they will begin tail flicking. If one does not retreat, a chase will ensue, but actual combat is rare. The fox squirrel scolds much as the gray does, but more hoarsely. In at least some areas, fox squirrel scolding consists of one or more long whines followed by five to 13 short grunts. A scolding gray squirrel, on the other hand, typically emits a series of short grunts and then one or more high-pitched whines (Moore 1957).

Relationship to People: The fox squirrel is an excellent game animal which, when prepared for the table, reveals another distinctive trait. Its bones, unlike the gray's, are pink rather than white (Levin and Flyger 1971). Its value as game and as an interesting animal to watch must be balanced against the occasional damage that it does to crops.

Additional References: J. M. Allen 1954; Baker 1944; Havera 1979; Havera et al. 1976; McCloskey and Shaw 1977; Moore 1968; Nixon and McClain 1969; Nixon et al. 1968, 1974, 1975; Packard 1956; Reichard 1976; Weeks and Kirkpatrick 1978; Zelley 1971.

Genus *Tamiasciurus*
American Red Squirrels

The genus *Tamiasciurus* occurs only in North America and includes two species, the red squirrel and the Douglas squirrel (*T. douglasii*). The latter species resides in the Pacific States. The well-known red squirrel of the Old World is a member of the genus *Sciurus*.

Tamiasciurus hudsonicus
Red Squirrel

Measurements: Total length 314 (284–343); tail 122 (113–130); hind foot 46 (42–50); ear from notch 23 (19–25); weight 205.1 (162.8–246.2).

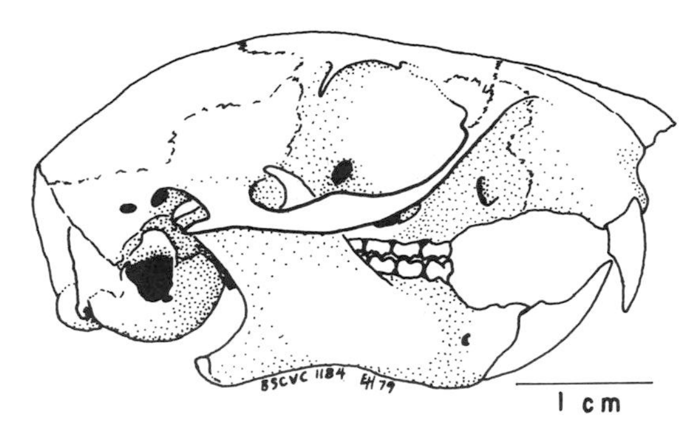

Figure 27. Skull of *Tamiasciurus hudsonicus*, red squirrel, BSCVC 1184.

Description: This small tree squirrel shows greater seasonal dimorphism than our other squirrels. The red squirrel's heavy winter coat is an orange-red on the back and sides, and in winter there are tufts of fur on the ears. In summer the ear tufts are gone, and a solid black line on the sides separates the white belly from the dorsal fur, which is now a dull grayish or greenish red. Both black and albino individuals occur but are rare. One Hubbard Co. female (BSCVC 3562) is nearly a solid gray except for an irregular white chest patch. The tooth formula is normally 1/1, 0/0, 1/1, 3/3 × 2 = 20, rarely 1/1, 0/0, 2/1, 3/3 × 2 = 22. P3, if present, is a tiny peg hidden by the crown of P4.

Range and Habitat: The red squirrel ranges from the northeastern United States and eastern Canada west to Alaska and south into the Rockies. It is the characteristic squirrel of Minnesota's coniferous forests and is also found in mixed forest and in some northern deciduous forest. The species is more at home in mature and continuous forests than in open woodlands or small isolated stands. In Minnesota, it is most common in the northern half of the state, but populations occur as far south as northern Iowa (Bowles 1975; Lynch and Folk 1968).

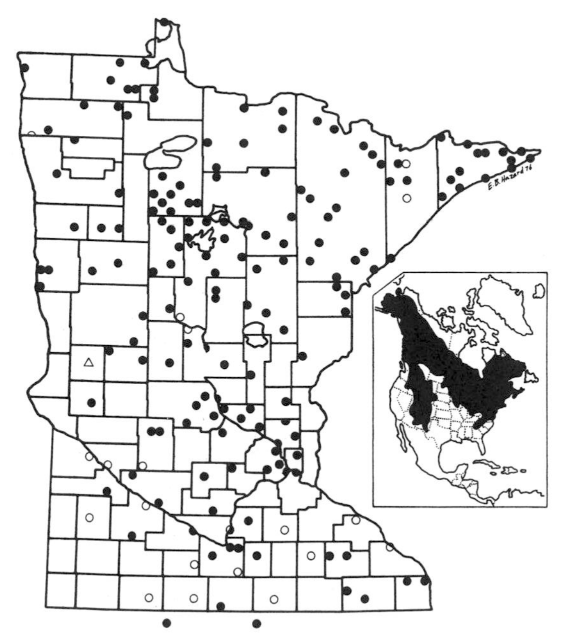

Map 30. Distribution of *Tamiasciurus hudsonicus*. ● = township specimens, selected locations. ○ = other township records, selected locations. △ = other county record. Iowa specimens reported in Bowles 1975. (Map produced by the Department of Biology, Bemidji State University.)

Natural History: The red squirrel's staple food is typically conifer seeds—of spruce, fir, pine, and cedar in the Upper Midwest and adjacent Canadian provinces—and its way of life is tied to this resource. Conifer seeds are too small to be of use if buried singly, as a gray squirrel buries acorns. The most economical way for a squirrel-sized mammal to live on conifer seeds is to harvest whole cones, before they ripen and shed their seeds, and then to store a winter's food supply of cones in one place. The only way to guarantee availability of the store or cache is to defend it. Red squirrels are strongly territorial in most of their range. A red squirrel defends not only its cache but its entire feeding area. This behavior is the most efficient way for each individual to insure itself a year-round supply of a food that is produced during one season (C. C. Smith 1968, 1970). This pattern breaks down in areas where the major squirrel food is acorns or other large nuts. An acorn is large enough to be economically stored and recovered singly. Furthermore, large caches of acorns would be vulnerable to predation by deer, which are common in mast-bearing woods (Smith and Follmer 1972). Gray and fox squirrels do best where mast is abundant, and the red squirrel, despite its aggressive behavior, cannot successfully defend its feeding territory or its cache against the larger, nonterritorial squirrels. In oak woods in Minnesota, red squirrels are relatively scarce, though they do well in some hardwood stands dominated by maple, birch, and basswood.

The commonest sign of territoriality in *Tamiasciurus* is the well-known drawn-out "cherr," a territorial advertising call. If an intruding red squirrel enters the territory (an area of up to 1 ha) despite this warning, a chase ensues, and the intruder is almost always driven out. Red squirrels also have an elaborate alarm display. The animal begins its response to a disturbance by piping or chirping, flicking the tail with each chirp. As the display proceeds, the chirps are replaced by explosive barks or coughs, and the squirrel flicks its tail more vigorously and stamps its hind feet, often leaping repeatedly from its perch as its fury heightens. It is perhaps the vigor of this display, plus the frequent ability of the red squirrel to chase off individual grays, that has given the red squirrel an underserved reputation for driving gray or fox squirrels from their habitats. The relative success of *Tamiasciurus* as opposed to *Sciurus*, however, is determined primarily by the major food sources available, not by temperament.

Although conifer (or other) seeds are their major food, red squirrels are gastronomic opportunists. They eat the buds and flowers of maple and elm, seeds of elm, maple, and basswood, various berries, any nuts that are available, mushrooms (including some poisonous to people), and whatever animal food they can get. They are reported to take insect larvae, birds, birds' eggs, and young gray squirrels and cottontails. They also lap sap from holes drilled by yellow-bellied sapsuckers (Coulter 1961; Hatfield 1937). In the spring, summer, and early fall, they forage primarily in the early morning and late afternoon; in the winter, they are more active at midday, thus conserving body heat.

Red squirrels prefer to nest in tree holes but often use leaf nests or even hollow logs or other shelters on the ground. They have one or two

Figure 28. *Tamiasciurus hudsonicus*, red squirrel.

litters per year, one litter being typical in the northern parts of the range (Dolbeer 1973; Kemp and Keith 1970; Kramm et al. 1975; Millar 1970). There are generally two to five young. When the young are able to fend for themselves, the mother excludes them from her territory (unless she moves to a new one), and each has to establish a territory of its own if it is to survive (C. C. Smith 1968).

Tamiasciurus is a preferred food of the pine marten, a mammal that has been largely extirpated in our area. It also falls prey to ermine, bobcats, lynx, and various hawks and owls.

Relationship to People: Red squirrels can do damage in cabins, resorts, and food storage areas, but they are seldom economically significant. They take an occasional songbird, but so do other songbirds. Red squirrels are good eating but are almost too small to be worth shooting. To me, this liveliest of tree squirrels personifies the north woods. It is a superb example of behavioral adaption to a particular environment, and this very behavior makes it a most interesting woodland neighbor.

Additional References: Ackerman and Weigl 1970; Bakko 1975, 1977; Ferron 1975, 1976; Ferron and Prescott 1977; Goudie 1978; Hatt 1929; Kirkland and Griffin 1974; Kramm 1975; Layne 1954a; Mohr 1965; Nelson 1945; Pauls 1978; Reichard 1976; Searing 1977; C. C. Smith 1978; M. C. Smith 1968; Zirul and Fuller 1970.

Genus *Glaucomys*
Flying Squirrels

The two species in this genus are restricted to the New World, but the taxonomic status of the genus needs further study. On the basis of chromosomal analysis, Schindler et al. (1973) conclude that these two species are closely related. However, the bacula of the two are strikingly different. That of *G. sabrinus* resembles the baculum of the Old World genus *Hylopetes* (Burt 1960), suggesting a more distant relationship (Weigl 1969). Both species have the tooth formula 1/1, 0/0, 2/1, 3/3 × 2 = 22.

Flying squirrels are the only rodents that glide and the only gliding mammals in North America (Gupta 1966a). Unlike bats, which are true flyers, flying squirrels cannot maintain or increase altitude. They do, however, maneuver skillfully during their descent from high in one tree, dodging obstacles, changing course, and, with a quick upward flip of the tail to brake speed, coming to a smooth upright stop at the base of a distant tree. It is pointless to ask how far they can glide, since this depends primarily on the vertical distance between the point of takeoff and landing. Gliding is used to escape enemies and to cover distances quickly.

Glaucomys volans
Southern Flying Squirrel

Measurements: Total length 228 (214–250); tail 95 (87–110); hind foot 30 (29–32); ear from notch 18 (16–21); weight 61.6 (42.5–74.5).

Description: The southern flying squirrel is small, weighing less than an eastern chipmunk. Its dense, silky fur is gray or brownish gray on the back and white or creamy white on the belly; the belly hairs are white to the base. Its tail is gray, and the tail fur extends to the sides, forming a horizontal plane. A loose fold of skin, the patagium, extends on each side of the body from the front to the hind legs. The patagium and the flattened tail constitute the animal's gliding surfaces. The front corner of the patagium is anchored to a spur of cartilage that extends outward from the wrist, thus increasing the gliding surface. The eyes are large.

Range and Habitat: The southern flying squirrel is found in the eastern United States, northeastern Mexico, peninsular Ontario between Lake Huron and Lake Ontario, southernmost Quebec, and Nova Scotia (Cameron 1976; Wood and Tessier 1974; Youngman and Gill 1968). It is a deciduous forest species, and in Minnesota its northern limit roughly coincides with the northern limit of forests dominated by oak (Weigl 1969). (See map 32). Earlier published species

Figure 29. *Glaucomys volans*, southern flying squirrel.

records from Clearwater and St. Louis counties in Minnesota apparently cannot be substantiated (Heaney and Birney 1975). Southern flying squirrels prefer heavy timber to young growth and are more often found in upland than in swamp woods. They do occur, however, in well-wooded towns and suburbs.

Natural History: Although abundant in many deciduous woodlands, *Glaucomys volans* is generally unfamiliar to us because it is strictly nocturnal. Occasionally people find southern flying squirrels in bird houses and attics, and they may be puzzled to find a flattened gray tail on the lawn, all that was left by a "domestic" cat. *G. volans* is relatively tame and will visit bird feeders at night. It has a sweet high-pitched chirp, so high that you may not realize you are hearing it. It is actually above the hearing range of many older people.

In the spring, summer, and fall, adult southern flying squirrels hole up individually, usually in an "abandoned" woodpecker hole in a tree or dead stub, but occasionally in a leaf nest (Landwer 1935). Females raise their litters of usually three to five young in these dens which are often only a meter or two above the ground and in small, tottering stubs that could not house a gray or fox squirrel. There are usually two litters, one in late March or April, the other in July. The gestation period is 40 days. Females with litters are often territorial (Madden 1974).

Southern flying squirrels eat much the same sorts of food as other deciduous forest squirrels. They feed on acorns, hickory nuts, and other seeds and nuts, and they also store these in tree holes, under bark, and in leaf litter (Muul 1965). They also eat buds, flowers, and whatever meat they can get, including a variety of insects and some birds (Stoddard 1920). The bird houses that flying squirrels sometimes occupy may provide board as well as lodging.

Southern flying squirrels are not true hibernators (Neumann 1967), but they do remain inactive and occasionally become torpid during periods of extreme cold. If appropriate holes (or attics) are available, a dozen or more will den together for warmth in the winter.

See the next account for a discussion of ecologic interactions between *Glaucomys volans* and *G. sabrinus*.

Relationship to People: Although common in some populated areas, southern flying squirrels are seldom of economic importance. To people who take the trouble to observe them, they are a source of much enjoyment. They do well in captivity and have been used successfully in the study of mechanisms by which animals regulate their daily activity and seasonal cycles (De Coursey 1960a, b, 1972; Muul 1965, 1969b) and in studies of the relationship of behavior and physiology to geographic distribution. In the state of Virginia, they are known to carry typhus, but there is no evidence that this has been a source of human infection (Bozeman et al. 1975).

Additional References: Bailey 1923; Dolan and Carter 1977; Giacalone-Madden 1977; Goertz et al. 1975; Heaney and Birney 1975; Johnson-Murray 1977; Jordan 1948, 1956; Layne 1954b, 1958b; Linzey and Linzey 1979; Muul 1968, 1970, 1974; Sollberger 1940, 1943.

Glaucomys sabrinus
Northern Flying Squirrel

Measurements: Total length 273 (248–300); tail 128 (109–148); hind foot 37 (34–41); ear from notch 22 (19–27); weight 89.6 (74.8–109.8).

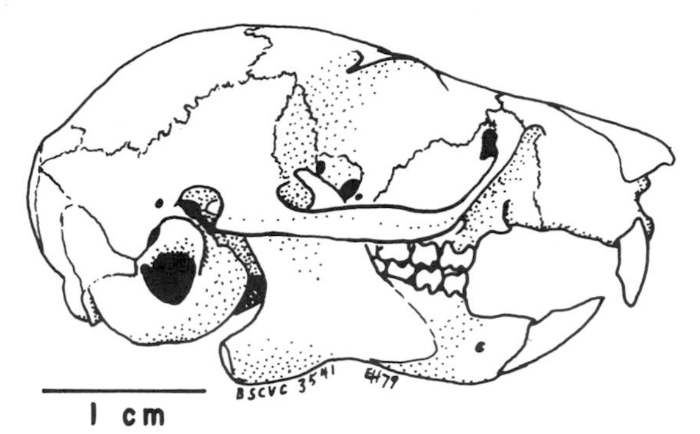

Figure 30. Skull of *Glaucomys sabrinus*, northern flying squirrel, BSCVC 3541.

Description: Glaucomys sabrinus weighs half again as much as *Glaucomys volans*. The fur on the back is generally a browner gray, and the hairs on the belly are white at the tips but lead-colored at the base.

Range and Habitat: The northern flying squirrel ranges across Canada and Alaska from tree line south into the northeastern and Great Lakes States and into the mountain forests of the western United States. There is a disjunct population in the Black Hills of South Dakota (Turner 1974). *G. sabrinus* is primarily an inhabitant of coniferous and mixed forests. In regions where both *G. volans* and *G. sabrinus* occur, *G. volans* often lives in the deciduous uplands and *G. sabrinus* in the coniferous swamps. The northern flying squirrel is confined to the northeastern half of Minnesota, with little overlap into the range of southern flying squirrel. *G. sabrinus* may well be a more recent immigrant into North America

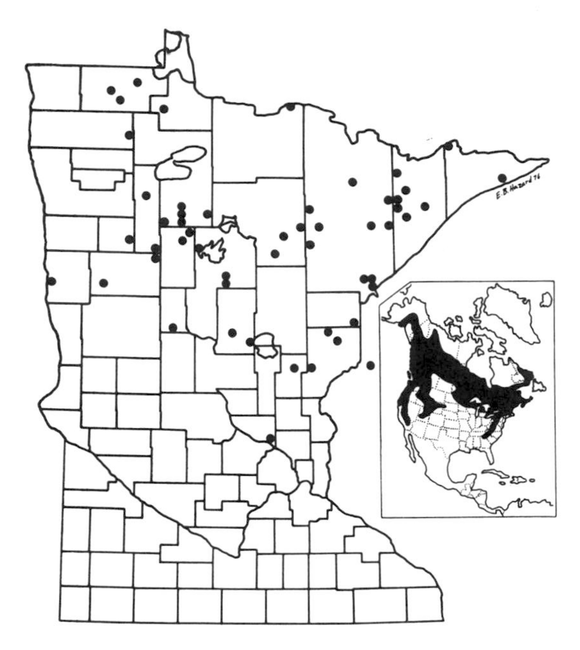

Map 31. Distribution of *Glaucomys sabrinus*. ● = township specimens. Wisconsin specimen reported in Jackson 1961. (Map produced by the Department of Biology, Bemidji State University.)

from Eurasia than was the ancestor of *G. volans*. The present overlap of the ranges of the two species in eastern North America may indicate the zone where the respective competitive advantages of one species over the other balance out (Muul 1969a; Weigl 1969).

Natural History: The northern flying squirrel, like its smaller relative, is a strictly nocturnal glider and is active throughout the year. The nest may be in a tree hole or outside, where it is built of such materials as cedar bark (Weigl and Osgood 1974). *G. sabrinus* may be less dependent on tree holes than *G. volans* (Weigl 1969). There is generally one litter a year, usually of three to six young. Northern flying squirrels are often common, and may take up residence in birdhouses, artificial nest boxes for ducks, and buildings. They have not been as extensively studied as the southern species, and some published reports of their food habits may essentially be extrapolations of the known food habits of *G. volans*. I have kept both species in captivity on sunflower seeds and dry dogfood. It may be that, to the south, *G. sabrinus* is limited by the presence of *G. volans*. *G. volans* females, which typically have two litters per year, breed earlier and apparently can successfully defend choicer nest sites not only from one another but from their larger but seemingly meeker cousins. To the north, on the other hand, *G. volans* may be limited not by *G. sabrinus* but its own smaller size, which may result in excessive heat loss in colder northern winters (Muul 1968). However, quite different factors may be involved. In northeastern California, the northern flying squirrel's diet consists almost totally of lichens and fungi (McKeever 1960). If this is true in eastern North America, the abundance of tree lichens and relative scarcity of acorns in the north woods may be important factors in limiting the distribution of the two flying squirrels (Weigl 1969).

Relationship to People: *G. sabrinus* generally inhabits less populated regions than *G. volans* and seldom has a direct effect on us. It plays a role both as prey and consumer in north woods ecosystems, and its unusual food preferences, as well as its docile behavior in captivity, suggest that it would be a fruitful subject for scientific study.

Additional References: Booth 1946; Brink and Dean 1966; Burt 1960; Coventry 1932; Guilday and Hamilton 1978; Gupta 1966a; Kittredge 1928.

Family Geomyidae

Pocket Gophers

Pocket gophers are restricted to North America and extreme northwestern Colombia. Fossil geomyids are known only from North America. Gophers are specialized for a subterranean existence, having strong large-clawed forelimbs, small eyes and ears, a short tail, and external (fur-lined) cheek pouches. The tooth formula is always 1/1, 0/0, 1/1, 3/3 × 2 = 20.

References: Burt 1958; Long 1976b; McLaughlin 1967; McNab 1966; Russell 1968; Scheffer 1958; Vaughan 1978.

Genus *Thomomys*

Thomomys talpoides
Northern Pocket Gopher

Measurements: Total length 235 (220–250); tail 64 (58–80); hind foot 30 (29–33); ear from notch 8 (7–9); weight 159.8 (114–195.4).

Description: The northern pocket gopher is not likely to be confused with any other local mammal except the plains pocket gopher, *Geomys bursarius*. *Thomomys* is not as big as *Geomys*, and its forelimbs are proportionately smaller and its claws lighter. Both species are generally brown, but the fur of the northern pocket gopher usually lacks the sheen typical of its larger relative. The surest distinguishing feature is on the anterior surfaces of the upper incisors. Those of *Thomomys* each have a faint groove near the inside edge; those of *Geomys* have two pronounced grooves.

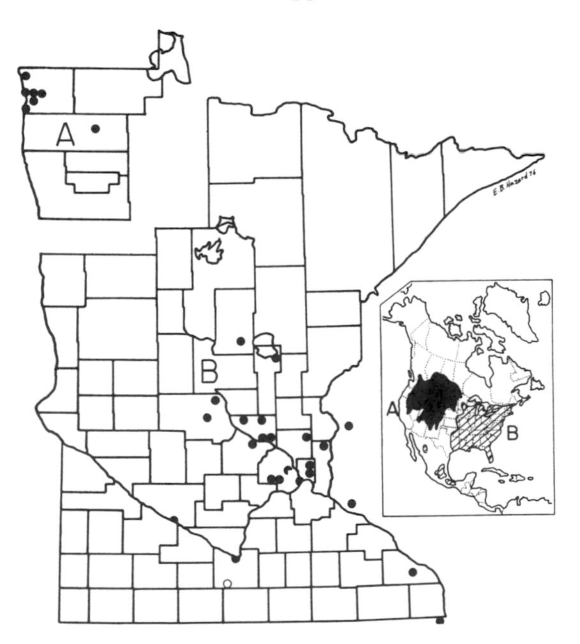

Map 32. A. Distribution of *Thomomys talpoides*. B. Distribution of *Glaucomys volans*. ● = township specimens. ○ = other township record. Iowa specimen reported in Bowles 1975. (Map produced by the Department of Biology, Bemidji State University.)

Range and Habitat: Thomomys talpoides ranges widely in open country in the western United States and the southern parts of the Prairie Provinces. It occurs across northwestern South Dakota and most of North Dakota, reaching its eastern limits in northwestern Minnesota.

Natural History: The northern pocket gopher is a burrowing, solitary, territorial herbivore, as are all pocket gophers. Except during the mating season, an adult gopher will commonly drive other adult gophers from its extensive burrow system—it may occasionally exclude pocket gophers of other, larger species. (See the Natural History section of the next account for a general discussion of pocket gopher burrows.) In general, the ranges of different species of pocket gophers are mutually exclusive (Bailey 1926; Baker 1974; Kennerly 1959; Vaughan and Hansen 1964). Miller (1964) found that most species prefer deep, light soils and that larger species usually can exclude smaller ones from choice habitats. In southeastern Manitoba, the ranges of *Thomomys talpoides* and *Geomys bursarius* apparently do not overlap, and *Geomys* appears to be less tolerant of wet or wooded habitats than is *Thomomys* (Wrigley and Dubois 1973). In Kittson County, Minnesota, on the other hand, Norum (1966) found both species in the heavy, clayey soils near the Red River, but only *Geomys* in the lighter soils to the east. *Geomys* ranges eastward well into partly wooded habitats in Minnesota, but *Thomomys* does not.

The northern pocket gopher is a strict vegetarian, generally preferring forbs to grasses, trees, and shrubs (Vaughan 1967b, 1974; Ward and Keith 1962). Gophers generally feed underground. Not only do they eat the underground parts of plants, but they also pull entire herbaceous plants down into the tunnels and feed on them. During the winter, gophers feed at the ground's surface under the snow. That they also sometimes feed on the surface when there is no snow is suggested by the presence of their remains in owl pellets (Fassler and Leavitt 1975). Gophers carry food in their cheek pouches and store food in their burrows.

Thomomys talpoides bears but one litter per year, usually in May or early June after a gestation of about 19 days (Banfield 1974). The average litter size in Minnesota is about four or five. When the young reach an age of six to eight weeks, the mother excludes them from her burrow, and they must establish territories of their own.

Relationship to People: The northern pocket gopher is frequently a pest on cultivated and grazing lands. The extent of damage, if any, depends on the particular land use and on the abundance of gophers. Pocket gophers are preyed upon by hawks and owls when they are above ground, and by coyotes, foxes, badgers, weasels, and snakes which can dig them out or follow them into their burrows. People persecute all these predators, often without reason. A more tolerant attitude toward predators might result in less excessive gopher populations.

In many areas, pocket gophers are important cultivators in their own right, loosening and aerating hard-packed soil, increasing soil moisture retention, and bringing nutrients to the surface. It has been estimated that, in some habitats, gophers annually bring to the surface as much as eight to 16 metric tons of soil per hectare (Banfield 1974).

Additional References: Andersen 1978; Hansen 1962b; Hansen and Miller 1959; Hansen and Morris 1968; Kalin 1964; Keith et al. 1959; Nevo et al. 1974; Patton et al. 1972; Quimby 1942a; Rusch et al. 1972; Thaeler 1968, 1974a, b, 1976, 1980; Vaughan 1961, 1963, 1967a.

Genus *Geomys*

Geomys bursarius
Plains Pocket Gopher

Measurements: Total length 260 (239–300); tail 66 (54–75); hind foot 33 (31–38); ear from notch 7 (5–9); weight 302.0 (190.8–429.7).

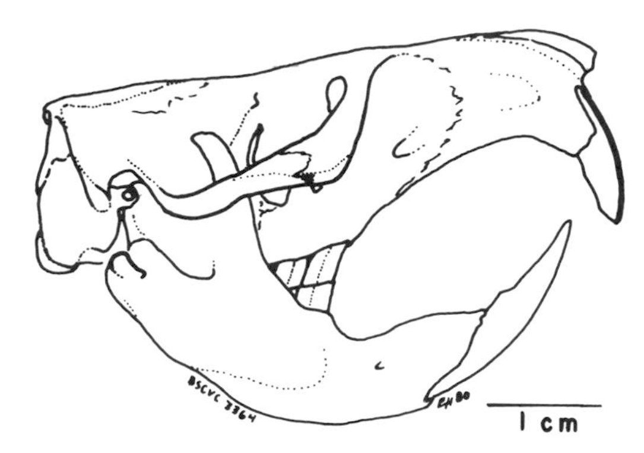

Figure 31. Skull of *Geomys bursarius*, plains pocket gopher, BSCVC 3364.

Description: The plains pocket gopher is larger than the northern pocket gopher, with more massive forelimbs, longer, heavier claws, a shiny brown coat, and two prominent grooves on the midline of the front of each upper incisor. The eyes and external ears are tiny, and the tail is short with short hair, as is generally true of pocket gophers. Albinos are relatively common. Unlike most mammals, males of this species grow larger throughout their lives.

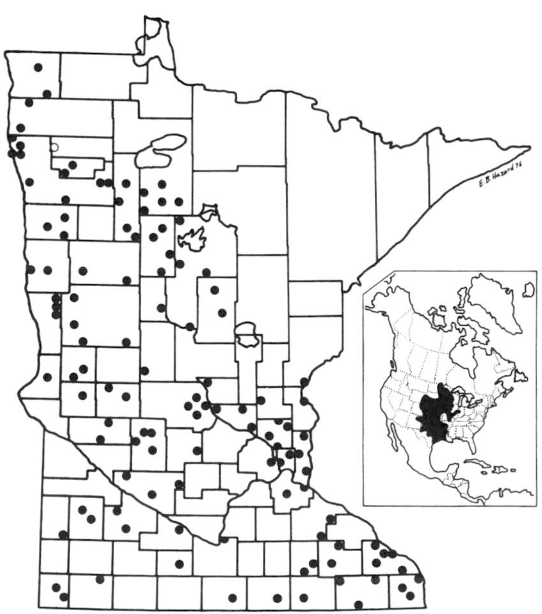

Map 33. Distribution of *Geomys bursarius*. ● = township specimens, selected locations. ○ = other township record. (Map produced by the Department of Biology, Bemidji State University.)

Range and Habitat: Geomys bursarius lives in the former tallgrass and midgrass prairie and associated open woodland from the Gulf Coast of Texas north into southern Canada, and from the western Great Lakes States to eastern Wyoming and New Mexico. It is at home in a variety of soils. The northern and plains pocket gophers occur together in parts of the Red River Valley in eastern North Dakota and northwestern Minnesota. Although both have been studied in this region of overlap (Kalin 1964; Norum 1966), no published research demonstrates conclusively whether their distribution is determined primarily by soil type or vegetation type, or by behavioral interaction. *Thomomys* has replaced *Geomys* in Emerson, Manitoba, where *Geomys* was abundant 50 years earlier (Wrigley and Dubois 1973). *Geomys* occurs in partly wooded areas of Minnesota but is absent from the coniferous forest of northeast Minnesota.

Natural History: Anatomically, the plains pocket gopher appears to be even more specialized for a fossorial niche than is the northern pocket gopher. The basic way of life of the two species seems to be much the same. *Geomys* is a strict vegetarian but has a greater tolerance for grasses than does *Thomomys* (Vaughan 1976b).

Each solitary, territorial gopher maintains an extensive burrow system which it uses for foraging, for food storage, for protection, and for reproduction. An important function of the burrows of pocket gophers (and of other burrowing mammals) is temperature and humidity control (Kennerly 1964). The burrows of many mammals have a higher humidity and more even temperature than the outside air, but gophers maintain particularly constant underground conditions by plugging the burrow opening with soil. The narrow range of burrow temperatures helps gophers avoid temperature stress. They can also respond to temperature stress by thermal lability, which permits considerable fluctuation of body temperature in response to fluctuations in burrow temperature (Bradley and Yousef 1975). During warm weather, however, the closed burrow with its high humidity makes loss of excess body heat difficult, especially for large individuals. This may place a maximum limit on the size of pocket gophers (McNab 1966).

Figure 32. *Geomys bursarius*, plains pocket gopher.

Breeding occurs in the spring; there is one litter, averaging three to four young. The young probably disperse from the parental burrows in July. Most young are solitary by late summer (Vaughan 1962).

Relationship to People: Plains pocket gophers are both pests and benefactors, depending on circumstances. When abundant, they can be very destructive on the farm and in the home garden. But they also cultivate and aerate the soil, and help prevent runoff of water. Gophers do less damage when predators are abundant enough to keep their numbers in check.

Additional References: Bailey 1926; Baker et al. 1973; Best 1973; Breckenridge 1929; Dalquest and Kilpatrick 1973; Hendricksen 1973; Miller 1964; Myers and Vaughan 1964; Pembleton and Baker 1978; Penny and Zimmerman 1976; Selander et al. 1974; Vaughan 1966.

Family Heteromyidae
Pocket Mice and Kangaroo Rats

Pocket mice and their relatives, like gophers, are primarily a North American group. The heteromyids, however, range somewhat more deeply into northwestern South America than do the geomyids. These rodents typically occupy arid and semi-arid habitats. Of the approximately 75 species in the family, only one occurs in Minnesota. Heteromyids typically have elongate hind legs and long tails and are, to varying degrees, specialized for jumping. Fossils in a preserved burrow in Nebraska indicate that pocket mice have been fossorial since the Pliocene (Voorhies 1974). Their auditory bullae are relatively large, and they have external, furlined cheek pouches. The tooth formula is always 1/1, 0/0, 1/1, 3/3 × 2 = 20.

Additional References: Bartholomew and Cary 1954; Burt 1958; Chiasson 1954; Duke 1957; Eisenberg 1963; McLaughlin 1967; Webster 1969; Webster and Webster 1975; Wondolleck 1978.

Genus *Perognathus*

Perognathus flavescens
Plains Pocket Mouse

Measurements (based on 48 adults measured by Hibbard and Beer 1960): *Body* length 120 (110–132); tail 54 (49–60); hind foot 16.8 (16–18); ear from notch 6 (5–7); weight males 9.2, females 8.8 (6.9–11.5).

Description: This small, long-tailed mouse is cinnamon buff with an admixture of blackish hairs on the upper sides and back. There are usually small white areas on the belly. Each upper incisor has a deep groove on the front. The auditory bullae are proportionately larger than those of any other mammal in Minnesota, and no other mouse in Minnesota has external (furlined) cheek pouches and four cheek teeth in each upper and lower jaw.

Range and Habitat: Perognathus flavescens ranges farther east than any other heteromyid in the United States, occurring from the shortgrass prairies of Montana and eastern Wyoming east into western Missouri, Iowa, and the former prairie areas of Minnesota. It also ranges south to northern Texas and eastern New Mexico. (See map 45.) The plains pocket mouse prefers open grassy habitats, with or without shrubs, but with a well-drained, sandy, partly bare soil (Brown and Metz 1966; Hibbard and Beer 1960).

Natural History: There have been many ecological studies of heteromyids, especially of communities that include more than one species of heteromyid. However, the ecology of *Perognathus flavescens* is practically unknown (Armstrong 1972; Cockrum 1952). The only extensive account of its natural history that I know of is Hibbard and Beer (1960), conducted in south-central Minnesota. The plains pocket mouse, like others of its genus, is a seed eater and a hibernator (Beer 1961a). Seeds commonly stored in its underground caches are those of sedges, spiderwort, wild buckwheat, puccoon, and foxtail.

Figure 33. *Perognathus flavescens*, plains pocket mouse.

Apparently, *P. flavescens* awakens periodically from hibernation to feed on its cache. There may be two or three litters per year. Six pregnant females examined by Hibbard and Beer (1960) were carrying three to five embryos each, with an average of four.

Relationship to People: The plains pocket mouse probably is of no practical significance to people. It is interesting to science as the only representative in Minnesota of a family primarily suited to arid conditions.

Additional References: Bowles 1975; Homan and Genoways 1978; Maxell and Brown 1968; Polderboer 1937.

Family Castoridae
Beaver

Beaver are large aquatic rodents which are found in both North America and Eurasia (McLaughlin 1967). There is only one living genus, but several genera of fossil beaver are known. *Castoroides ohioensis*, which lived in Minnesota during the Pleistocene, was as big as a black bear. A skeleton of *C. ohioensis* is on exhibit at the Minnesota Science Museum in St. Paul.

Genus *Castor*

Castor canadensis
Beaver

Measurements (based on 69 adults from Wyoming [Osborn 1953]): Total length 1082 (1000–1200); tail vertebrae 478 (440–523); scaly tail 282 (258–318); hind foot 178 (156–190); weight: usual adult range 20–27 kg, with maximum reports of 50 kg (Jackson 1961).

Description: The beaver is the largest rodent in North America. It is unmistakable because of its stout build, luxurious brown fur, and broad, flat, scaly tail. The powerful hind legs have webbed feet, and the second toe has a divided claw which

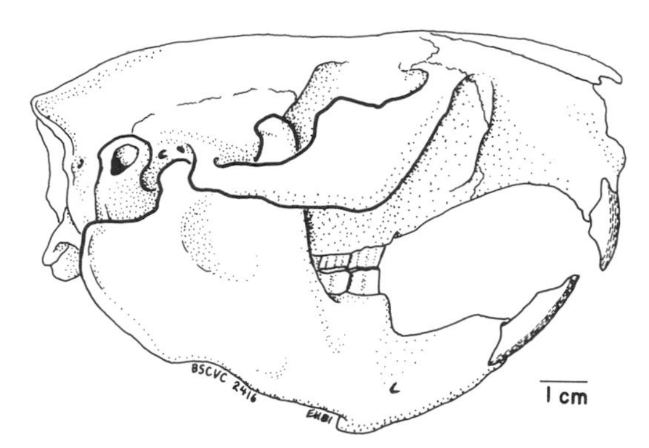

Figure 34. Skull of *Castor canadensis*, beaver (young adult), BSCVC 2416.

is used as a comb. The skull of a beaver is massive, with sturdy incisors faced with deep orange enamel, and relatively high-crowned cheek teeth with parallel sides. The flattened grinding surfaces of the cheek teeth consist of a series of loops of enamel separated by dentine. The tooth formula is 1/1, 0/0, 1/1, 3/3 × 2 = 20.

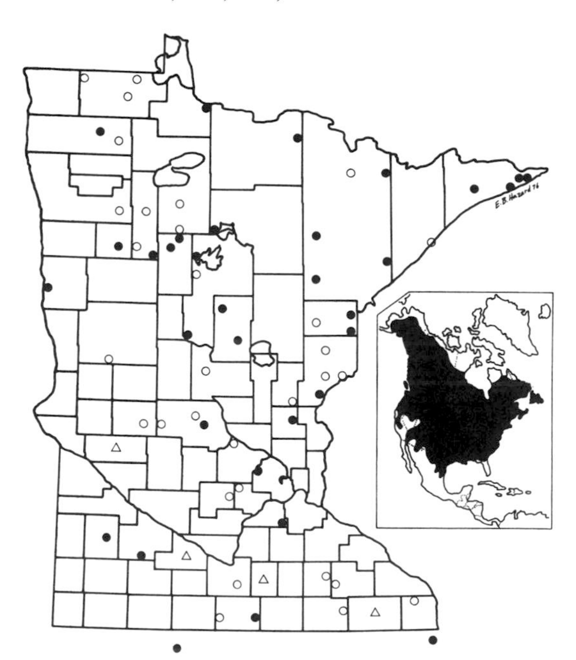

Map 34. Distribution of *Castor canadensis*. ● = township specimens. ○ = other township records, selected locations. △ = other county records. Iowa specimens reported in Bowles 1975. (Map produced by the Department of Biology, Bemidji State University.)

Range and Habitat: Castor canadensis and *Castor fiber*, the Old World beaver, are generally considered to be separate species (Youngman 1975). The former originally ranged over almost all of North America north of Mexico, except for the High Arctic, extreme southeastern United States, and parts of the southwestern deserts. Beaver occurred in both open and wooded country, wherever there were suitable streams, lakes, or even marshes and swamps. They were reduced or extirpated over much of their range by uncontrolled trapping, but have now reoccupied many areas (sometimes being reintroduced by people) and are generally prospering under good management practices. Currently they occur throughout Minnesota (Hiner 1938; Longley and Moyle 1963; Roberts 1935, 1941; Wechsler 1974).

Natural History: If technology is the production of new objects or structures by the manipulation of the natural environment, then, among mammals, only people rival beaver as technologists. Beaver are not as efficient hydraulic engineers as we are, but their dams do a better job of flood control than do many of ours.

In the North, beaver face the same problem as red squirrels—storing food for the winter (see p. 69). But the food and habitat of beaver are such that they function best in a small colony, rather than as individuals. The summer food of beaver comprises a variety of succulent aquatic plants as well as the bark of trees that grow near water. The winter store consists of tree branches and trunks: primarily aspen, willow, and alder. A stable body of water provides a safe means for reaching both summer and winter food supplies, a medium for transporting cut wood, an underwater site for food storage, a refuge from predators, and underwater protection for the entrance to a reasonably secure lodge or bank burrow. The dam or dams provide the stable water level upon which the whole system depends. In habitats where the water level is stable and food is at hand, beaver may build no dam, just as they may build no lodge if a suitable bank is available for a burrow.

Figure 35. *Castor canadensis*, beaver.

The colony generally comprises four to 10 individuals and usually centers on a dominant mature female (Hodgdon and Larson 1973). Her mates may stay for one to a few years, and kits often disperse during their second year, but she remains for many years. Gestation has been estimated to be 100 to 128 days. Mature females generally have four or more young about the first of May; the average in Minnesota is five (Wechsler 1974). Females may first breed in their second year, but some do not breed until their third year (Brenner 1964; Larson 1967; Longley and Moyle 1963; Osborn 1953). All but the young of the year are active during the summer and fall, cutting trees and building and repairing dams, mostly at night. The whole colony generally lives in a single lodge or bank burrow, thus conserving body heat in winter. The interior lodge temperature seldom drops below 0° C, especially if the snow cover is deep (Stephenson 1969).

Beaver have many anatomical and physiological adaptations that make them well suited to an aquatic life. They keep their thick fur dry by oiling it with the secretions of abdominal skin glands. The large webbed hind feet provide propulsion, and the tail is used primarily as a rudder and sometimes as a scull—it is not used to carry mud. When a beaver dives rapidly, its tail may slap the water, the noise alerting the colony to potential danger. Valves can shut water out of both the nose and the ears. The lips can close behind the gnawing incisors. The internal nares (the openings at the rear of the nasal passages) form a tube which lies within the glottis (the opening of the trachea or windpipe). This arrangement provides a continuous air passage, functionally separate from the mouth and pharynx, which are often full of water (Coles 1970). Hence, a beaver can feed or carry wood and still breathe by bringing its nostrils above water. Beaver, like other diving mammals, tolerate a high blood level of carbon dioxide and have other physiological mechanisms that permit prolonged submersion.

Relationship to People: The beaver has perhaps affected human history in North America more than any other wild mammal. The fur trade, which influenced political events and the pattern of settlement from coast to coast in both the United States and Canada, was primarily the beaver trade. The role of the beaver in history is recognized in hundreds of place-names and in the official seals of various levels of government.

At first, beaver were trapped year-round, primarily in the summer. There were no trapping seasons and no limits. The fur is prime only in late winter and early spring, and only the best furs were used for fur coats. Nonprime or damaged furs were made into felt hats. The excessive uncontrolled harvest of the early years was the major factor that caused widespread extirpation of the beaver. Early trappers, like early loggers, simply moved farther west as their resource dwindled. Although it is difficult to overharvest some species, the beaver is a classic example of a species that can be exterminated by overexploitation. In most states with substantial beaver populations, the species is now managed in order to provide a reliable annual harvest and a relatively stable population. Management primarily involves regulating the total catch, restricting the trapping season to the early spring, and managing selected habitats. An early spring trapping season accomplishes three things. It insures that the adult male will not be removed before he has bred the adult female, that the young of the previous year will be old enough to survive if the mother is trapped, and that the fur will be prime, thus giving maximum return on the restricted catch (Wechsler 1974). Beaver manage their own habitat, but our land use practices may or may not encourage them in this regard. Replacement of aspen by conifers, stream channelization, and wetland drainage are all detrimental to beaver.

Beaver ponds may benefit ducks and other wildlife by providing nesting and feeding sites but, by replacing running water with relatively still water and by silting the bottom, they may harm trout streams (Knudsen 1962; Renouf 1972). The ponds also conserve spring runoff, thus insuring more constant stream flow, diminishing floods, conserving soil, and helping maintain the water table. Silted-in beaver ponds may become profitable farm- or forestland. Inconveniently located dams may cause problems, such as washed-out roads or destruction of valuable bottomland timber (Vesall et al. 1947; Wechsler 1974). Conservation officers are authorized to issue permits for removal of beaver that cause such damage.

Additional References: Aldous 1938; Aleksiuk 1968; Beer 1955b; Bradt 1938; Erickson 1939; Hansen et al. 1973; Henderson 1960; Hibbard 1958; Highby 1940; Jenkins and Busher 1979; Kittredge 1938; Marshall and Miquelle 1978; Novak 1977; Novakowski 1967; Provost 1962; Slough 1978; Stenlund 1953; Svendsen 1978; Tevis 1950; Townsend 1953; Warren 1927; Wilde et al. 1950.

Family Cricetidae

New World Rats and Mice, and Voles and Lemmings

This large family of small to medium-sized rodents is distributed over most of the major land masses of the world, except for southeast Asia, the East Indies, and the Australian region. Of the several subfamilies, two occur in Minnesota. The subfamily Cricetinae includes the relatively long-tailed, large-eared, large-eyed harvest mice, deer mice, and grasshopper mice. The subfamily Microtinae (treated by some as a separate family, the Arvicolidae) includes the relatively short-eared, beady-eyed voles and lemmings. The latter group, except for the muskrat, generally have short to medium-length tails. All members of the family have the tooth formula 1/1, 0/0, 0/0, 3/3 $\times$ 2 = 16. Some mammalogists classify cricetids in the family Muridae (Hooper and Musser 1964; Lawlor 1979).

Additional References: Arata 1967; Hartung and Dewsbury 1978; Reig 1977; Vaughan 1978.

Subfamily Cricetinae
New World Rats and Mice

Genus *Reithrodontomys*
Harvest Mice

The genus *Reithrodontomys* comprises about 18 species, most of which inhabit the western United States and Latin America. One species is isolated from the others in the southeastern United States, and one western species reaches Minnesota.

Reithrodontomys megalotis
Western Harvest Mouse

Measurements: Total length 136 (123–153); tail 62 (55–74); hind foot 16 (15–18); ear from notch 12.5 (12–14); weight 11.8 (six specimens, 10.0–12.9).

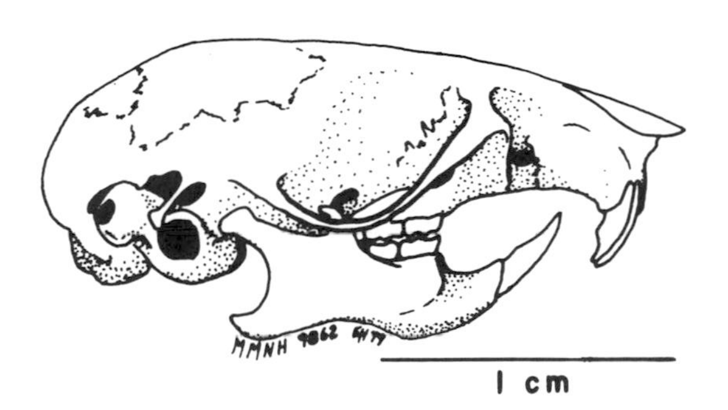

Figure 36. Skull of *Reithrodontomys megalotis*, western harvest mouse, MMNH 9862.

Description: Western harvest mice are among the smallest rodents in Minnesota. They are brown above and whitish below. The upper incisors are grooved. House mice (*Mus musculus*) average larger than *R. megalotis*, and are more uniformly colored above and below. Pocket mice are a lighter brown, have external cheek pouches, and have four cheek teeth in each jaw. Harvest mice, like all cricetids, have no premolars.

Range and Habitat: Reithrodontomys megalotis has a wide range through much of the western United States and Mexico, the central Rockies, and the central and Upper Midwest. In recent decades it has apparently extended its range through Illinois to eastern Indiana (Whitaker and Mumford 1972b; Whitaker and Sly 1970). In Minnesota, harvest mice occur in the southern part of the state only and seem to be uncommon there. (See map 36.) They are most likely to be found in grassy, overgrown fields, in fencerows near fields, and on unmown roadsides (Stains and Turner 1963).

Natural History: The habits of harvest mice are well known in other areas (e.g. California and Indiana), but not in Minnesota. The gestation is 23 or 24 days, the maximum known litter size is 8, and up to 14 litters in one year have been recorded in captivity (Egoscue et al. 1970). I would be surprised if there were more than four litters a year in Minnesota. In warm, dry parts of the range, breeding occurs in late spring and early fall, with a hiatus during the hottest part of the summer. Harvest mice often build their nests aboveground: in a shrub or attached to several stems of forbs or grasses, or on the surface of the ground (Shump 1975). In some areas, at least, males are territorial and are more abundant than females (Fisler 1971). The diet consists largely of the seeds of grasses and of what many people choose to call weeds. Invertebrates, especially lepidopterous larvae, also make up a large part of the diet.

Relationship to People: Harvest mice have expanded their range eastward and possibly northward as the land has been cleared, but in much of their former prairie habitat, cultivation of the land has restricted them to roadsides and other "waste" habitats. They are probably not abundant enough in Minnesota to be economically important. Their effects, if any, tend to be positive, with the small amount of grain they eat being overbalanced by their consumption of weed seeds and caterpillars.

Additional References: Beer and MacLeod 1955; Blaustein and Rothstein 1978; Carleton and Myers 1979; Fisler 1965,

1970; Ford 1977; Frydendall 1969; D. R. Johnson 1961; Jones and Mursaloğlu 1961; Kaufman and Fleharty 1974; Long 1962; Meserve 1977; Pearson 1964; Smith 1936; Tanner 1976.

Genus *Peromyscus*
White-footed Mice

Mice of the genus *Peromyscus* are common in most habitats of North America. (Several authors review the biology of the genus in King 1968.) There are two species of *Peromyscus* in Minnesota. One is *P. leucopus*, the white-footed mouse, a woodland species. The other, *P. maniculatus*, the deer mouse, occurs here as two distinct subspecies: *P. m. gracilis*, the woodland deer mouse, and *P. m. bairdii*, the prairie deer mouse. In this book I have not generally considered subspecies because they usually intergrade from one subspecies to another of the same species as you cross the state. In most parts of its range, subspecies of *P. maniculatus* also intergrade from region to region, but in the upper Great Lakes states *P. m. gracilis* and *P. m. bairdii* overlap with relatively little intergradation. These two subspecies are often found in distinct habitats within the same area, usually with little or no interbreeding. I therefore treat these forms of *P. maniculatus* separately.

Additional References: Greenbaum and Baker 1978; Greenbaum et al. 1978; Hooper 1958; King, Maas, and Weismann 1964; King et al. 1967; Lawlor 1979; McNab and Morrison 1963; Miller and Getz 1977a; Osgood 1909; Terman and Sassaman 1967; Zimmerman et al. 1978.

Peromyscus maniculatus bairdii
Prairie Deer Mouse

Measurements: Total length 152 (138–168); tail 62 (56–69); hind foot 18 (16–19.5); ear from notch 15 (14–18?); weight 21.5 (19.4–24.8).

Description: The prairie deer mouse and the other two forms of *Peromyscus* in our area are one shade or another of brown on the back and sides, and white below. Harvest mice are smaller and more delicate and have grooved incisors; grasshopper mice are stockier; and house mice are buff below, not white. Of the three *Peromyscus*, the prairie deer mouse has the shortest tail, averaging about 40 percent of the total length, and the shortest ears. The tail is usually distinctly bicolored, the dark upper surface sharply separated from the white lower portion. This mouse is generally the grayest of the three *Peromyscus*. Subadults of all three forms are gray and may be indistinguishable.

Range and Habitat: P. m. bairdii is an upland, grassland mammal which ranges from the Dakotas through much of western and southern Minnesota and eastward south of the Great Lakes and into the Appalachians. It has spread eastward and northward as the eastern forests have been replaced by farmland. In Minnesota, Wisconsin, and Michigan it now occurs within the range of the woodland deer mouse, *P. m. gracilis*, and the two subspecies are generally reported to not

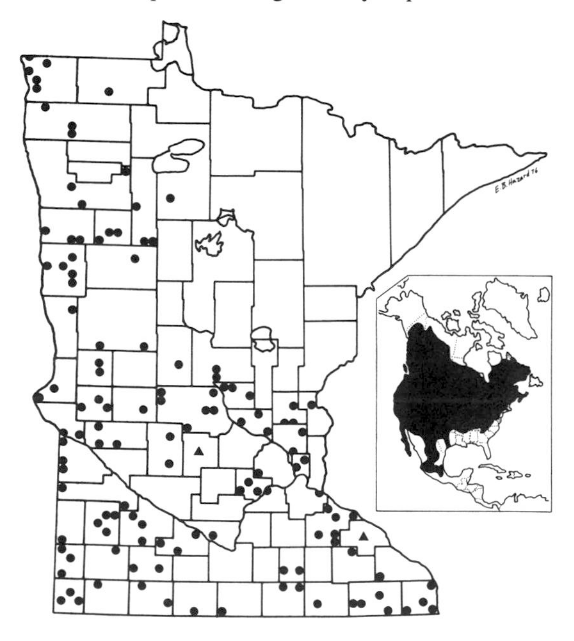

Map 35. Distribution of *Peromyscus maniculatus bairdii*. ● = township specimens, selected locations. ▲ = county specimens. Inset = continental distribution of *all* subspecies of *P. maniculatus*. (Map produced by the Department of Biology, Bemidji State University.)

interbreed in nature, even where both may be trapped at the same forest-grassland edge. However, at least in northwest Minnesota, intermediate individuals have been found (Iverson et al. 1967). In Wisconsin, *P. m. bairdii* is more abundant on cultivated land than the meadow vole (*Microtus pennsylvanicus*), which prefers the grassy edges (LoBue and Darnell 1959).

Natural History: Mice of the genus *Peromyscus* are nocturnal and omnivorous. They feed extensively on seeds, and the prairie deer mouse stores seeds of such plants as bush clover in its burrows (Barry 1976; Howard and Evans 1961). Prairie deer mice also eat insects (Flake 1973). Females can breed at an age of about seven weeks. Breeding goes on from April through September, gestation lasts about 23 days, and one female may raise several litters in a season (Beer and MacLeod 1966). Weaning occurs at about 18 days of age (King et al. 1964). During the breeding season, at least, adults are generally territorial (Stickel 1968; Terman 1961, 1962). Males are reported to reside occasionally in a nest with a female and young, and two litters with two mothers sometimes occur in one nest (Howard 1949b). Prairie deer mice are active throughout the winter, when several may aggregate in a single nest.

Relationship to People: P. m. bairdii eats both weed seeds and forage plant seeds, as well as insects. It probably seldom eats a large enough proportion of the seed crop to hamper reproduction of any desirable species.

Additional References: Baker 1971; Beer et al. 1957; Blair 1940, 1948; Criddle 1950; Crowner and Barrett 1979; Drickamer 1970, 1972, 1976; Grant and Birney 1979; Harris 1952, 1954; Hart and King 1966; Houtcooper 1978; King 1958; Savidge 1974a, b; Terman 1973; Tester and Marshall 1961; Thomas and Terman 1975; Verts 1957; Wecker 1963.

Peromyscus maniculatus gracilis
Woodland Deer Mouse

Measurements: Total length 181 (162–200); tail 89 (80–104); hind foot 21 (20–22); ear from notch 20 (17–22); weight 20.6 (14.9?–25.1).

Description: In Minnesota, *P. m. gracilis* is the *Peromyscus* with the longest tail (up to half of the total length) and the largest ears. (However, in northeastern Minnesota where it intergrades with *P. m. maniculatus* of eastern Canada, the ears may be somewhat shorter.) The fur is brown above and white below, and the tail, as in *P. m. bairdii*, is sharply bicolor. Except for occasional hybrids (Iverson et al. 1967), *P. m. bairdii* and *P. m. gracilis* can readily be separated on measurements or proportions alone. It is more difficult to distinguish the woodland deer mouse from the white-footed mouse (*P. leucopus*), which may inhabit the same woods, than from the prairie deer mouse.

Range and Habitat: This subspecies is the characteristic long-tailed mouse of northern and

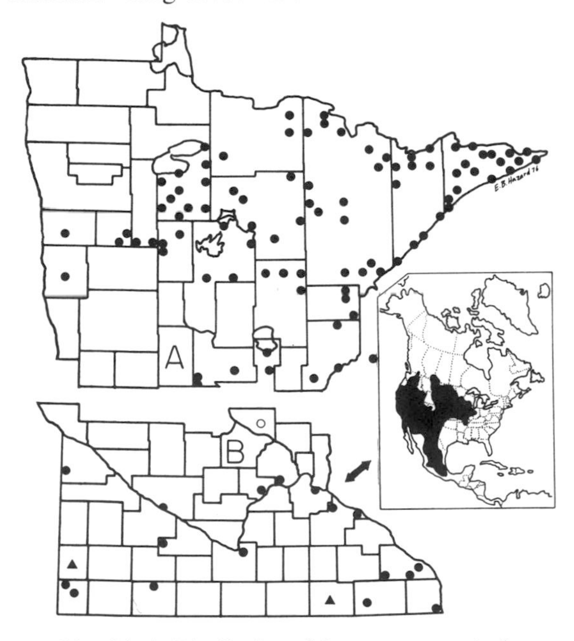

Map 36. A. Distribution of *Peromyscus maniculatus gracilis*. ● = township specimens, selected locations. B. Distribution of *Reithrodontomys megalotis*. ● = township specimens. ○ = other township records, selected locations. ▲ = county specimens. Inset = continental distribution of *R. megalotis*. (Map produced by the Department of Biology, Bemidji State University.)

northeastern Minnesota, of the northern parts of the Great Lakes States, and of parts of southeastern Canada and New England. It is largely restricted to woodlands and is commonest in relatively deep woods with at least some conifers. In northern Michigan, however, some populations favor lichen-grass fields (Fitch 1979), and *P. m. gracilis* often invades recently burned or clear cut areas in large numbers (Sims and Buckner 1973).

Figure 37. *Peromyscus maniculatus gracilis*, woodland deer mouse.

Natural History: In contrast to the prairie deer mouse, the woodland deer mouse is partly arboreal; its long tail is an adaptation for balancing as it climbs about in shrubs and trees (Horner 1954). The woodland deer mouse is omnivorous, feeding on nuts and other seeds, buds, berries, and whatever meat, mostly invertebrate, is available. It readily eats vertebrate carrion, as do many rodents. Females can breed at seven weeks of age and may have several litters per year. Average litter size varies from 4.0 to 5.5 (Flake 1974). Weaning takes place at about 24 days of age (King et al. 1964). The nest may be above ground in a hollow stub, or under a log or some other shelter. Like the other members of the genus, the woodland deer mouse is strictly nocturnal and will not usually be seen in the daytime unless it has been disturbed. The breeding females, at least, seem to be territorial. In captivity, males have been observed helping to care for the young (Horner 1947).

P. m. gracilis is often apparently the most abundant mammal in a woods, but the population may decrease markedly every four years or so, then gradually increase to peak numbers again. Terman (1968) reports that some populations of *Peromyscus* maintain relatively low but stable densities. The population of *P. m. gracilis* that I know best, in southeastern Beltrami County, seems to fluctuate noticeably. Manville (1949) found an inverse relationship between numbers of *P. m. gracilis* and of *Clethrionomys gapperi* (the southern red-backed vole).

Relationship to People: The woodland deer mouse benefits us by helping to prevent outbreaks of defoliating insects. On the other hand, rapid influxes into burned or cutover areas may hamper forest reseeding (Sims and Buckner 1973). It can be a pest in summer cabins, where it may find shelter, nest sites and nest material, and food for the winter months. In towns and in farm buildings, the house mouse (*Mus musculus*) is generally more successful. The best way to control mice in houses is to set mouse traps indoors near potential entrances and along walls.

Additional References: Beer et al. 1957; Blair 1942; Dice and Clark 1962; Harris 1952, 1954; Hart and King 1966; Hooper 1942; King 1958; Kirkland 1976; Kirkland and Griffin 1974; Klein 1960; Miner 1966; Rasmussen 1964; Smith and Speller 1970; Tadlock and Klein 1979.

Peromyscus leucopus
White-footed Mouse, Wood Mouse

Measurements: Total length 166 (153–178); tail 75 (65–84); hind foot 20 (19–22); ear from notch 16 (15–17); weight 27.3 (19.0–35.4).

Description: It is sometimes difficult to tell this mouse from *P. m. gracilis*, the woodland deer mouse. On the average, the tail is shorter (about 45 percent of the total length), the ears are

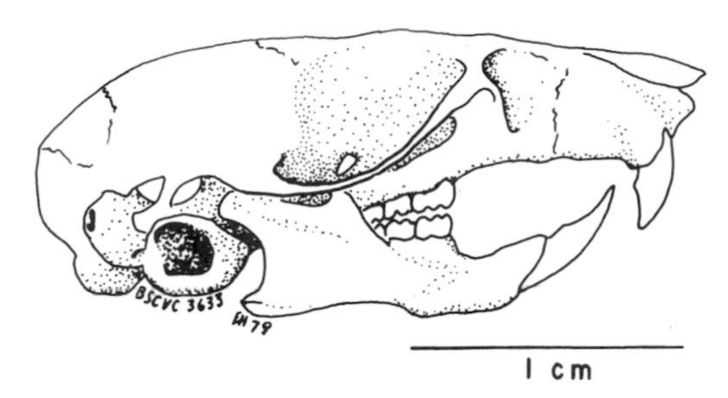

Figure 38. Skull of *Peromyscus leucopus*, white-footed mouse, BSCVC 3633.

shorter, and the sides and back a more reddish brown. The upper lip is usually dusky, and the tail is not sharply bicolored, although it is darker above than below. It is not safe to try to distinguish juveniles or subadults. Additional useful characters are mentioned in the skull key.

Range and Habitat: Peromyscus leucopus has a wide range, from eastern Mexico through the central states, to extreme southern Alberta and Saskatchewan, and east to the Atlantic. It is absent from the extreme southeastern United States and apparently also absent from extreme northeastern Minnesota. The white-footed mouse is primarily a species of deciduous woodlands and is generally more tolerant of open conditions and early stages in ecologic succession than is the woodland deer mouse (Bowker and Pearson 1975; Hirth 1959; Iverson et al. 1967; Klein 1960; M'Closkey 1975a; Pearson 1959; Verts 1957; Wetzel 1958).

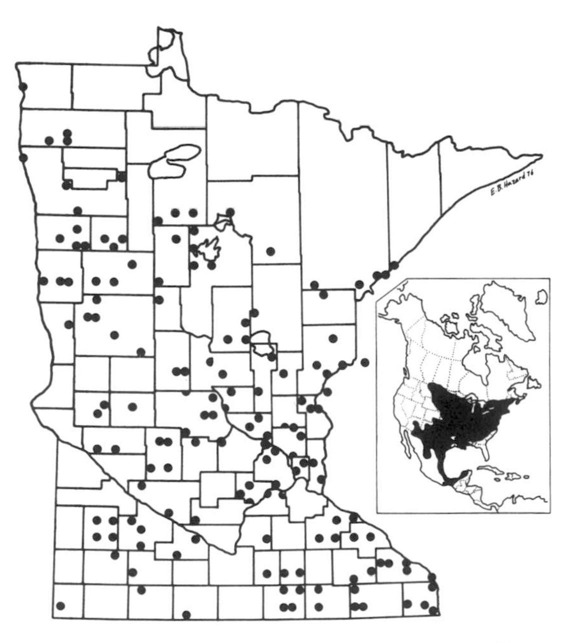

Map 37. Distribution of *Peromyscus leucopus*. ● = township specimens, selected locations. (Map produced by the Department of Biology, Bemidji State University.)

Natural History: Like the other members of its genus, *P. leucopus* is nocturnal and omnivorous. It feeds on a variety of seeds, nuts, and berries, but is primarily insectivorous when suitable insects are abundant (M'Closkey 1975b). White-footed mice are at home in trees and shrubs as well as on the ground. They may nest in burrows or other cavities at or below ground level, or above ground in a stump or hollow tree. They reputedly do not dig their own burrows.

Females first breed at eight to 10 weeks of age; gestation generally lasts 23 days; and the number of young in the litters of older females is usually three to six, with four being most common (Jackson 1961). A female may have four or more litters in a season, but there may be a lull in breeding during midsummer (Sheppe 1965c). *P. leucopus* is typically promiscuous, and males generally play no role in raising the young.

There is little overlap between the home ranges of adults of the same sex, especially among females (Metzgar 1971). Female territoriality may help prevent overpopulation among wood mice, as may a delay in sexual maturation in young animals under crowded conditions (Rogers and Beauchamp 1976). Nonetheless, substantial fluctuations of numbers occur in *P. leucopus*, as they do in other members of the genus. The species spreads readily into unoccupied areas because adults as well as young are distinctly exploratory and because individuals move readily across areas of unsuitable habitat, including water barriers as wide as 125 m (Sheppe 1965b, 1966).

Relationship to People: White-footed mice are unlikely to be of economic significance to people except as pests in forest reseeding projects, or as possible benefactors in checking populations of some harmful insects. Their abundance in heavily populated areas of North America has made them ideal subjects for research in population dynamics, habitat selection, food habits, and reproductive biology.

Additional References: Baker 1971; Barry 1976; Beer 1961b; Brown 1964; Browne 1977; Burt 1940, 1943; Christianson 1977; Cook and Terman 1977; Cornish and Bradshaw 1978; Dewsbury 1975; Dewsbury et al. 1977; Drickamer 1970, 1972, 1976; Fleming and Rauscher 1978; Getz 1961d, 1968a; Grant and Birney 1979; Hansen and Batzli 1979; Hill 1972; Long 1973; M'Closkey and Fieldwick 1975; M'Closkey and Lajoie 1975; Metzgar 1973; Millar 1975; Myton 1974; Nicholson 1941; Orr 1966; Sheppe 1965a; Smith and Speller 1970; Snyder 1956; Vestal and Hellack 1978.

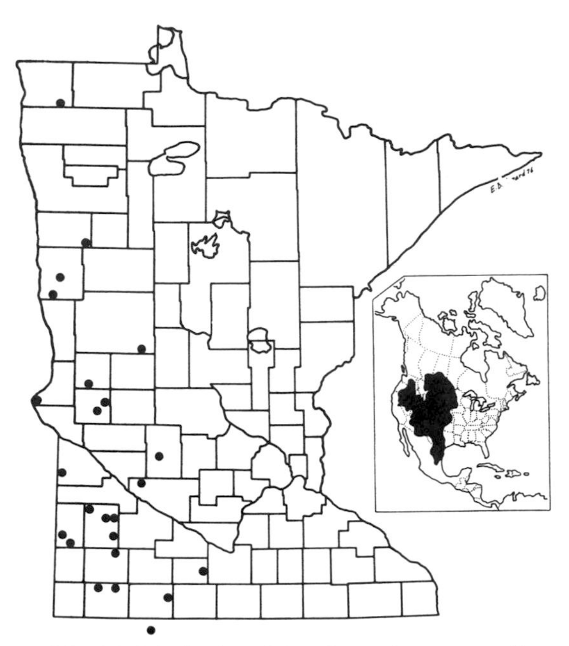

Map 38. Distribution of *Onychomys leucogaster.* ● = township specimens. Iowa specimen reported in Bowles 1975. (Map produced by the Department of Biology, Bemidji State University.)

Genus *Onychomys*
Grasshopper Mice

This genus is restricted to North America, and there are only two or three species (R. J. Baker et al. 1979; Hollister 1914; Van Cura and Hoffmeister 1966).

Onychomys leucogaster
Northern Grasshopper Mouse

Measurements: Total length 151 (139–166); tail 40 (34–45); hind foot 22 (19–23); ear from notch 16 (14–18); weight 42.1 (34.8–50.6).

Description: This relatively rare mouse resembles *Peromyscus*, but it is stouter and about twice as heavy, and distinctly gray above (juvenile *Peromyscus* are gray also); moreover, its tail is less than one-third of the total length, and it has a distinct white patch in front of each ear. Its underparts are white.

Range and Habitat: The northern grasshopper mouse inhabits grasslands from northern Mexico to the southern Prairie Provinces, and from the Great Basin to the eastern edge of the Great Plains. It is most common in sparsely vegetated areas (Choate and Terry 1974), though it is generally rare in comparison to other resident mice. In Minnesota, it is restricted to the prairie counties.

Natural History: Onychomys leucogaster is an unusual mouse, both in its food habits and in its social organization. It feeds almost wholly on animal matter, vertebrate as well as invertebrate. As much as 25 percent of the stomach contents of grasshopper mice is vegetation (Flake 1973), but much of this comes from the intestines of the herbivorous insects that they eat (Hansen 1975). Male grasshopper mice in spacious laboratory enclosures invariably killed other species of rodents housed with them, including species three times their weight (Ruffer 1968). In the southern grasshopper mouse, *O. torridus*, the male kills all mammalian prey, at least under laboratory conditions (O'Farrell and Cosgrove 1975). Aside

Figure 39. *Onychomys leucogaster*, northern grasshopper mouse.

from its stocky, powerful build, the grasshopper mouse has no obvious external anatomical adaptations for a carnivorous diet. Like most carnivores, it is considerably less abundant than herbivores of comparable size in the same habitat.

Male *O. leucogaster* are strongly territorial, and adults live in male-female pairs. The pair stores food underground, presumably for short periods. Breeding in Colorado lasts from March until August or later; the season may well be shorter in Minnesota. Gestation lasts 33 days, longer than in most other cricetines, and may be delayed to 52 days or more if the female is nursing an earlier litter (O'Farrell 1975). In Colorado, litter size varies from three to seven, with an average of about 4.6 (Flake 1974).

Relationship to People: Grasshopper mice are probably not abundant enough to be of great practical importance to people, but they can only be beneficial, since they feed primarily on herbivorous insects and rodents which pose a potential threat to crops. They also make good laboratory subjects, taking well to captivity, having a greater median life-span than white-footed mice (O'Farrell and Cosgrove 1975), and permitting study of a social organization unusual for rodents.

Additional References: Anderson and Jones 1971; Carleton and Eshelman 1979; Dewsbury et al. 1977; Dickerman and Tester 1957; Egoscue 1960; Engstrom and Choate 1979; Frydendall 1969; Grant et al. 1977; Harriman 1973; Jahoda 1973; Kaufman and Fleharty 1974; Kleiman 1977; Lanier and Dewsbury 1977; McCarty 1978; Pinter 1970; Ruffer 1965a, c.

Subfamily Microtinae
Voles and Lemmings

Microtines are typically chunky rodents with small eyes and short ears. Many species have relatively short tails, but some, including the muskrat, are long-tailed. In contrast to cricetines, most of which are omnivores, microtines feed primarily on green vegetation. Their molars are typically high-crowned, with grinding surfaces consisting of loops and triangles (fig. 86). Some species have evergrowing molars (Oxberry 1975). Because the morphology of voles and lemmings is distinctive, Guilday and Hamilton (1978), van der Meulen (1978), and others recognize them as a separate family, the Arvicolidae.

Microtines (or arvicolids) are common in the Northern Hemisphere, in both the Old and the New Worlds.

Additional References: Anderson 1959; Arata 1967; Elton 1942; Freeland 1974; Freeland and Janzen 1974; Hooper 1968; Innes 1978; Nadler et al. 1978; Richmond and Stehn 1976; Schlesinger 1976; Vaughan 1978.

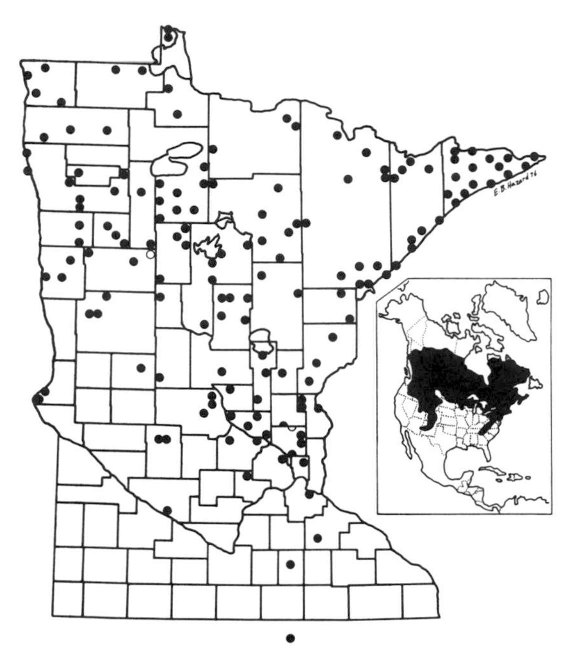

Map 39. Distribution of *Clethrionomys gapperi*. ● = township specimens, selected locations. ○ = other township records. (Map produced by the Department of Biology, Bemidji State University.) Iowa specimen reported in Bowles 1975.

Genus *Clethrionomys*

Clethrionomys gapperi
Southern Red-backed Vole

Measurements: Total length 134 (114–150); tail 37 (30–45); hind foot 18 (17–19.2); ear from notch 14 (11.5–17); weight 24.3 (14.8–34.3).

Description: The southern red-backed vole has the stocky build, short ears, small eyes, and short tail typical of voles. It has gray underparts, brown sides, and a rich, reddish brown back. In winter, the pelage is distinctly longer and lighter colored than in summer.

Range and Habitat: Clethrionomys gapperi is primarily an inhabitant of the taiga. It occurs in Canadian forests from the Pacific to the Atlantic, and in coniferous and mixed forests of the northern United States, including the eastern Dakotas, Minnesota, and northern Iowa. It ranges far to the south in the Rockies and Appalachians. *C. gapperi* occurs in forests throughout most of Minnesota but is most common in moist coniferous and mixed forests in the north. It seems to be at least as partial to humid forest conditions as to the conifers themselves (Getz 1962a; Hirth 1959; Iverson et al. 1967; Kirkland and Griffin 1974).

There are three recognized species of *Clethrionomys* in North America: *C. occidentalis* in the coastal forests of the Pacific Northwest; *C. rutilus* in Alaska and the Canadian Arctic; and *C. gapperi*. *C. rutilus* also occurs across northern Eurasia. Grant (1974) has shown that *C. gapperi* is partly interfertile with *C. glareolus*, the bank vole, which occurs only in Eurasia. Youngman (1975), on the other hand, suggests that *C. gapperi* is conspecific with *C. rutilus*. Canham and Cameron (1972) have investigated variation in the serum proteins of *C. rutilus* and *C. gapperi*.

Natural History: In moist coniferous and mixed forests and in cedar, tamarack, and black spruce swamps, the red-backed vole is generally our most common rodent. It is primarily terrestrial but seems to be in part a competitor of the semiarboreal woodland deer mouse which is often common in the same forests (Kirkland and Griffin 1974; Manville 1949). It also interacts with the meadow vole where suitable grassland and woodland habitats meet; the two species seem to be able to coexist in either habitat, but not during the breeding season (Clough 1964; Crowell 1973; Grant 1969; Iverson and Turner 1972; Morris 1969; Turner et al. 1975). *C. gapperi* may exclude *Napaeozapus insignis*, the woodland jumping mouse, from mutually suitable habitat (Lovejoy 1973).

Clethrionomys, like some other voles, is active both day and night, in contrast to the more strictly nocturnal cricetines. Red-backed voles often construct burrows but do not maintain well-groomed surface runways to the extent that meadow voles do. The nest is either in a burrow under a shelter such as a rotten log, or among tree roots (Jackson 1961). These voles are less likely to move into houses and cabins than deermice are. Gestation lasts 17 to 19 days, and the two to eight (most often four or five) young are weaned and on their own in less than three weeks. There may be several litters in a single season, which in Minnesota probably extends from early May through late September. The young are mature and able to reproduce in four or five months. *C. gapperi* is typically a forest dweller, but, when populations are high, it is likely to invade adjacent grasslands or disturbed areas such as burns or blowdowns. The dispersing animals are usually subadults (Powell 1972).

The red-backed vole feeds primarily on green vegetation. In addition to herbaceous leaves and shoots, it eats the bark of young trees and shrubs, and a variety of fruits and nuts. It also eats some insects (Hamilton 1941b; Jackson 1961). It, in turn, is a staple item in the diet of various mammalian and avian predators. Masked shrews and short-tailed shrews inhabit the same forests as do red-backed voles (Timm 1975), and may be significant predators, especially on young voles.

Relationship to People: Red-backed voles are most common in relatively wild country and seldom have much effect on people. They occasionally damage young deciduous trees in winter; however, they may help control populations of defoliating insects.

Additional References: Beer 1953b, 1961b; Beer et al. 1957; Beuch, Siderits, et al. 1977; Cameron 1964; Fuller 1977a, b; Getz 1968a, b; Gunderson 1959; Heaney and Birney 1975; McManus 1974a; Merritt and Merritt 1978; Mihok 1976; Miller and Getz 1977a; Murie and Dickinson 1973; Schloyer 1977.

Genus *Phenacomys*

Phenacomys intermedius
Heather Vole

Measurements (four specimens from Churchill, Manitoba): Total length 135–154; tail 24–31; hind foot 18–20; weight 39.0–47.

Description: This small, brownish gray vole has a yellowish nose and a tail less than 42 mm long (see rock vole, below). The back of the one specimen known from Minnesota "is bright yellowish brown" (Handley 1954). The belly fur is dark gray with white tips.

Range and Habitat: Phenacomys intermedius is the most widespread of the four species of this Nearctic genus. It ranges across the boreal forests of Canada from Labrador to the southern Yukon, and into the Cascades, northern Sierras, and northern Rockies in the western United States. In Minnesota, its southern limit is the north shore of Lake Superior and the coniferous forests of the Canada-Minnesota boundary. (See map 41.) Shaler Aldous trapped one specimen in 1940 near Ely, but the exact location, and therefore the county, is unclear (Handley 1954; Swanson et al. 1945; Timm 1975). Within its wide range, the heather vole occupies a variety of wooded and open habitats. It probably does not occupy stands vegetated only by grass, and it is generally associated with heaths (family Ericaceae).

During the Pleistocene, *Phenacomys intermedius* or a close relative occurred in various sites south of the glaciated region of the United States. This supports the hypothesis that a belt of boreal forest occurred south of the glaciers (Guilday et al. 1971; Guilday and Parmalee 1972).

Natural History: Phenacomys intermedius inhabits sparsely settled areas and is not commonly trapped. Most of our knowledge of its biology comes from the research of Foster (1961) in Aulthier-Nord, Quebec, and Fort Churchill, Manitoba. The heather vole is apparently entirely herbivorous, feeding on the bark of willow, birch, and heaths, on the fruit and leaves of such heaths as blueberry and bearberry, on hypogeous fungi (Williams and Finney 1964), and perhaps on lichens. It caches both berries and foliage. Heather voles are active to some extent during the day but seem to be more crespuscular and nocturnal than *Clethrionomys*.

Phenacomys intermedius builds its summer nest underground in the shelter of rocks or roots but constructs its winter nest on the surface where it generally is protected by snow. The gestation period is about 19 days, and the first litter is born in mid-June. Litter size ranges from two to eight and averages about five. The young leave the nest when about three weeks old, and young females can breed at an age of four to six weeks. Primiparous females may weigh about half as much as fully grown females (Vaughan 1969). Males are not sexually mature during their first summer.

Relationship to People: The heather vole probably is of no economic significance to people. It is at the edge of its range in northeastern Minnesota, where it may well occur in disjunct rather than continuous populations and therefore be vulnerable to human alteration of the habitat. Despite intensive efforts, no specimen has been collected since the first one in 1940. If *P. intermedius* is rediscovered in the state, overcollecting should be avoided.

Additional References: Brown 1967b; Foster and Peterson 1961; Graham 1976; Guilday and Hamilton 1973, 1978; M. L. Johnson 1967, 1973; Martell and Radvanyi 1977; Peterson 1966; Youngman 1975.

Genus *Microtus*
Voles

Voles of this genus occur in North America, Eurasia, and Africa. All the species in Minnesota are confined to North America.

Microtus pennsylvanicus
Meadow Vole

Measurements: Total length 155 (137–171); tail 39 (32–43); hind foot 19 (17–20); ear from notch 12 (10–15); weight 40.0 (31.3–52.2).

Description: This is the largest vole in Minnesota. It has a dark brown back and somewhat lighter sides. Its gray-brown belly fur is tipped with silver. The tail in adults normally measures 37 mm or more. The ears are more deeply hidden in the fur than are those of the red-backed vole. There is no distinct red or chestnut band down the back, and neither the nose nor the underparts are yellowish. The pattern of the lower molars is distinctive (see fig. 86). Females have eight teats, all of them functional.

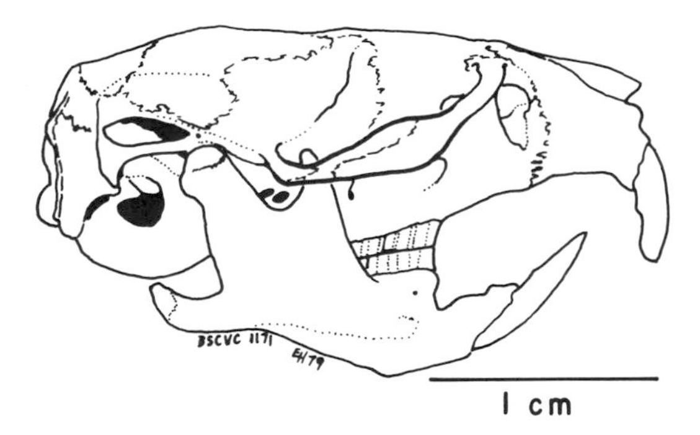

Figure 40. Skull of *Microtus pennsylvanicus*, meadow vole, BSCVC 1171.

Range and Habitat: The meadow vole ranges from Alaska to Labrador and as far south as the southern Rockies and the coast of Georgia. It is absent from a large area of the central plains states. The species occurs throughout Minnesota but is uncommon in the extreme northeast (Timm 1975). *M. pennsylvanicus* prefers open habitats

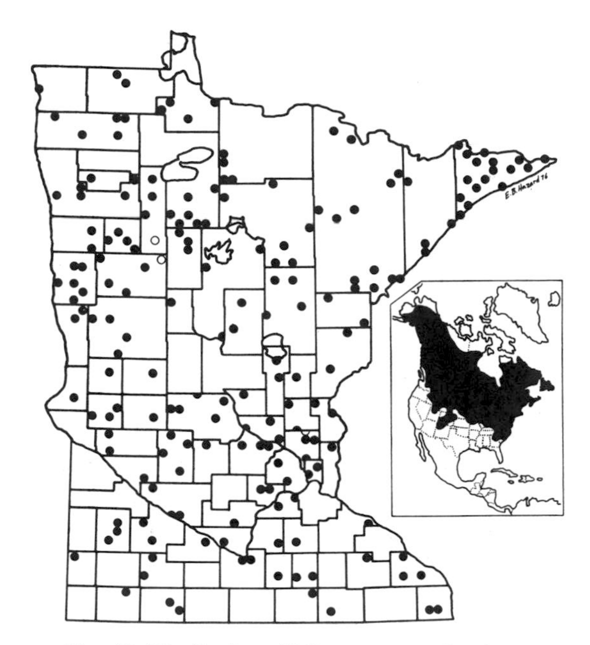

Map 40. Distribution of *Microtus pennsylvanicus*. ● = township specimens, selected locations. ○ = other township records. (Map produced by the Department of Biology, Bemidji State University.)

with a dense grass cover (Bowker and Pearson 1975; Eadie 1953; Hodgson 1972; Iverson et al. 1967). When populations are dense, it may move into adjacent woods, especially if there is a grass understory. It can survive and reproduce in woodland in the absence of competitors (Grant 1971, 1975). In cropland, meadow voles may be confined to grassy edges, with prairie deer mice occupying the cultivated fields (LoBue and Darnell 1959).

Natural History: The meadow vole is Minnesota's most prolific rodent. There are as many as 11 young in a litter, though five and six are most common (Hamilton 1941a). Gestation lasts 21 days or less, the young leave the nest at three weeks of age or less; and females can breed for the first time when 25 days old. The minimum generation time is thus under seven weeks.

M. pennsylvanicus is capable of rapid population increase under favorable conditions. During periods of increase, many individuals disperse, and these may establish populations in unoccupied areas. *M. pennsylvanicus* is a relatively successful colonizer (Crowell 1973). Apparently the voles that disperse differ genetically, on the average, from those that remain behind. Those that disperse possess a high reproductive potential but are subordinate in intraspecific conflicts; those that remain are successful defenders of territories (Bridsall 1974; Keller and Krebs 1970; Krebs et al. 1973; Krebs et al. 1969; Krebs and Myers 1974; Myers and Krebs 1974).

Meadow voles forage both during the day and at night. They are less active on brightly moonlit nights, partly in response to the light, but also in response to an internal timing mechanism related to lunar phases (Doucet and Bider 1969). For orientation in the daytime, they use the sun as a compass; this indicates that they have an internal biological clock (Fluharty et al. 1976). Meadow voles feed on a variety of green plants, both monocots and dicots, some of them introduced species (see Relationship to People, below). They also eat some insects, though not to the extent that *Peromyscus* does (Fish 1974; M'Closkey and Fieldwick 1975).

Over its wide range, *M. pennsylvanicus* interacts with many other small mammals. It tends to avoid *Blarina brevicauda*, which may prey upon it (Barbehenn 1958; Eadie 1944, 1952, 1953; Fulk 1972). It does not usually succeed in occupying woodlands inhabited by *Peromyscus leucopus* or *Clethrionomys gapperi* (M'Closkey and Fieldwick 1975; Morris 1969). Farther west, it successfully excludes *Microtus montanus*, the montane vole, from mesic habitats which the latter would otherwise utilize (Koplin and Hoffmann 1968).

Relationship to People: Meadow voles are common in agricultural areas and may do significant damage to forage crops, garden vegetables, and fruit trees. They have adapted to our alteration of the environment as well as have any native rodents. With their high reproductive potential and their colonizing ability, they can adapt well to frequent changes in agricultural land use. To the extent that food preferences are genetically determined, *Microtus pennsylvanicus* may be said to have been preadapted to the introduction of European agriculture into this continent. In food preference tests at the University of Minnesota Field Biology Station at Lake Itasca, Thompson (1965) found that, of the 10 plants most often selected by meadow voles, at least six and perhaps eight were introduced from Europe. Furthermore, the grasses among these preferred foods dominate previously tilled fields which have been allowed to go ungrazed and unmowed, habitats in which the meadow vole thrives. It does not do as well in native tallgrass prairie.

Perhaps the best insurance against excessive damage by meadow voles is an abundance of predatory mammals and birds. Unfortunately for us, we tend to persecute rather than encourage such predators.

Additional References: Baker 1971; Beer 1961c; Beer et al. 1957; Getz 1961a, b, 1963, 1970; Grant and Birney 1979; Gray and Dewsbury 1975; Guthrie 1965; Iverson and Turner 1974; Kirkland and Griffin 1974; Kohn and Tamarin 1978; Krebs 1977; Madison 1978; Mallory and Clulow 1977; Murie 1971; Myers and Krebs 1971; Rose 1979; Seabloom et al. 1978; Tamarin 1977, 1978; Tester and Marshall 1961; Turner and Iverson 1973; Van Vleck 1968; Zimmerman 1965.

Microtus chrotorrhinus
Rock Vole, Yellownosed Vole

Measurements: Total length 153 (132–172); tail 42 (38–49); hind foot 20 (19–21); ear from notch 15 (12–16); weight 36.3 (21.8–46.0).

Description: The rock vole is dark grayish brown on the back and sides, and silvery gray below. The nose is orange or yellowish orange, a richer yellow than that of the heather vole. The tail is longer than 38 mm. Meadow voles are generally larger, their nose is not yellowish, and their ears are shorter.

Range and Habitat: Microtus chrotorrhinus is found in the Appalachian Mountains from North Carolina to Quebec and in the coniferous forest belt from Labrador to the Quetico-Superior area of Ontario and northeastern Minnesota. (See map 42.) In much of its range it seems to be restricted to areas of glacial boulders and rocky outcrops. Hall and Kelson (1959) suggested that *M. chrotorrhinus* might be conspecific with *M. xanthognathus*, the yellow-cheeked vole of northwestern Canada and interior Alaska, but the two species can be consistently separated morphologically and cytologically (Banfield 1974; Youngman 1975).

Figure 41. *Microtus chrotorrhinus*, rock vole.

Natural History: The biology of this species is poorly known. Its distribution is generally spotty, perhaps because its habitat is discontinuous. *M. chrotorrhinus* is often associated with boulder fields or other rocky habitats, but not always. It also occurs in Cook County in at least one upland shrub community without rocks, with a sparse overstory, and with an abundance of organic litter and dead wood, possibly the result of a blowdown. If organic litter is important to the species, shrubby, sparsely wooded communities which result from fire would probably be poor habitat (Buech, Siderits, et al. 1977; Buech, Timm, et al. 1977; Timm 1980). Isolated pockets of suitable habitat may often be unoccupied, because of the difficulty of dispersal through intervening biotic communities. Collection at known localities should probably be restrained to avoid local extermination. Success of rock voles seems to be sometimes related to the absence of meadow voles. In at least one locality in Minnesota, rock voles seem to be compatible with shrews, least chipmunks, woodland deer mice, red-backed voles, woodland jumping mice, and red squirrels (Timm 1974; Timm et al. 1977).

Breeding occurs at least from June through August. A typical litter size in Canada is four. Presumably, like voles in general, *M. chrotorrhinus* is primarily a vegetarian.

Relationship to People: Except for its intrinsic interest as a relatively rare and little-known mammal, the rock vole probably has little direct relationship to people. Such a rare animal, like a rare wildflower, should be treasured for the variety it adds to the wilderness.

Additional References: Graham 1976; Guilday and Hamilton 1978; Kirkland 1977a, b; Komarek 1932; Martell and Radvanyi 1977; Martin 1971a, b, 1972, 1973a, b; Peterson 1962; Roscoe and Majka 1976; Whitaker and Martin 1977.

Microtus ochrogaster
Prairie Vole

Measurements: Total length 141 (119–166); tail 34 (27–46); hind foot 18 (15.5–22); ear from notch 12 (9–14); weight 27.9 (16.8–49.1).

Description: The prairie vole is smaller on the average than the meadow vole, but the two species overlap in size and proportions. There is a distinct yellowish cast to the fur of the prairie vole, especially on the belly. The ears are almost hidden in the fur. There are six teats in females, and the anterior pair is generally nonfunctional.

Range and Habitat: This is truly a prairie species, ranging from the southern Prairie Provinces in a broad band southeastward across the grasslands of the midwestern states to northern Arkansas and the bluegrass region of Kentucky. *Microtus ochrogaster* is typically found in grassy and treeless locations (Martin 1956). In regions where *M. pennsylvanicus* is absent, the prairie vole may occupy a variety of open habitats, but in the Upper Midwest it is largely restricted to grasslands that are relatively dry and relatively

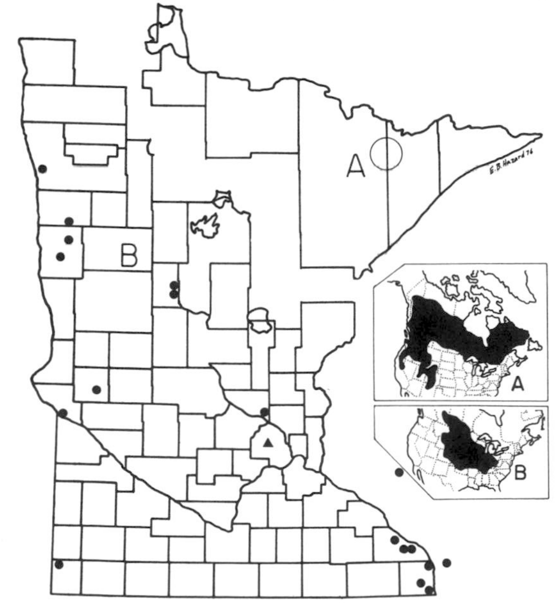

Map 41. A. Distribution of *Phenacomys intermedius*. Circle = approximate trap site of the only Minnesota specimen. B. Distribution of *Microtus ochrogaster*. ● = township specimens. ▲ county specimen. Wisconsin specimens reported in Jackson 1961. (Map produced by the Department of Biology, Bemidji State University.)

undisturbed. Consequently, it is rare, and may be absent today from areas for which there are earlier valid locality records (Bowles 1975; Timm 1980). Areas maintained as pasture and meadow are commonly preempted by the meadow vole.

Natural History: Field studies of the biology of the prairie vole have been done in Kansas, Indiana, and Nebraska (Fitch 1957; Jameson 1947; Krebs et al. 1969; Martin 1956; Meserve 1971). Its natural history in Minnesota may be somewhat different. On the prairie, the vegetation is dominated by grasses and the climate is characterized by occasional prolonged drought. *Microtus ochrogaster* is highly adapted to such an environment. The extensive runways of prairie voles are sheltered by the dense grass, which also retards surface evaporation, maintaining a high relative humidity in the immediate surroundings of the voles. Prairie voles nest in burrows which open into the runway system. The gestation period, as in meadow voles, is three weeks or less, but prairie vole litters are smaller, averaging just over three young (Fitch 1957). During a drought, relatively few litters are born; the females apparently do not ovulate then. But when food, especially young plant growth, is abundant, prairie vole populations increase rapidly. Females can breed when a month old, though most are not pregnant before the end of their second month. Breeding occurs through the winter farther south (Meserve 1971), but this is unlikely in Minnesota. *Microtus ochrogaster* populations fluctuate irregularly, with population density depending largely on the availability of succulent plant food and thus ultimately on rainfall.

Relationship to People: In regions where it is abundant, *M. ochrogaster* can cause significant losses to forage crops and to orchards. In Minnesota, on the other hand, we have adversely affected the prairie vole by so altering its habitat that it commonly is displaced by the meadow vole.

Additional References: Allan 1936; Batzli et al. 1977; Calhoun 1945; Carrol and Getz 1976; Flattum 1962; Gaines et al. 1978; Getz 1962c, 1963; Grant et al. 1977; Gray and Dewsbury 1973; Heaney and Birney 1975; Hoffmeister and Getz 1968; Kaufman and Fleharty 1974; Keller and Krebs 1970; Krebs 1970, 1977; Krebs et al. 1973; Myers and Krebs 1971; Rose and Gaines 1976; Stehn and Richmond 1975; Tamarin and Krebs 1969.

Microtus pinetorum
Woodland Vole

Measurements (two Houston Co. specimens): Total length (one) 122; tail 18, 18; hind foot 15, 17; ear from notch 10, 11; no weight taken.

Description: The soft pelage of the small woodland vole is auburn above and on the sides, and buffy below. The ears are hidden in the fur, and the tail is less than 25 mm long. The southern bog lemming, with which it might be confused, has grizzled fur and grooved upper incisors.

Range and Habitat: Microtus pinetorum ranges

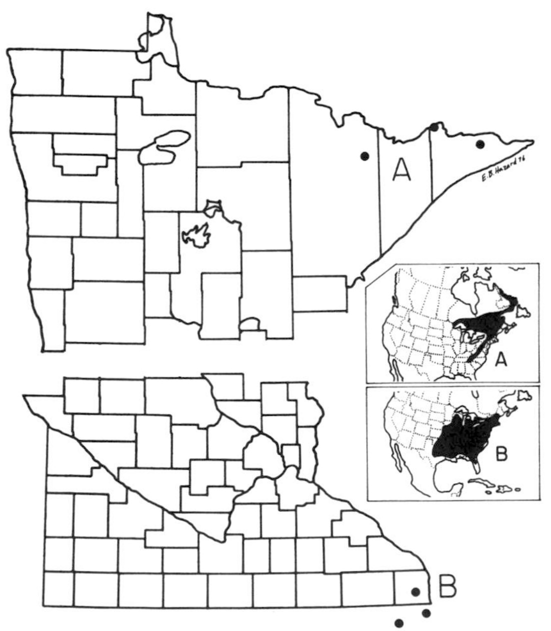

Map 42. A. Distribution of *Microtus chrotorrhinus*. B. Distribution of *Microtus pinetorum*. ● = township specimens. Iowa specimens reported in Bowles 1975. (Map produced by the Department of Biology, Bemidji State University.)

throughout the eastern United States except for Maine and southern Florida, barely reaching southeastern Minnesota (Hatfield 1939c; Heaney and Birney 1975). Its Latin name suggests a coniferous habitat, and it was formerly known as the pine vole. This was unfortunate, for it characteristically resides in upland deciduous forests, especially those with a grassy understory (Benton 1955; Goertz 1971). It occurs in pine woods only in the southeastern states. Woodland voles thrive in many orchards (Benton 1955; Cengel et al. 1978; Hamilton 1938).

Natural History: Fleshy roots, rhizomes, and grasses are staple foods of the woodland vole. Minnesota's most fossorial vole, this species spends most of its time underground in relatively shallow burrows in the loose soil and duff of the forest floor. The burrow system is used for active feeding and for food storage (Hamilton 1938). The nest is usually under a shelter such as a log or in a relatively deep burrow. Goertz (1971) found the average litter size to be 2.6 (there are only four teats). As many as four litters are born per year. The reproductive potential of the woodland vole is not as great as that of the meadow vole, but it is probably less vulnerable to predation in its subterranean habitat.

Relationship to People: Where populations are dense, *Microtus pinetorum* often seriously damages orchards by girdling trees. Orchardists in the eastern states have used endrin extensively as a rodenticide, and this has resulted in natural selection for resistance to it in some orchard-dwelling populations of the woodland vole (Hartgrove and Webb 1973; Webb and Horsefall 1967).

Additional References: Beck and Mahan 1978; Briese and Smith 1974; Cole and Batzli 1978; Dewsbury 1976; Fish and Whitaker 1971; Graham 1976; Hartgrove and Webb 1973; Hatt 1930; Hiaasen et al. 1978; Jameson 1949; Kirkpatrick and Valentine 1970; Layne 1958b; Miller and Getz 1969, 1977b; Novak and Getz 1969; Paul 1970; Pearson 1959; Schadler and Butterstein 1979; Staples and Terman 1977; Valentine and Kirkpatrick 1970; Wertheim and Giles 1971; Wrigley 1969.

Genus *Ondatra*

Ondatra zibethicus
Muskrat

Measurements: Total length 556 (487–576); tail 247 (203–277); hind foot 76 (66–83); ear from notch 19 (15–25); weight, 1,014 (595–1424).

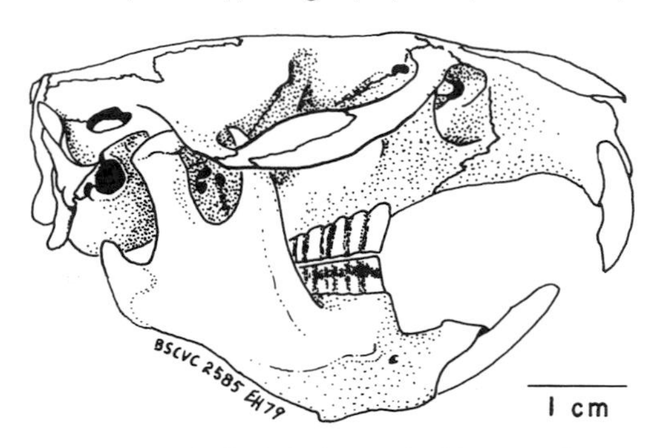

Figure 42. Skull of *Ondatra zibethicus*, muskrat, BSCVC 2585.

Description: The muskrat is a large, brown, aquatic vole, with a long, black, compressed (vertically flattened) tail. Its belly fur is silvery, and its dense, luxurious underfur is overlain by shiny guard hairs. The large hind feet, the main source of propulsion in swimming (Mizelle 1935), have stiff hairs fringing the toes. The name muskrat is derived from the musky secretion of scent glands near the external genitalia of both sexes.

Range and Habitat: Ondatra is a monotypic genus, native only to the New World. Muskrats occur in extreme northern Mexico, most of the contiguous United States, and all of Canada and Alaska except for much of the Arctic coast. They are absent on the southeastern coastal plain from western Florida to southeastern North Carolina (Errington 1963a). The species has been introduced into Europe (Mohr 1933; Storer 1937; van den Brink 1968), where, in only 65 years, it has spread and differentiated into geographic races that are comparable in diversity to those in North America (Pietsch 1970).

Except during seasons of dispersal, muskrats

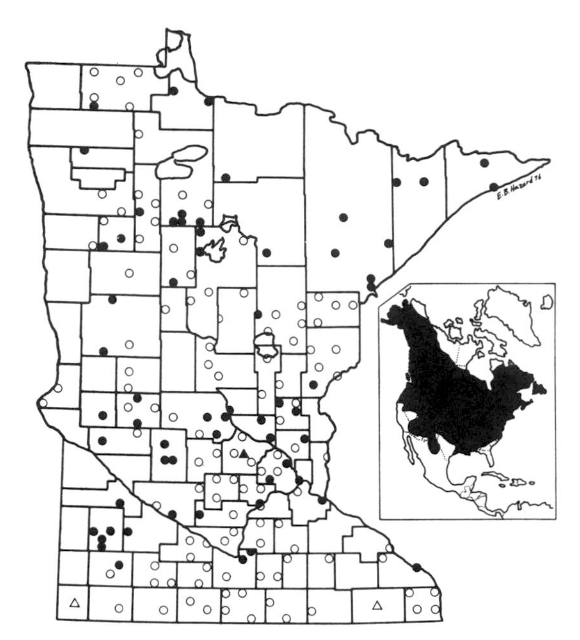

Map 43. Distribution of *Ondatra zibethicus*. ● = township specimens. ○ other township records, selected locations. ▲ = county specimen. △ = other county records. (Map produced by the Department of Biology, Bemidji State University.)

always live near water. They prefer still or slow-flowing waters with at least some protection from the sweep of the wind and with a growth of edible marsh plants. They also favor bodies of water that are not subject to abrupt changes in level, but they may be found in and near waters that are far from optimal.

Natural History: The biology of this species is well known as a result of studies by many scientists, particularly the late Paul Errington of Iowa State University. The most extensive single work on the muskrat is his *Muskrat Populations*, published posthumously in 1963. Muskrats live in groups, partly because suitable habitat occurs in discrete units, and partly because transients are attracted to occupied waters if populations are not too dense. There is, however, no complex social structure; each individual acts on its own behalf most of the time. During the breeding season, primarily in April through June, adults generally avoid close contact, except when mating. Lodges or burrows containing females with young are generally 18–36 m apart, but occasionally two litters are raised in a single lodge. Most females do not bear young during the year of their own birth. Gestation lasts three to four weeks. There may be a dozen young in a litter, but the average for a particular year and area is generally six to eight. The number of litters per year varies from five in the south to one or two in the extreme north. In Minnesota and adjacent areas, few females have over three litters per year, and two may be more common (Errington 1963a; Stewart and Bider 1974). Males typically remain as companions to particular females for some time, in a more or less monogamous relationship (Sather 1958).

Muskrats are primarily vegetarian. Although they may have individual or regional preferences for particular plants such as cattails, bulrushes, sedges, duck potatoes, and dandelions (Christian 1964; Errington 1963a; Northcott 1974), they are versatile and adaptable feeders (Errington 1941). In some populations, they store food such as duck potatoes, and they feed on and store ear corn. In winter they eat stored foods, rootstocks below the freeze line, and sometimes stalks of plants above the ice or ground, or even some material from the walls of lodges themselves. When dissolved oxygen in ponds has decreased to levels that are critically low for fish, bullheads will crowd into the plunge holes of muskrat lodges for air, and the muskrats will eat them.

The lodge is a mass of vegetation built up on the shallow bottom, on the shore or an island, or on top of the ice. After the mound is formed, the muskrat hollows out a chamber and plunge holes from below. Small lodges may contain a single chamber, but large ones, up to 1.8 m high and over 2.5 m across, may contain several entries and chambers. These may or may not be interconnected. Lodge building is most intense in the fall. From late summer through late winter, most muskrats tolerate close contact with one another, and it is common for several to huddle together in one chamber for warmth. The concentration of oxygen apparently remains adequate at such times, but that of carbon dioxide and other waste gases becomes surprisingly high (Huenecke et al. 1958). Abandoned lodges are useful to other animals. Raccoon and mink use them as feeding platforms, mink build their own nests in them; and coots and Forster's terns nest on them (Bergman et al. 1970; Frederickson 1970).

Muskrats are also effective burrowers. Those inhabiting stream banks often live in burrows rather than lodges, and, even in marshes with abundant lodges, there may be extensive burrow systems (Errington 1963a).

All mammals are subject to seasonal crises. For the muskrat, these relate primarily to the marshes or other bodies of water in which they live (Errington 1957b, 1963a). Unusually deep freezes may seal off underwater foods, and floods may sweep away lodges. Droughts may deprive muskrats of food, cover, and water itself (Errington 1939a). Muskrats can do without liquid water for a short time, but they require more water per gram of body weight per day than do most other rodents (Holleman and Dieterich 1973). Human modification of water levels, by dams, drainage, or channelization, can profoundly after muskrat habitat (Choate 1972; Hazard 1973).

Physical factors like cold and drought affect dense and sparse muskrat populations more or less equally. Predation and disease, however, typically operate more intensely on dense populations. The mink is probably the commonest vertebrate predator on muskrats. Red fox, raccoon, and raptorial birds also take their share. Diseases, including tularemia (Walker and Moore 1971) and a hemorrhagic disease (Errington 1963a), spread through dense populations much more rapidly than through sparse ones. Fluctuations in muskrat populations are probably more the result of disease and physical adversity than of predation (Elton and Nicholson 1942; Errington 1940, 1951, 1954a, b). Predation by mink is heaviest on individuals made vulnerable by disease, excessive cold, or other adversity (Errington 1954b, 1963a). Muskrat populations may decrease drastically at times, and occasionally whole populations in particular marshes are exterminated. Muskrats, however, are prolific (Errington 1951, 1954a), and vacant habitats are soon replenished by dispersing individuals (Errington 1940).

Relationship to People: Ondatra zibethicus is Minnesota's most important furbearer (Errington 1948; Johnson 1969). It has also been an important source of human food in some areas (Errington 1963a). Although it is fond of corn and may be a problem where cornfields are near marshes (Errington 1938), the muskrat must be rated as an overall economic asset. For me, the major value of the muskrat is aesthetic. It is the only microtine that can be readily observed in nature. Although primarily nocturnal, it is often abroad at dawn or dusk, or even in broad daylight. A quiet observer can spend many a pleasant hour by a pond or marsh watching muskrats feed, quarrel, carry material for lodge building, and otherwise engage in the business of "staying alive" (Errington 1963a).

Additional References: Aleksiuk and Frohlinger 1971; Beer 1950b; Beer and Meyer 1951; Beer and Truax 1950; Bellrose 1950; Berner 1980; Boyce 1978; Danell 1978; Errington 1937b, c, 1939b, 1943; Highby 1941; MacArthur 1978; MacArthur and Aleksiuk 1979; Marshall 1937; McCann 1944; Neal 1968, 1977; Olsen 1959; Stewart and Bider 1977.

Genus *Synaptomys*
Bog Lemmings

Two other genera of lemmings occur in both Eurasia and North America, but the bog lemmings (*Synaptomys*) are restricted to North America.

Synaptomys cooperi
Southern Bog Lemming

Measurements: Total length 112 (100–123); tail 16 (11–21); hind foot 17 (15.5–19); ear from notch 12 (10–13); weight 24.4 (19.1–27.5).

Description: This lemming superficially resembles a small meadow vole, but its grizzled gray pelage is longer, and its tail is only about as long as its hind foot. The belly fur is tipped with silver. The skulls of lemmings are broad, and the short rostrum gives the southern bog lemming an abrupt profile. The incisors are broad, heavy, and more strongly curved than those of voles, and the upper incisors are grooved.

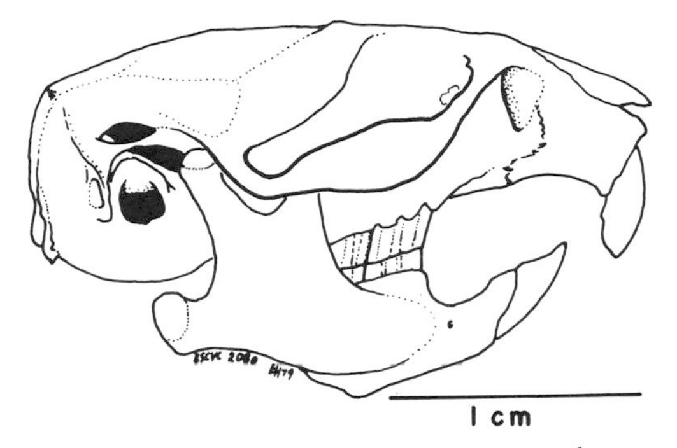

Figure 43. Skull of *Synaptomys cooperi*, southern bog lemming, BSCVC 2080.

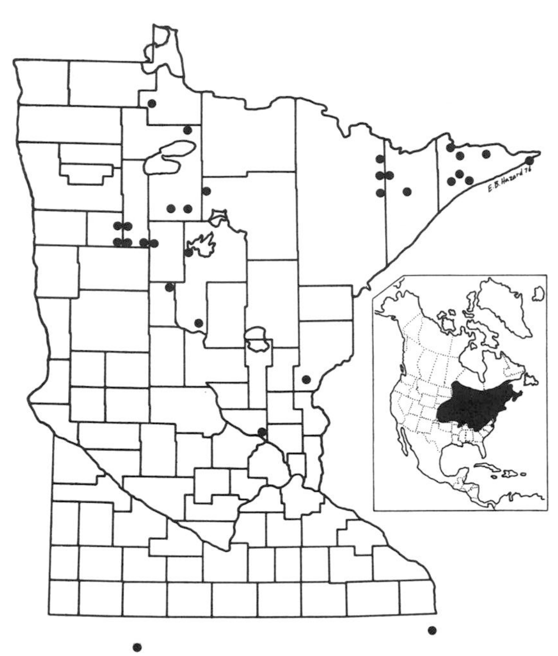

Map 44. Distribution of *Synaptomys cooperi*. ● = township specimens. Iowa specimens reported in Bowles 1975. (Map produced by the Department of Biology, Bemidji State University.)

Range and Habitat: The southern bog lemming is found in roughly the northeastern third of the United States and, in Canada, from southern Manitoba east to the mouth of the St. Lawrence. It inhabits a variety of wooded and open moist areas, including low fields with matted grass or sedge cover, shrubby fields, and sphagnum bogs with ericaceous shrubs and an overstory of tamarack or spruce (Burt 1940; Connor 1959; Getz 1961a; Pruitt 1959). It sometimes occurs in dry upland woods (Iverson et al. 1967); bog lemmings are most likely to appear in upland habitats when populations are high (Burt 1940). The species ranged as far south as Texas during the Pleistocene (Roth 1972). In Minnesota, it presently occurs in the northern half of the state.

Natural History: The southern bog lemming is not frequently encountered near human dwellings, and is seldom of economic significance. Perhaps for these reasons few mammalogists have studied it thoroughly. Connor (1959) studied it in the pine barrens of New Jersey. The biology of the species in Minnesota is probably much the same.

Bog lemmings, like the related voles, are primarily if not exclusively herbivorous. They feed on the leafy parts of sedges and grasses; on the leaves and tender twigs of leatherleaf and other ericaceous shrubs; on blueberries and other fruit; and on starchy rootstocks of various herbs. They cut and cache leaves and stems, piles of which can be found in their numerous runways. They are active both day and night.

S. cooperi typically builds nests high in a sphagnum clump, well above water level. In southern Michigan, Getz (1961a) found that, despite its preference for moist habitats, the southern bog lemming avoided standing open water whereas the meadow vole did not. The nest is a 9–15 cm ball of shredded sedge or grass, typically with an inner chamber about 6 cm across.

Figure 44. *Synaptomys cooperi*, southern bog lemming.

Presumably the species is promiscuous. Gestation in lactating females lasts 23 days. The number of young per litter ranges from one to five, with a mode of three and a mean of 3.0. There may be several litters per season. In the pine barrens, home ranges were small (0.04–0.10 ha), and densities ranged from a high in October of 2.0 lemmings per ha to a low in May of 0.9. Since lemming populations fluctuate markedly, home range and density figures are likely to vary from year to year. In some situations, *S. cooperi* seems to be competitively excluded by *Microtus pennsylvanicus*, whereas in others it is not (Baker 1971; Blair 1948; Getz 1961a).

Relationship to People: When they are abundant, southern bog lemmings may make significant inroads on forage crops, but they are generally of little economic importance.

Additional References: Gaines et al. 1977; Graham 1976; Guilday and Hamilton 1978; Heaney and Birney 1975; Kirkland 1977b; Martell and Radvanyi 1977; Wetzel 1955, 1958.

Synaptomys borealis
Northern Bog Lemming

Measurements (three specimens): Total length 110–122; tail 18–37; hind foot 18–19; ear from notch (one) 13.

Description: This species cannot readily be distinguished from the southern bog lemming by pelage alone. In the first lower molar of the northern bog lemming, the outer (labial) border is relatively smooth, without deep indentations (see fig. 86). In *S. borealis* there is a backward-directed spine at the rear of the bony palate, which is absent in *S. cooperi*. Any specimens of *Synaptomys* taken near the Minnesota-Canada border should be referred to a professional mammalogist for positive identification.

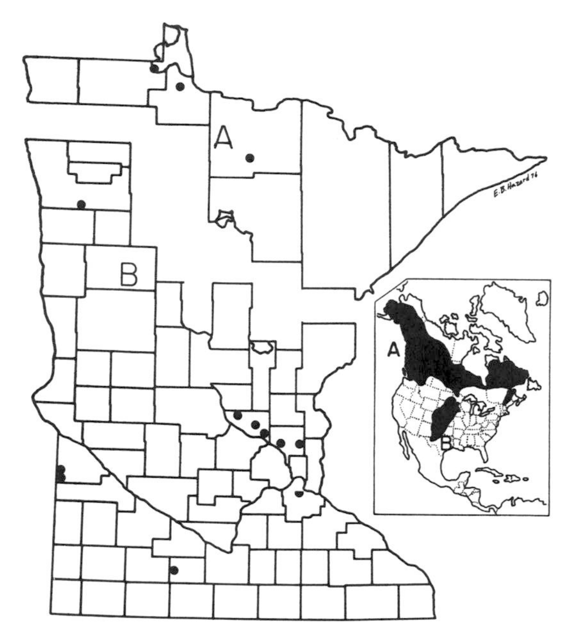

Map 45. A. Distribution of *Synaptomys borealis*. B. Distribution of *Perognathus flavescens*. ● = township specimens. (Map produced by the Department of Biology, Bemidji State University.)

Range and Habitat: The northern bog lemming ranges from southern Alaska across central Canada to Labrador. It is not known how broadly its range overlaps that of the southern bog lemming, nor is it known if the three Minnesota specimens represent the southern limit of a more or less continuous population or a population that has become isolated from the rest of the species.

Natural History: Little is known of the natural history of *Synaptomys borealis*. Presumably it is similar to that of *S. cooperi*. It has been collected in moist habitats similar to those of *S. cooperi*, but specimens have also been collected in dry open woodland in northern Quebec.

Relationship to People: The northern bog lemming is of no known economic importance.

References: Graham 1976; Guilday and Hamilton 1973, 1978; Heaney and Birney 1975; Martell 1974; Wetzel and Gunderson 1949; Youngman 1975.

Family Muridae
Old World Rats and Mice

Murids are among the world's most successful mammals. They have spread, more or less naturally, from central and southeast Asia to most of the Old World, including Australia and many oceanic islands. They seem to have been prevented from reaching North America only by the severe winters of Beringia. Several murids have become parasitic on human populations and, with our help, have spread throughout the world (Arata 1967).

The murids now found in North America are superficially similar to cricetines, with long tails and prominent ears. The tail, however, is scaly and nearly hairless. The molar teeth have three rows of tubercles, rather than the two rows that are found in *Peromyscus* and *Onychomys*. The murid tooth formula is 1/1, 0/0, 0/0, 3/3 × 2 = 16.

References: Arata 1967; Elton 1958.

Genus *Rattus*

Rattus norvegicus
Norway Rat

Measurements: Total length 390 (360–415); tail 175 (160–183); hind foot 41 (40–43); ear from notch 20 (19–21); weight about 200–270.

Description: Wild specimens of this large rat are more or less uniformly brown or, more rarely, black. The belly is not appreciably lighter than the back and sides. The long, tapering, scaly tail is less than half the total length. *R. norvegicus* is often called the brown rat—it reached Norway only a few hundred years ago.

Range and Habitat: Cosmopolitan; occurs wherever there is either a large concentration of

Figure 45. *Rattus norvegicus*, Norway rat.

people or an exploitable farming enterprise. The smaller black rat, or roof rat (*R. rattus*), the major host of the flea that carries bubonic plague, is more vulnerable to cold and has apparently not reached Minnesota. It can be told from young brown rats by its longer tail (over half the total length).

Natural History: The natural environment of *Rattus norvegicus* is an artificial one—our cities, farmsteads, and town dumps.

Both people and rats (as well as a variety of other species symbiotic with us) have become adapted to the peculiar environments that people create, to the point where it becomes difficult to use the term "natural" with meaning. The Norway rat has found us to be fine providers of food, often in large amounts and in concentrated form, and of shelter. Rats maintain much smaller territories than do native North American mice and are thus able to maintain high densities where

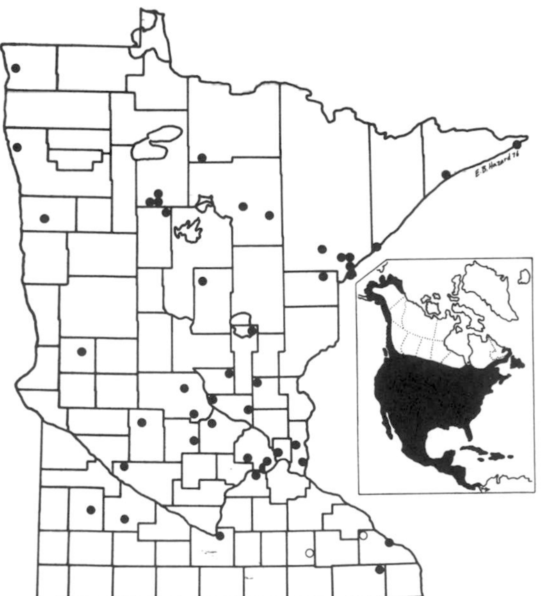

Map 46. Distribution of *Rattus norvegicus*. ● = township specimens. ○ = other township records. (Map produced by the Department of Biology, Bemidji State University.)

food is abundant (Barnett 1963, 1967; Lore and Flannelly 1977). They have a short gestation (about 21 days), typically bear six to nine young, and may have eight or more litters per year. Rats are omnivorous; they can live on a variety of plant and animal matter, but their cheek teeth are not suited to a diet of coarse vegetation alone. They also actively prey on invertebrates, small reptiles and birds, and small mammals, including house mice (O'Boyle 1974).

Relationship to People: Rattus norvegicus causes more economic loss than any other mammal, except for people themselves (Canby 1977). Rats destroy or spoil food and fabrics, are vectors for many diseases, prey on small stock such as baby chicks, and may attack people. They are difficult to control, partly because, as in other species, natural selection favors those who are inherently resistant to controls. Just as most houseflies are now resistant to DDT, there are now some rats that are immune to the anticoagulant Warfarin (Brooks and Bowerman 1973; Brothers 1972; Greaves and Ayres 1977). A recent approach is to control rats with chemosterilants (Garrison and Johns 1975; Gwynn 1972a, b; Gwynn and Kurtz 1970; Mischler et al. 1971). Another proven method is to build structures that are rat-proof to begin with.

By a long process of artificial selection, people have produced various strains of domesticated rats that are useful in research. The social behavior of wild and domestic rats has been studied in considerable detail (Boice 1972; Boreman 1972; Calhoun 1962). Price and Loomis (1973) found a number of differences in the responses of wild and domestic Norway rats to novel environments. They then tested the offspring of reciprocal crosses (male wild × female domestic; male domestic × female wild). These hybrids were genetically alike but raised in different social situations—the first by domestic mothers, the second by wild mothers. The two groups of hybrid offspring behaved alike, suggesting that the observed behavioral differences in wild and domestic rats are largely determined by hereditary differences, not environmental differences.

Additional References: Andrews et al. 1971; Barfield and Geyer 1972; Chesness et al. 1968; Christian and Davis 1956; Elton 1958; Gunderson 1944; Margulis 1977; Nusetti and Aleksiuk 1975; Strecker 1954; Taylor et al. 1974; Taylor and Quy 1978; Whitaker 1977.

Genus *Mus*

Mus musculus
House Mouse

Measurements: Total length 160 (150–177); tail 75 (67–85); hind foot 18 (17–20); ear from notch 12 (11.5–13); weight 19.2 (13.5–35.3).

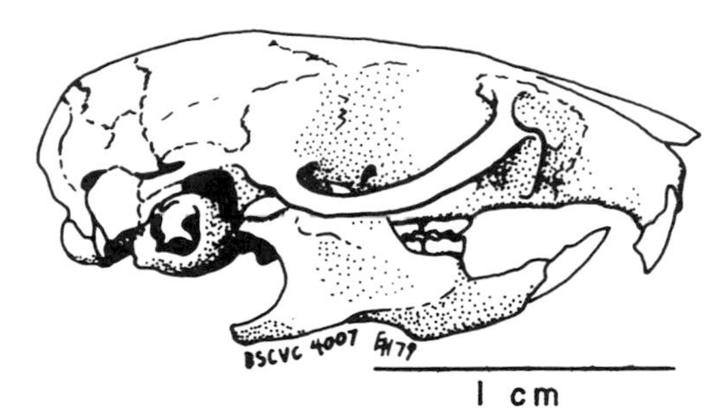

Figure 46. Skull of *Mus musculus*, house mouse, BSCVC 4007.

Description: Wild house mice are large-eared, long-tailed, short-haired mice, smaller than *Peromyscus*, and smellier. Their eyes are not as prominent as those of deer mice, and the pelage is generally a uniform grayish brown, only slightly lighter below than above. Occasional black individuals occur in the wild. Laboratory strains are most often albinos, but many other domestic color variants exist.

Range and Habitat: See Norway rat. In the southern states, *Mus musculus* can maintain field populations over the winter, but, in the northern United States and Canada, it apparently is restricted to areas where people provide shelter. The house mouse has become cosmopolitan as a symbiont of humans, but it has also established feral populations on many islands (Bellamy et al. 1973; Berry and Jakobson 1975a, b; Myers

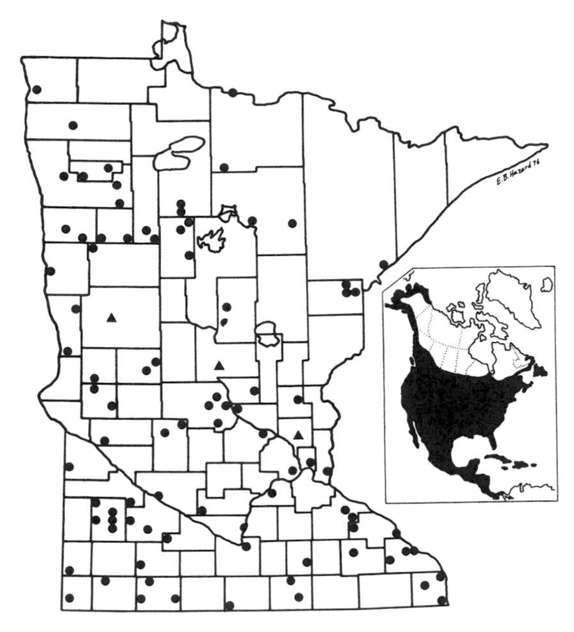

Map 47. Distribution of *Mus musculus*. ● = township specimens. ▲ = county specimens. (Map produced by the Department of Biology, Bemidji State University.)

1974a, b) and in some mainland habitats, especially in disturbed ones (Baker 1971; Breakey 1963; Briese and Smith 1973; Caldwell and Gentry 1965). This species is descended from central Asian ancestors that were adapted for arid conditions (Anderson 1978); it is well suited to living in human structures which provide food and shelter but little water.

Natural History: Like Norway rats, house mice are omnivores with a high tolerance for crowding. In some environments they are territorial (Crowcroft and Rowe 1963), but their territories are small. I once maintained a colony of two dozen adults in an enclosure of less than two m², in which several litters were raised within a few dm of each other, and in which two females raised their litters jointly in a single nest. An earlier attempt to raise a colony of *Peromyscus* in the same enclosure failed, despite abundant food and water, because adults killed both nest young and other adults. Because their reproductive capacity is high, house mice can quickly reach the carrying capacity of a habitat. They respond to such overcrowding in several ways, including inhibition of sexual maturation, increased mortality of infants and young adults, lower fecundity, reduced litter size, embryonic reabsorption, and dispersal (Drickamer 1974; Lloyd 1975; Rowe et al. 1964; Strecker 1954). *Mus musculus* can compete effectively with native mice and voles (Baker 1971; Briese and Smith 1973; Caldwell and Gentry 1965; Delong 1966), especially in disturbed habitats. In Minnesota, the vulnerability of *Mus* to cold winter conditions prevents year-round competition with native mammals except in shelters provided by people. However, house mice (and perhaps other mice not generally considered hibernators) sometimes conserve energy by becoming torpid during daily periods of inactivity (Hudson 1973; Morton 1978).

Relationship to People: Only the Norway rat damages food and property more than the house mouse. Mice are as hard to control as rats are. Perhaps both are best controlled by construction of vermin-proof storage facilities.

Domesticated varieties of house mice serve as standard laboratory animals, and some people fancy them as pets. Wild urban, farm, and feral populations have been the subjects of research in population dynamics and social behavior, and particularly in the operation of natural selection in relation to population and social biology.

Additional References: Berry 1978; Crowcroft 1966; Crowner and Barrett 1979; Ebert and Hyde 1976; Fertig and Edmonds 1969; Goehring 1971b; Grant and Birney 1979; Houtcooper 1978; Leamy 1977; Lynch 1977; Selander et al. 1969.

Family Zapodidae
Jumping Mice

Jumping mice and their relatives live in North America and Eurasia. The genera *Zapus* and *Napaeozapus* occur only in North America (Arata 1967).

Genus *Zapus*

Zapus hudsonius
Meadow Jumping Mouse

Measurements: Total length 216 (205–226); tail 131 (122–135); hind foot 30 (28–31); ear from notch 13 (12–14); weight 18.8 (13.2–25.1).

Description: No other Minnesota mice have tails as long as those of jumping mice, which are over 57 percent of the total length. The hind legs are much longer than the forelegs, the ears and eyes rather prominent, more mouselike than volelike in proportions. The coarse coat of the meadow jumping mouse is yellowish brown to olive brown above, yellowish on the sides, and white below. The belly hairs, unlike those of *Peromyscus* which are gray at the base, are white for their full length. There is a dark stripe down the back where there is a concentration of black hairs (Krutzsch 1954). The tip of the tail is not normally white-tipped. Unfortunately the rare occurrence of *Z. hudsonius* with white-tipped tails (Schorger 1951) precludes distinguishing this species in the field from the woodland jumping mouse, *Napaeozapus insignis*, by this character alone.

The upper incisors of *Z. hudsonius* are grooved, and a small premolar precedes the three upper molars. *Zapus* is the only North American genus with the tooth formula 1/1, 0/0, 1/0, 3/3 × 2 = 18.

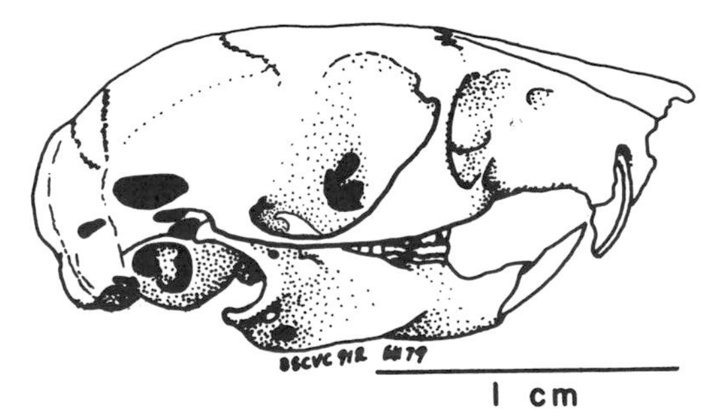

Figure 47. Skull of *Zapus hudsonius*, meadow jumping mouse, BSCVC 912.

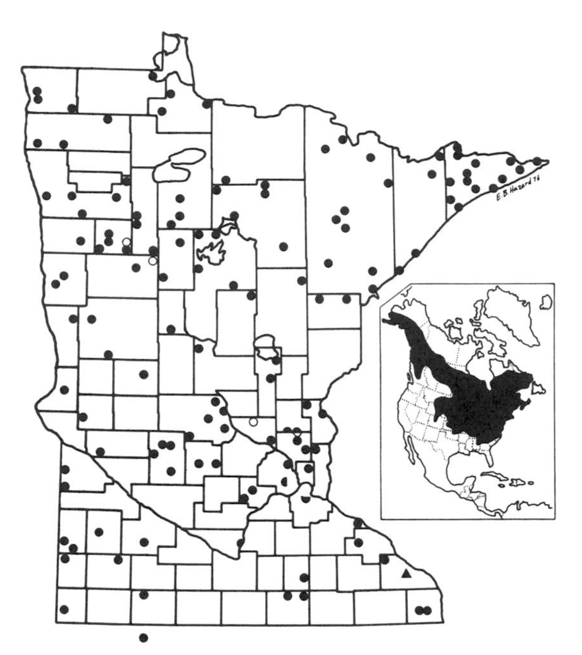

Map 48. Distribution of *Zapus hudsonius*. ● = township specimens, selected locations. ○ = other township records. ▲ = county specimen. (Map produced by the Department of Biology, Bemidji State University.)

Range and Habitat: Zapus hudsonius ranges from southern Alaska across Canada to Labrador and south to the central and mid-Atlantic states. It occupies a variety of semi-wooded and open habitats, often moist ones (Getz 1961b). Whitaker (1963a) suggests that the meadow jumping mouse is less attracted to soil moisture as such than to the rank herbaceous vegetation often characteristic of moist habitats. *Z. princeps*, the western jumping mouse, occurs in the Dakotas but apparently does not reach Minnesota (Davis and Ernst 1971; Seabloom 1978; Wiehe and Cassel 1978).

Natural History: Despite its common name, the meadow jumping mouse typically moves on all fours. It may also move along rapidly in a series of short leaps of a few cm, and will leap distances of up to a meter to avoid danger.

Zapus hudsonius feeds on a variety of plant and animal matter. Insects comprise up to 50

Figure 48. *Zapus hudsonius*, meadow jumping mouse.

percent of the diet. Jumping mice feed primarily on fruits and seeds, rather than on the vegetative tissues of plants. Hypogeous fungi are important in the diet (Baker 1971; Quimby 1951; Whitaker 1963a). Although the meadow jumping mouse hibernates, it apparently does not store food. Different individuals enter hibernation at different times over a period of several weeks in late summer and early fall; each adds massive amounts of fat (up to slightly over 100 percent of the normal weight) for a few weeks before hibernating (Whitaker 1972b). Whitaker (1963a) found that more than half of the population that entered hibernation did not appear the following spring and that only the heavier animals survived. Apparently, accumulation of adequate fat reserves is critical to overwinter survival. The hibernaculum may be underground, under a log, or protected in some other way. In the Waubun area, *Zapus* is prevalent on and near mima mounds, and presumably prefers these raised mounds on the prairie as hibernation sites (Ross et al. 1968).

Gestation ranges from 17 to 19 days in nonlactating females. Lactation will lengthen the gestation period to 20 or 21 days. Most births in Minnesota occur between 15 June and 30 August; Minnesota litter size ranges from four to seven, with an average of 5.7 (Quimby 1951).

Relationship to People: Meadow jumping mice are not common on tilled land and probably do little damage to seed crops. If they have any economic significance, it is probably as a check on insect populations. As experimental animals, they have been useful in studies of the physiology of hibernation.

Additional References: Beer et al. 1954; Elwell et al. 1973; Grant and Birney 1979; Hamilton 1935; Iverson et al. 1967; Kirkland and Kirkland 1979; Klingener 1963, 1964; Lyman 1963; Meierotto 1967; Muchlinski and Rybak 1978; Sheldon 1934; Sims and Buckner 1973; Wetzel 1958; Whitaker 1966.

Genus *Napaeozapus*

Napaeozapus insignis
Woodland Jumping Mouse

Measurements: Total length 241 (231–250); tail 147 (143–153); hind foot 31 (29–32); ear from notch 15 (11–19); weight 25.6 (17.6–29.9).

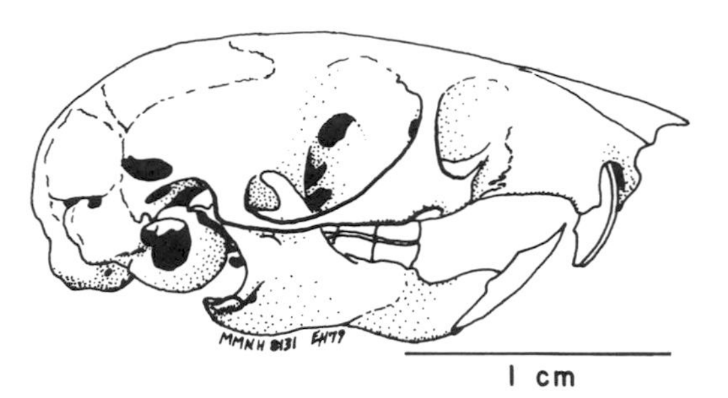

Figure 49. Skull of *Napaeozapus insignis*, woodland jumping mouse, MMNH 8131.

Description: The woodland jumping mouse averages larger than the meadow jumping mouse, and is more distinctly tricolor, having yellow or orange-yellow sides. The tail is sharply bicolor, with a white tip that may measure over 40 mm. The dorsal stripe is dark brown, the belly hairs white to the base. The upper incisors are grooved. The molars have an intricate pattern of folds and islands; there is no premolar. The tooth formula is 1/1, 0/0, 0/0, 3/3 × 2 = 16.

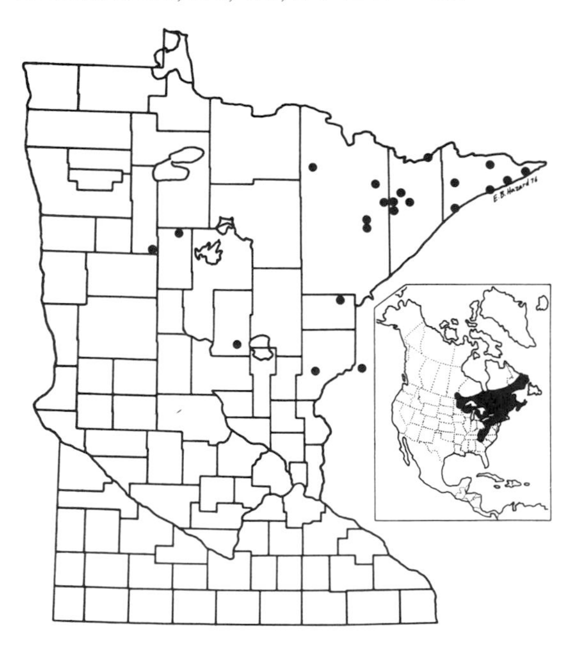

Map 49. Distribution of *Napaeozapus insignis*. ● = township specimens. Wisconsin specimen reported in Jackson 1961. (Map produced by the Department of Biology, Bemidji State University.)

Range and Habitat: Napaeozapus insignis ranges from southeastern Manitoba and northern Minnesota through southeastern Canada and the northeastern United States, and south through the Appalachians to northernmost Georgia. Unlike *Zapus hudsonius*, which prefers somewhat open habitats and early successional stages, *Napaeozapus* is characteristic of closed, relatively mature spruce-fir or hemlock-hardwoods forest (Whitaker and Wrigley 1972). It typically occurs where there is substantial shrubby or herbaceous ground cover (Brower and Cade 1966; Whitaker 1963b).

Natural History: Like the meadow jumping mouse, the woodland jumping mouse is a hibernator. It stores no food but subsists on stored fat for a hibernation period typically longer than six months. Immatures are usually the last to fatten up and enter hibernation, and immature males are the first to emerge in the late spring (Preble 1956). In Minnesota, hibernation probably lasts from September or early October to May. Females probably emerge in late May and are bred then. Gestation lasts 23 to 29 days (Sheldon 1934, 1938). Numbers of embryos and of placental scars range from two to seven (Whitaker and Wrigley 1972). Female *Napaeozapus* have only four teats, in contrast with *Zapus* females who have eight.

Woodland jumping mice eat diverse seeds and fruit, insects and other invertebrates, and hypogeous fungi. Lovejoy (1973) characterized optimum *Napaeozapus* habitat as that having "low woody vegetative cover, high soil moisture, and abundant invertebrate food." Whitaker (1962, 1963b) found that fungi, especially the genus *Endogone*, comprised over 35 percent of the diet in late summer and fall in New York. In form, function, and diet, *Napaeozapus* seems more like woodland forms of *Peromyscus* than like other small mammals. Nevertheless, there is no evidence that either interacts adversely with the other. There does, however, seem to be an inverse relationship between the abundance of *Na-*

paeozapus and *Clethrionomys* in habitats suitable for both (Brower and Cade 1966).

Relationship to People: Woodland jumping mice probably are of little economic importance, unless they act as a check on outbreaks of foliage-feeding insects. They reach their western limit in northwestern Minnesota, and probably occur there in well separated islands of suitable habitat, rather than as a more or less continuous population. Their biology in such areas merits further study.

Additional References: Beer 1953b; Bider 1968; Blair 1941; Clough 1959; Hamilton 1935, 1941b; Harper 1929; Kirkland and Griffin 1974; Kirkland and Kirkland 1979; Klein 1957; Klingener 1964; Layne and Hamilton 1954; Manville 1949; Meierotto 1967; Snyder 1924; Thibault 1969; Whitaker 1966; Wrigley 1969, 1972.

Family Erethizontidae

Arboreal, or New World, Porcupines

This New World family is Neotropical in origin. It has existed from the Oligocene to Recent in South America but has only occurred in North America since the late Pliocene. Of the eight living species, only one occurs in North America.

References: Kleiman 1974; Moody and Doniger 1956; Starrett 1967; Wood 1950, 1955, 1972; Woods 1972, 1973.

Genus *Erethizon*

Erethizon dorsatum

Porcupine

Measurements: Total length 686 (612–740); tail 180 (160–223); hind foot 93 (75–110); ear from notch 32 (30–35); weight 6.8 kg (5.3–9.1).

Description: Porcupines are large, stout, brownish black rodents. In the western states, most porcupines are yellowish. The coat consists of short underfur, long guard hairs (some of them white-tipped), and, on the back and sides, sharp spines or quills, which are most abundant on the

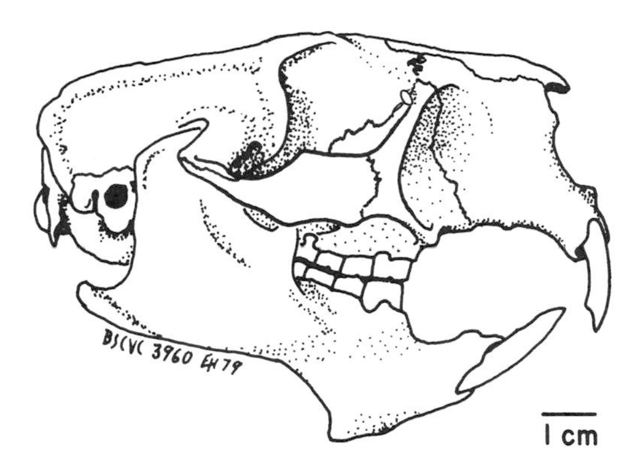

Figure 50. Skull of *Erethizon dorsatum*, porcupine, BSCVC 3960.

rump and tail. The quills are off-white with black tips. Microscopic scales at the tip of the quills overlap, their free edges pointing away from the tip. These edges have the effect of barbs, so that the quill works into the flesh that it penetrates and is extremely difficult to remove (Quick 1953; Shadle 1947, 1955a, b; Shadle and Po-Chedley 1949). The rostrum is deep and the infraorbital

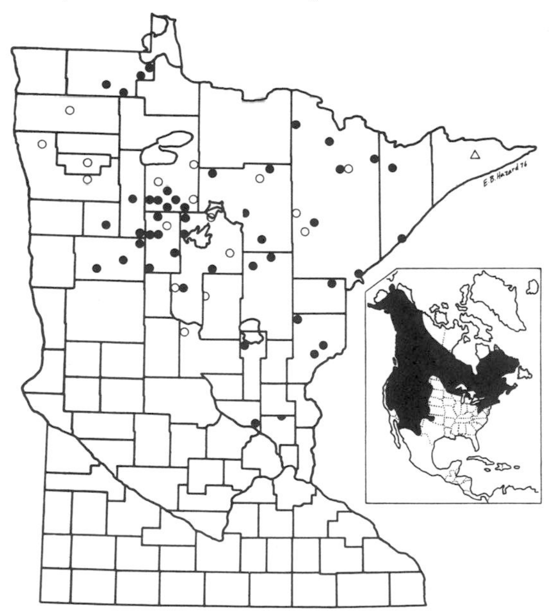

Map 50. Distribution of *Erethizon dorsatum*. ● = township specimens. ○ = other township records. △ = other county record. (Map produced by the Department of Biology, Bemidji State University.)

foramen is large, so large that mammalogy students commonly misidentify it as the orbit. The grinding surfaces of the cheek teeth have ridges of enamel that superficially resemble those of beaver, an instance of convergent evolution in two distantly related bark eaters. The tooth formula is 1/1, 0/0, 1/1, 3/3 × 2 = 20. Porcupines move on the ground with an awkward waddle and climb trees efficiently, if slowly.

Range and Habitat: Erethizon ranges northward from northern Mexico through most of the western United States and Canada to Alaska, and eastward across Canada and the Great Lakes states to Labrador and the northeastern states. It most commonly inhabits boreal forests but also occurs north of tree line in parts of Canada and Alaska, in various scrub habitats in the arid Southwest, and in habitats devoid of woody plants in parts of Texas (Kasper and Parrish 1974). During the Pleistocene, it occurred in West Virginia, Arkansas, Missouri, and Florida (Guilday and Hamilton 1973; Martin 1958; Parmalee 1971; White 1970).

Natural History: Most rodents have a high rate of reproduction, correlated with a high rate of mortality. The porcupine has neither. Gestation lasts over 200 days (Shadle 1948, 1951), so females can have only one litter per year, normally comprising but one pup. Not all adult females bear each year (Brander 1973). In Minnesota, young porcupines arrive most often in March or April. They are born in an unusually mature state, fully furred and with eyes open, and are soon able to follow the mother. The quills harden shortly after birth (Tyron 1947). Young porcupines may nurse for three months, but they can eat solid food at nine days (Shadle and Ploss 1943) and can subsist on vegetation alone at about two weeks (Woods 1973). That porcupines persist despite such a low reproductive rate attests to a low rate of mortality. Mortality is probably highest among the young; once a porcupine reaches adulthood, it is likely to live a long time. Maximum longevity in the wild is not known, but captives have lived longer than 10 years (Brander 1971). Although a porcupine's spines afford considerable protection, various predators do take porcupines. The best known of these is the fisher (Cook and Hamilton 1957; de Vos 1951; Schoonmaker 1938b). Other predators that at least attempt to take porcupines are the lynx, bobcat, coyote, red fox, and marten (de Vos 1953; Keller 1937; Pollack 1951; Quick 1953; Rollings 1945; Westfall 1956; Woods 1973).

Porcupines are noted for their diet of bark, and, indeed, most of their winter food is the inner bark of coniferous trees. In Minnesota, unfortunately, white pine is a favorite food. In the summer, porcupines eat a great variety of woody and herbaceous vegetation, much of it at ground level (Shapiro 1949; Woods 1973). They even eat water lily pads (Schoonmaker 1930). Porcupines gnaw ax handles and other wooden objects, supposedly for the salt left by human sweat (Jackson 1961). Surprisingly, Bloom et al. (1973) found that porcupines would not drink weak salt solutions tolerated by other mammals. Porcupines are also known to gnaw glass, aluminum, rubber tires, and plastic tubing (Cook and Hamilton 1957; Preston 1948; Scott 1941). Over a period of three years, I collected several porcupines that gnawed the plywood, but not the two-by-two pine framework, of a wooden shelter for a gasoline-driven electric generator. Most of them were collected at night, the time when porcupines are most active. In an early summer radio-telemetry study in Carlton County, Marshall et al. (1962) found that porcupines were largely but not entirely nocturnal.

Porcupines are generally solitary, except during mating and for the few months that the pup may remain with its mother. In the winter, however, several may den together in a hollow log or tree, abandoned house, or other shelter (Brander 1973). In the second-growth woods around Bemidji, porcupines often share winter dens under the large, decaying stumps left by the early loggers. Porcupines are active throughout the year, but in winter an individual's home range may be

Figure 51.
Erethizon dorsatum,
porcupine.

restricted for some time to the immediate area of a single food tree (Schoonmaker 1938a). Shapiro (1949) found winter feeding areas to encompass about 5 ha in the Adirondacks. By contrast, in a 30-day period during early summer, Marshall et al. (1962) found the home ranges of two adults in Carlton County to be 13 and 14.6 ha, respectively.

Relationship to People: Porcupines are often considered pests, because they damage timber (Cook and Hamilton 1957; Faulkner and Dodge 1962; Gill and Cordes 1972). Actually the dollar-per-hectare loss of timber is generally too low to warrant intensive control programs (Woods 1973). The porcupine also will take garden crops (Costello 1966; Gabrielson 1928). The offending porcupine is usually easy to find and is best removed by shooting. The density of porcupine populations in some areas might decrease if fishers were to become abundant again.

Professional biologists differ on the subject of wild animals as pets. Some species should not be kept by private individuals under any circumstances, and certainly no species should be kept if it is not properly cared for, which implies, among other things, the supervision of a responsible and knowledgeable adult. Porcupines take well to captivity (Crandall 1964), can learn to recognize individuals (probably by voice and smell), and might even be termed affectionate. Newborns are so advanced at birth that they can be hand-reared successfully; in fact, pups removed during the last week or two of gestation from freshly killed mothers often survive and grow normally. MacDuff II, removed from his mother in April 1962, thrived in my home and then in the animal room at Bemidji State University, living primarily on commercial dog chow, with occasional melon rinds, apples, poplar branches, and probably other things I never found out about. He was a source of instruction and entertainment for college students and visiting school children until 3 May 1971, when he died at just over nine years of age.

Additional References: Curtis and Kozicky 1944; Ferguson and Merriam 1978; Gupta 1966b; Harder 1979; Heaney and Birney 1975; Marshall 1951b; Mirand and Shadle 1953; Po-Chedley and Shadle 1955; Shadle 1946, 1952; Shadle et al. 1946; Smith 1977; Speer and Dilworth 1978.

ORDER CARNIVORA
Carnivores

The order Carnivora includes some of our best-known mammals. Although "carnivore" means meat eater, few carnivores eat only meat, and some (like the giant panda) are strict vegetarians. All Minnesota carnivores eat at least some animal matter. Our species range in size from the least weasel, which is about as heavy as a meadow vole (40 or 50 g), to the black bear, which may reach 270 kg.

Wild carnivores occur on all continents except Antarctica, but the ancestors of the wild dog of Australia, the dingo, were probably introduced into that continent by people. Dogs are the most common and varied of the domesticated carnivores, but people have also domesticated other carnivores, including various cats and a weasel, the Old World ferret. Carnivores occupy a variety of habitats—arboreal, terrestrial, aquatic, and marine—and are found in locations ranging from deep forests to dry deserts, and from the Arctic to the equator. Many single species are ecologically tolerant, ranging widely through many habitats, but others are quite restricted.

All Minnesota Carnivora have prominent upper and lower canines, and three upper and lower incisors. The cheek teeth vary in number and form, largely in relation to the degree of specialization for a meat diet. In predominantly meat-eating carnivores, P4 and m1 are carnassials—large teeth modified for shearing connective tissue and for cracking bone. The P4/m1 carnassial arrangement first appeared during the Paleocene in the Miacidae, a group which can reasonably be classified as the earliest family of the order Carnivora, or as one of several families of the extinct order Creodonta (Simpson 1945).

The mandibular condyle in carnivores is approximately a horizontal cylinder, and it fits closely into the mandibular fossa of the skull, giving strong support to the action of the jaw, but permitting little lateral or fore-and-aft motion. The relatively large carnivore brain and the three upper and lower incisors are features that make it easy to tell any carnivore skull from that of an opossum, the only other mammal in Minnesota whose skull is apt to be mistaken for that of a carnivore. For other distinguishing features of the order, consult Stains (1967), and for a broad overview see Ewer (1973).

Additional References (including several on predator control and predator-prey interactions): Balser 1964; Barbehenn 1973; Berryman 1972; Ehrlich and Birch 1967; Errington 1946, 1956, 1957a, 1963b, 1967; Griffiths 1975; Hibbard 1956b; Hornocker 1972; Kruuk 1972; Lockie 1966; McCabe and Kozicky 1972; Mech 1967; Munkel and Fremling 1967; Radinsky 1969, 1978; Rosenzweig 1968; Rosenzweig and MacArthur 1963; Savage 1977; Tanner 1975; Vaughan 1978.

Family Canidae
Dogs

Canids are generally moderate-sized terrestrial carnivores that eat meat when they can, but utilize many other foods. They are digitigrade and often run long distances in pursuit of prey. Many canids are relatively social mammals, running in packs or at least remaining in mated pairs for a considerable part of the year. Species that typically live in packs have an elaborate repertoire of dominance-subordinance behavior patterns, and Fox (1972a) has argued that natural selection favors the maintenance of a high degree of variation in genetically influenced social roles in such species.

This family is the most widespread in the order Carnivora, inhabiting all the continents but Antarctica. The tooth formula of all canids in Minnesota is 3/3, 1/1, 4/4, 2/3 × 2 = 42, and carnassials are present. Van Gelder (1978) maintains that most living canids, including the three genera in the accounts that follow, should be included in the single genus *Canis*.

Additional References: Anisko 1976; Bekoff 1977b; Christian 1970; Elder and Hayden 1977; Ewer 1973; Fox 1969a, b, 1970, 1971a, 1975; Hildebrand 1952; Kleiman 1966, 1967; Kleiman and Eisenberg 1973; Lawrence and Bossert 1975; Nudds 1978; Radinsky 1973; Schmidt-Nielsen et al. 1970; Scott 1967; Seal 1975; Stains 1975; Vaughan 1978.

Genus *Canis*
Dogs

In Minnesota, there are two wild species of this genus, *Canis latrans* (the coyote) and *C. lupus* (the wolf). There is also the widespread and highly varied domestic dog, *C. familiaris*. (See the section on Domestic and Feral Mammals for a discussion of the application of Latin names to domestic forms.) Coyote × dog, wolf × dog, and coyote × wolf hybrids occur in some parts of the nation (Bekoff et al. 1975; Gipson et al. 1975; Kennelly and Roberts 1969; Kolenosky 1971; Lawrence and Bossert 1969; Mengel 1971; Richens and Hugie 1974; Silver and Silver 1969). Such hybrids are unlikely to be as fertile as any of the parent species, and there is no evidence at present that populations of such hybrids exist in Minnesota.

Canis latrans
Coyote

Measurements: Total length 1,226 (1,177–1,330); tail 328 (300–356); hind foot 199 (191–218); ear from notch 103 (97–112.5); weight 13.04 kg (10.97–17.24).

Description: The coyote is a medium-sized, slender dog, with a bushy tail, pointed ears, and a pointed nose. The back and sides are usually a yellowish (sometimes reddish) gray. The feet and legs are yellowish, the underside and throat white. There is a black wash down the middle of the back. Considerable variation occurs; a few coyotes are almost white, others almost black. Adults in Minnesota weigh from about 11 to 14 kg. The largest on record, a male from Pine Co. (Chesness 1974), is 19 kg. Coyotes, which are

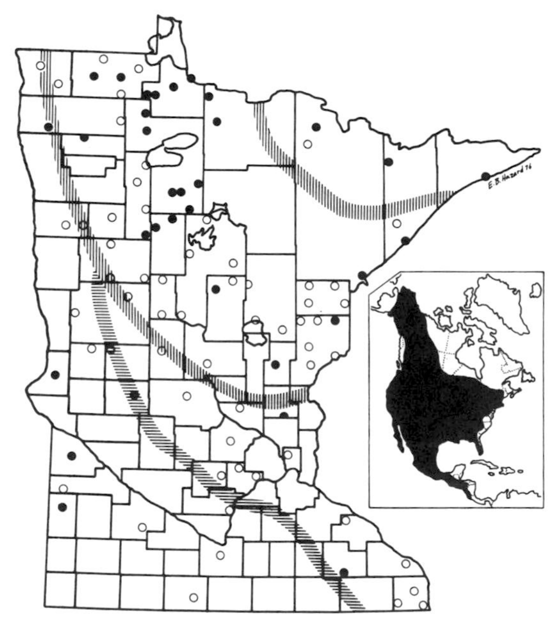

Map 51. Distribution of *Canis latrans*. ● = township specimens. ○ = other township records, selected location. Area enclosed by vertical hatching = main coyote range; area to southwest delimited by horizontal hatching = peripheral range (both after Berg and Chesness 1968). (Map produced by the Department of Biology, Bemidji State University.)

Figure 52. *Canis latrans*, coyote.

often referred to as "brush wolves," can thus be easily distinguished from adult gray wolves (or timber wolves), which weigh from about 26 to 44 kg in Minnesota. There are no authentic records of coyote × wolf hybrids in Minnesota. In the field, the coyote can be told from the wolf by its longer tail, which it holds between its legs when it runs. At a distance, some domestic dogs can be confused with either species.

Range and Habitat: Canis latrans occurs only in the New World. It is adapted to a wide variety of open and brushy habitats throughout western North America from Alaska to Central America. It originally ranged east to the western edge of the eastern deciduous forest, perhaps being limited in the north and east by the presence of the timber wolf (Gier 1975). As people have replaced the continuous forest with a mixture of habitats, and have eliminated the timber wolf, the coyote has dispersed into the northeastern states and southeastern Canada (Georges 1976; Gier 1975; Hamilton 1974; Richens and Hugie 1974; Severinghaus 1974a, b). In Minnesota, it originally occurred only in the prairie counties, but now it is most abundant in the northern part of the state and has become rarer in the intensively farmed prairie. Despite persecution, coyotes do well in many settled areas.

Natural History: Coyotes are primarily meat eaters, but, like most dogs, they eat a variety of fruits when available. In Minnesota, their diet includes various small and medium-sized mammals (mice, voles, ground squirrels, hares and rabbits, and porcupines), ground-nesting birds, various invertebrates, deer, and livestock (Chesness 1974; Hamilton 1974; Mathwig 1973). Careful studies show that much of the flesh of deer and livestock that *Canis latrans* eats is carrion (Gier 1975; Nellis and Keith 1976; Ozoga and Harger 1966; Richens and Hugie 1974). Coyotes live singly or associate in pairs or family groups, depending on the nature of the food supply (Bekoff and Wells 1980; Gier 1975). They do not hunt in packs as consistently as wolves do.

Females generally bear five to seven young in April, after a 60- to 63-day gestation. Both parents help care for the young, and the family will usually hunt as a group until the young disperse in midwinter. In Minnesota, yearling females commonly do not bear a litter. Coyotes dig their own dens, or enlarge the dens of foxes or badgers. The home ranges of adult females are about 15 km^2, and these overlap but little. The ranges of adult males generally are much larger and overlapping (Chesness 1974).

Relationship to People: Coyotes are disliked and persecuted by many stockmen and hunters because of their alleged depredations on cattle, sheep, and deer. Some claims of coyote killing of stock and game in Minnesota have been validated, but many such claims have been shown to be erroneous. Inaccurate reports of the slaughter of large mammals by *Canis latrans* arise from at least two sources. First, much of the meat taken is carrion—most dogs are opportunists and take what comes easiest, and coyotes are no exception. Today the coyote scavenges largely from people, as it once scavenged from the wolf (Gier 1975). In central Alberta, one-half of the diet is carrion (Nellis and Keith 1976). Second, observers often cannot, or do not bother to, distinguish between coyotes and free-running dogs, or between coyotes and wolves. Stock management practices also may encourage or discourage coyote predation. Fewer problems occur if sheep are penned at lambing time, and if dead sheep are not left lying out where they will attract coyotes (Chesness 1974). There is some evidence that we can condition coyotes against taking domestic stock by "lacing" carcasses with appropriate repellents (Conover et al. 1977, 1979; Ellins et al. 1977; Gustavson 1979; Olsen and Lehner 1978).

Coyotes feed largely on smaller herbivorous mammals which themselves can be serious agricultural pests if their populations are too dense. The relative economic effects of coyotes in an area will depend on local circumstances, and decisions on the need for control of coyote numbers should be based on careful professional studies of local situations. *Canis latrans* is by no means an endangered species, but neither is it totally harmful. Coyotes play a role in the regulation of natural populations, and they are a clever and colorful part of the Minnesota countryside. The state would be poorer without them.

Additional References: Andelt et al. 1979; Andelt and Gipson 1979a, b; Andelt and Mahan 1977; Bekoff 1977a, 1978; Bekoff and Diamond 1976; Bekoff and Jamieson 1975; Boggess et al. 1978; Connolly et al. 1976; Fichter et al. 1955; Gier 1968; Howard 1949a; Kennelly and Johns 1976; Knowlton 1972; Krefting 1969; Lawrence and Bossert 1967; Lehner 1976; Linhart and Knowlton 1967; Linhart and Robinson 1972; McCarley 1975; MacConnell-Yount and Smith 1978; Robinson 1961; Robinson and Grand 1958; Wells and Lehner 1978; Young and Jackson 1951.

Canis lupus
Wolf, Timber Wolf, Gray Wolf

Measurements (six specimens): Total length 1,511–1,581; tail 394–457; hind foot 250–267; ear from notch (one) 122; weight 26.31–44.45 kg (Mech 1974b lists weights—including all subspecies—of adult males as 20–80 kg; adult females 18–55 kg).

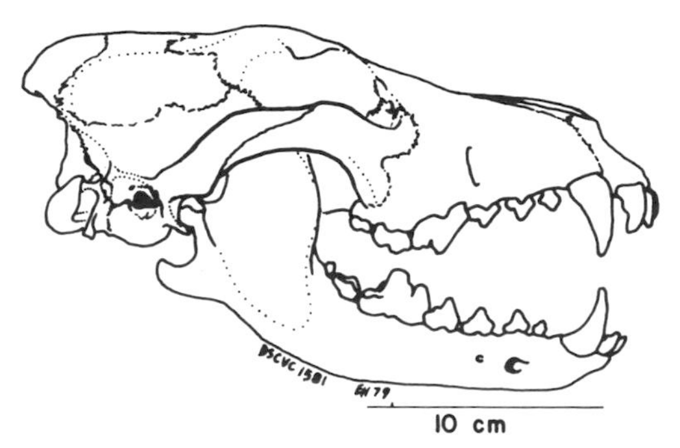

Figure 53. Skull of *Canis lupus*, wolf, BSCVC 1581.

Description: The wolf is a large, densely furred dog with a bushy tail which it holds straight out behind as it runs. Wolves vary greatly in color, but most are predominantly gray, with prominent facial markings, and generally lack the yellowish or reddish tones of coyotes. The muzzle is relatively broad. As with coyotes, the skull lacks the distinct brow or forehead which is characteristic of many domestic dogs. Adult wolves generally are twice as heavy as adult coyotes.

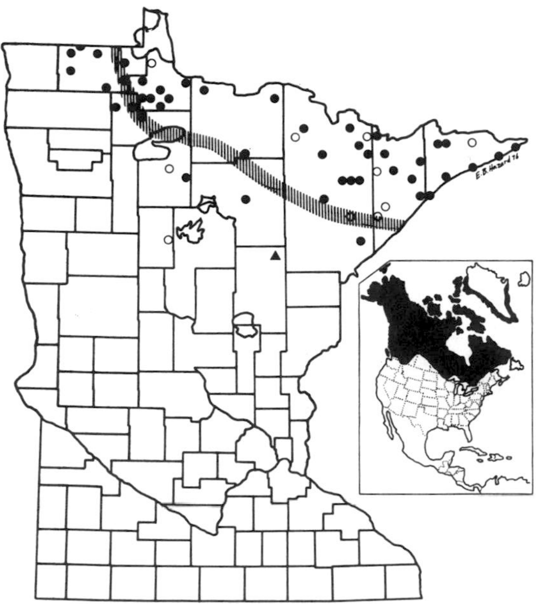

Map 52. Distribution of *Canis lupus*. ● = township specimen, selected locations. ○ = other township records, selected locations. ▲ = county specimen. Area north and east of hatching = primary wolf range (after Berg and Chesness 1968). (Map produced by the Department of Biology, Bemidji State University.)

Range and Habitat: Canis lupus once occurred throughout most of North America and was also widespread in Eurasia. The red wolf of the lower Mississippi Valley and Gulf Coast, now nearly extinct, has been named a separate species, *Canis niger*, but may be a subspecies of *C. lupus* (Lawrence and Bossert 1967, 1975; Kolenosky 1971, 1972; Mech 1974b; Riley and McBride 1975). Gray wolves or timber wolves are now gone from most of the contiguous United States and occur only in the sparsely settled parts of Alaska and Canada, and in northeastern Minnesota. They are relatively common in Cook, Lake, and northern St. Louis counties, and occur sporadically in remote parts of Itasca, Koochiching, Beltrami, and Lake of the Woods counties. Wolves have probably disappeared from northern Wisconsin and Upper Michigan (Robinson and Smith 1977; Thompson 1952). Individuals can move such great distances that it is not possible to delimit the range of the wolf (or comparable large mammals) as sharply as one can the range of a vole or shrew.

Natural History: A number of wildlife biologists have studied the wolf in northeastern Minnesota and elsewhere (see especially Frenzel, Fritts, Mech, S. Olson, Stenlund, Van Ballenberghe, and their coauthors). Although much remains to be learned, the ecology of *Canis lupus* in Minnesota is comparatively well known. Wolves are relatively sociable, long-distance hunters that feed primarily on deer, especially in the winter. The typical timber wolf pack consists of a dominant adult pair and their subadult and juvenile offspring, although there may be other subordinate adults (Pimlott 1975; Rabb et al. 1967; Schenkel 1967; Woolpy 1968; Zimen 1975). Packs normally consist of fewer than 10 adult and subadult individuals. The pack ranges over a territory of 125 to 310 km^2. There is typically a "no-man's-land" between pack territories, and the survival of deer may be relatively high in such areas (Hoskinson and Mech 1976; Mech 1972, 1979; Van Ballenberghe and Erickson 1973). Wolves mark their territories with urine and feces, and also howl to announce their presence in their territories (Harrington and Mech 1978a, b, 1979). Olfactory and auditory signals thus probably enable wolf packs to maintain territories and avoid the territories of other packs, and presumably enable young adults that have left the territory of the parent pack to recognize unoccupied areas. Commonly only a single pair in the pack breeds, which prevents the production of an unduly high density of wolves. The social organization of wolves tends to prevent overutilization of the prey resource, at the same time insuring that all areas of suitable habitat will become occupied.

Wolves mate about the end of February and produce four to seven young 63 days later. The young are generally reared in an underground den. They usually remain with the parents for at least one year and do not disperse and establish their own families until two or three years of age. During the summer when one female is rearing a new litter, individual subadults may leave the pack and hunt alone within the pack's territory. In wolves, as in most wild canids (but not usually in domestic dogs), both males and females care for and bring food to the young.

In Minnesota, the principal food of the wolf is the white-tailed deer (Mech 1966b, 1970b; Mech and Karns 1977). In summer, wolves supplement their diet with beaver and other prey and with some wild fruits, but in winter they subsist almost entirely on deer. Therefore, the condition

Figure 54. *Canis lupus*, wolf.

of the northeastern Minnesota timber wolf population is directly dependent upon the condition of the local white-tailed deer herd (Mech 1970a; Mech and Karns 1977). Although northeastern Minnesota was not originally white-tail country, it became relatively good deer habitat when the primeval coniferous forest was destroyed by lumbering and fire and was replaced by open, early successional stages of poplar and brush. (See Odum, 1971, for a discussion of ecologic succession.) The frequent occurrence of severe winters, however, prevents northern Minnesota from being consistently good deer habitat. Moreover, in recent decades conifers have seeded in and replaced these early stages in many areas, the deer population has declined, and so has the wolf population (Mech 1975b, 1977b; Mech and Karns 1977; Seal et al. 1975; Van Ballenberghe and Mech 1975). The major threat to the white-tailed deer within the range of the wolf is neither human nor natural predation but the decrease in available browse due to ecologic succession in a marginal habitat, coupled with deep snow and extreme cold in winter (Mech 1970a; Mech and Karns 1977). The major threat to the timber wolf in northeastern Minnesota is neither hunting nor trapping but the decrease in deer. The timber wolf has been extirpated from most of its former range in the contiguous 48 states. In northeastern Minnesota, however, the wolf is probably at the maximum population density that its natural prey species will support.

The wolves on Isle Royale, Michigan, some 30 km from the north shore of Lake Superior, subsist on moose. The two species are in balance, but continued survival of the moose population depends on substantial areas of the island returning to early successional stages, generally as a result of fire (Allen and Mech 1963; Hansen et al. 1973; Jordan et al. 1967; Mech 1966a; Peterson 1977; Wolfe and Allen 1973).

Relationship to People: The prevailing attitudes toward the timber wolf are prime examples of the persistence of fallacy in the face of facts. Indeed, the wolf takes second place to no other mammal when it comes to being misjudged by people. It is true that wolves, like some other large mammals, are incompatible with certain intensive human activities and that they are quite capable of killing and eating livestock. However, the documented percentage of livestock taken annually within the range of timber wolves in Minnesota is very low (Fritts 1980). Despite persistent tales to the contrary, there are no confirmed instances of wolves killing or injuring people in Minnesota. *Canis lupus* generally avoids people. There is nonetheless a deep-seated, culturally based fear of wolves. Although wolves have lived in a balanced predator-prey relation with deer for millennia, many people who live in wolf country remain convinced that wolves must be eliminated to save the deer. Some other people, who often have little familiarity with northeastern Minnesota and even less understanding of the mechanisms of wildlife ecology, believe that the wolf will disappear forever if it is not afforded absolute protection from all hunting and trapping. Neither group seems interested in the careful ecological studies that have shown that habitat management to insure a continued supply of suitable browse and cover for deer is the key to both a thriving deer herd and a healthy wolf population. If there is good deer habitat, the wolf population will be in no danger and will be able to sustain a moderate, controlled level of hunting and trapping (Minnesota Volunteer Staff 1977; Stenlund 1974).

Additional References: Chapman 1978; Colinvaux and Barnett 1979; Floyd et al. 1978; Fox 1971b, 1972a, b, 1973; Fox and Andrews 1973; Frenzel 1974; Frenzel and Mech 1967; Fritts 1979; Jolicoeur 1959, 1975; Klinghammer 1978; Mech 1973a, 1974a, 1975a, 1977a, c, d, e, f; Medjo and Mech 1976; Murie 1944; Olson 1938a, b; Pimlott 1967; Pimlott et al. 1969; Stenlund 1955b, 1965; Theberge and Falls 1967; Van Ballenberghe 1972; Van Ballenberghe et al. 1975; J. F. Wolfe 1977; Young 1946; Young and Goldman 1944.

Genus *Vulpes*
Foxes

Vulpes vulpes
Red Fox

Measurements (a series of adults from northern Iowa, reported by Storm et al. 1976): *Males*—total length 998 (954–1,045); tail 359 (320–390); hind foot 157 (140–170); ear 85 (75–93); weight 4,822 (4,131–5,675); *females*—total length 946 (842–1,020); tail 337 (294–368); hind foot 145 (127–156); ear 79 (66–85); weight 3,938 (2,951–4,585). Eleven males from Minnesota average somewhat larger: total length 1,055 (980–1,115); tail 396 (356–420); hind foot 169 (162–190); ear from notch 93 (87–97); weight 5.14 kg (4.31–6.12 kg).

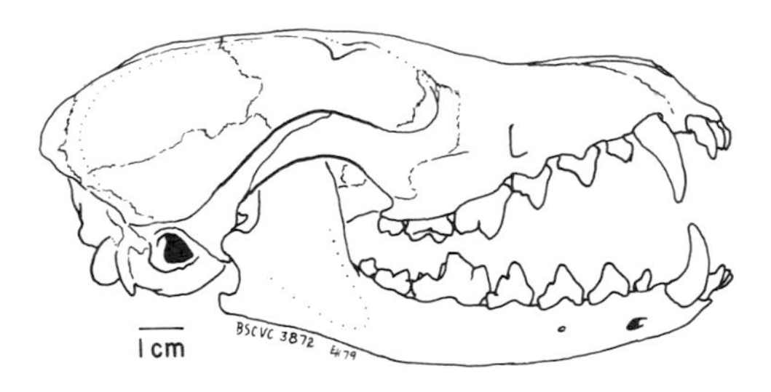

Figure 55. Skull of *Vulpes vulpes*, red fox. BSCVC 3872.

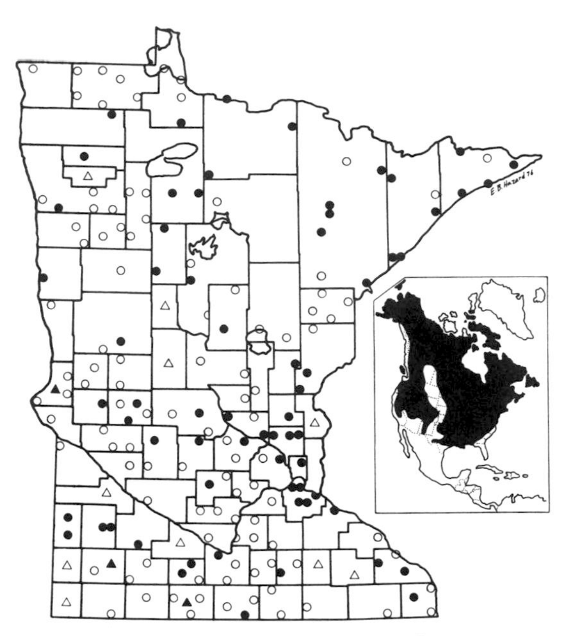

Map 53. Distribution of *Vulpes vulpes*. ● = township specimens. ○ = other township records, selected locations. ▲ = county specimens. △ = other county records. (Map produced by the Department of Biology, Bemidji State University.)

Description: Most red foxes are some shade of yellowish red. There is usually a trace of black in the fur of the middle of the back and in the bushy tail, and the feet and outer side of the tips of the ears are black. The throat, underside of the body, and tip of the tail are white. In winter, the coat is thick and luxurious. A few red foxes are black or silver (black with white-tipped hairs) and others are "cross" foxes—brownish with a cross-shaped darker area over the shoulders. These rarer forms are simply striking genetic variants of the red fox and may occur in otherwise typical litters.

Range and Habitat: Vulpes vulpes ranges across the northern hemisphere in Eurasia and North America through a variety of habitats, and from the Arctic to warm temperate zones. The New World form may not have ranged as far east as the Atlantic coast in colonial times, and was once thought to be a distinct species, *Vulpes fulva*. However, it has interbred freely and merged completely with the red foxes which the colonial aristocracy imported from England for sport, so there seems to be no basis for naming separate species (Churcher 1959). Red foxes occupy a variety of habitats throughout Minnesota, but prefer semi-open country to continuous forest.

Natural History: Foxes are relatively solitary animals. Pairs with their young typically stay together into late summer, but family groups often do not persist through the fall and winter. Red foxes are probably monogamous for one season at a time (Ables 1975). They breed in January or February, and the females bear four to 10 (usually four to six) young after a 51- to 53-day gestation (Schofield 1958; Switzenberg 1950). The family occupies a burrow which the foxes dig them-

Figure 56. *Vulpes vulpes*, red fox, and *Philohela minor*, American woodcock.

selves or remodel from the den of another mammal. The den usually has two or more entrances, and there may be alternative dens within the territory. Red foxes are probably territorial; they may exclude other families within an area of a few km² (Ables 1969b; Preston 1975; Sargeant 1972), although the behavioral mechanisms are not fully documented (Ables 1975). The young usually disperse in late September or early October, males typically traveling farther than females (Phillips et al. 1972; Storm et al. 1976; Storm and Montgomery 1975).

Vulpes vulpes is primarily carnivorous, especially in winter. The diet is varied, including woodchucks, other ground and tree squirrels, rabbits and hares, muskrats, mice and voles, song birds and game birds, and snakes, turtles, and frogs (Ables 1975; Brickner 1953; Errington 1935; Findley 1956a; Hatfield 1939b; Korschgen 1959; MacGregor 1942; Scott 1943). In the spring, summer, and fall, foxes add a variety of insects, fruits, and other items to their diet. In fall and winter they often eat wounded game birds and mammals. During the first month after the young are born, the male may bring food back to the den, and later both parents do this if both are present (Ables 1975). Since foxes bring back only large items, studies based on remains at dens emphasize larger prey items, like muskrats and cottontails. Actually, the bulk of the animal matter eaten during much of the year, as indicated by scat analysis, is mice and voles. Foxes may take shrews but often will not eat a shrew they have killed, because of the strong odor (Murie 1936). Foxes will often cache larger kills and visit them over a period of several days, as they will dead deer or other carrion (Henry 1977).

Like other wild dogs, red foxes hunt primarily at night, but they are sometimes abroad in daylight, especially in early morning and late afternoon (Ables 1969a, 1975). At midday, they are usually inside in their dens, or perhaps sunning themselves while keeping a watchful eye (or sensitive ear) out for intruders. You may surprise them, but they will likely see or hear you first.

Relationship to People: Some people trap red foxes for their fur, which fluctuates in value with changes in style, and a few hunt them for sport with horses and hounds, or just hounds. Most people, however, seem to dislike foxes, thinking of them primarily as predators on poultry or pheasants. Foxes can make serious inroads in a hen house, especially a badly maintained one. However, it has been demonstrated that upland game bird and waterfowl populations usually do well in the presence of both foxes and well-regulated human hunting, as long as food and cover are adequate (Errington 1933). By territorial behavior, foxes probably keep their own populations at a level that generally will not deplete prey populations. The major economic impact, if any, of most foxes is probably as a curb on the numbers of various small herbivorous rodents.

Red foxes, as well as other carnivores, are potential vectors for rabies and other diseases of people and other mammals (Gier 1948; Pryor 1956; Schnurrenberger et al. 1970; Verts and Storm 1966). They are not carriers of rabies as often as skunks are, but foxes do interact with skunks (Houseknecht and Huempfner 1970) and may thus be exposed to rabies.

Additional References: Barash 1974c; Gingerich and Winkler 1979; Isley and Gysel 1975; Jorgenson et al. 1978; Layne and McKeon 1956a, b; Longley 1962; Richards and Hine 1953; Sargeant 1978, 1980; Scott and Klimstra 1955; Stanley 1963; Storm 1965.

Vulpes velox
Swift Fox

This small fox is generally cream-colored and is about the size of a house cat. It formerly lived in the prairies of southwestern Minnesota (Egoscue 1979; Hawley 1974). Over the first several decades of this century, the swift fox was apparently extirpated from the northern prairie states and southern Prairie Provinces, but it may be returning. Specimens have recently been taken in Nebraska, western and central South Dakota, and North Dakota (Van Ballenberghe 1975a, b). The return of the swift fox into its former range may be a result of the ban on predator poisoning, and the species will be in danger again if poisoning is resumed. It probably has not yet moved as far east as Minnesota. The swift fox is certainly not numerous enough to be of economic consequence, but, as a unique remnant of the original prairie fauna, it should be protected.

Additional References: Bowles 1975; Chambers 1978; Jones 1964; Kilgore 1969; Long 1965; Rohwer and Kilgore 1973; Thornton and Creel 1975; Thornton et al. 1971.

Genus *Urocyon*
Gray Fox

Urocyon cinereoargenteus
Gray Fox

Measurements (seven specimens): Total length 905–1,040; tail 293–378; hind foot 127–143; ear from notch 65–81; weight 3.86–5.17 kg.

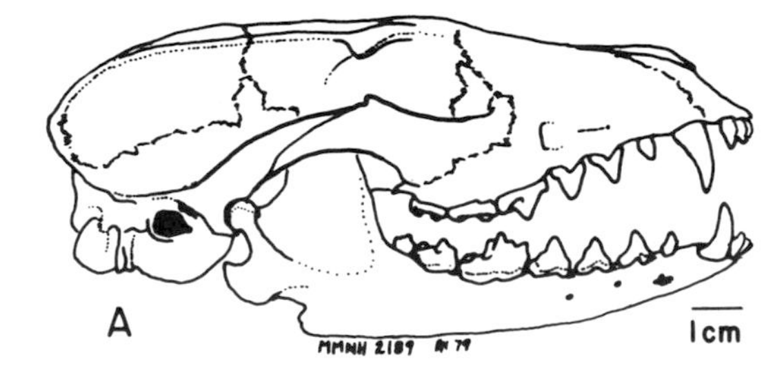

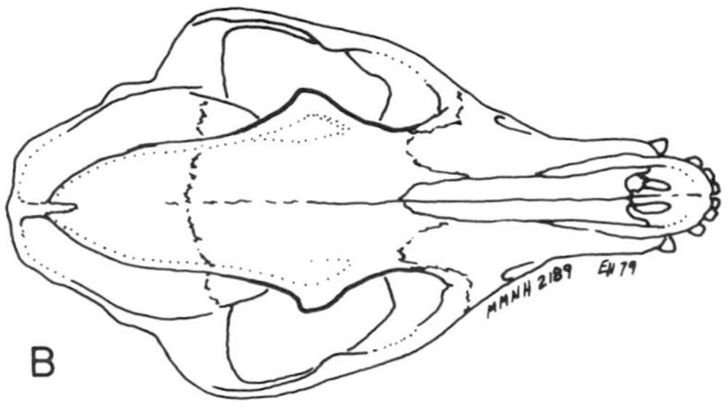

Figure 57. Skull of *Urocyon cinereoargenteus*, gray fox, MMNH 2189. A. Lateral view: note notch below angular process. B. Dorsal view: note U-shaped pattern of parietal ridges.

Description: The gray fox is about the same size as the red fox but is not generally as heavy. The overall gray color results from guard hairs that have white bands and black tips. A distinct black stripe runs down the middle of the back and continues onto the tail, which has a black tip. The underfur is a yellowish buff, as are the legs and feet, the back of the ears, and the sides of the neck. Although the animal is beautifully marked, its fur is not as luxurious as that of the red fox and has little economic value.

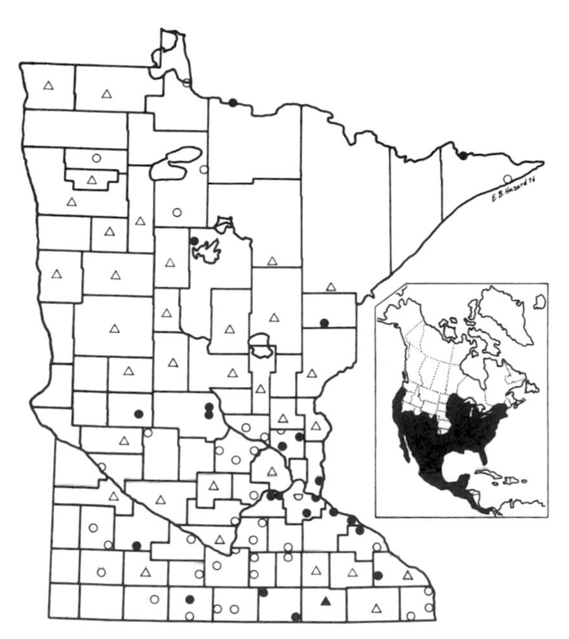

Map 54. Distribution of *Urocyon cinereoargenteus*. ● = township specimens. ○ = other township records, selected locations. ▲ = county specimen. △ = other county records. (Map produced by the Department of Biology, Bemidji State University.)

Range and Habitat: Urocyon cinereoargenteus occurs only in the New World. It ranges from Latin America to the Pacific coastal states, the southern Rockies, most of the eastern United States, and extreme southern Ontario and Quebec. There are occasional records from most of Minnesota, but the gray fox is not common except in the southeastern counties. It prefers wooded areas or heavy brush, in contrast to the red fox, which favors more open habitats (Choate and Krause 1976; Errington 1935; Follmann 1974; Trapp and Hallberg 1975).

Natural History: The biology of the gray fox is not nearly as well known as that of the red fox. Trapp and Hallberg (1975) have reviewed the literature through the early 1970s. Like most canids, *Urocyon* is an opportunist, feeding on whatever vertebrate or invertebrate meat is available and eating various fruits in season. Gray foxes feed on cottontails more often than on meadow voles; the reverse is true of red foxes (Errington 1935). When abundant, insects may make up nearly half the diet of the gray fox (Wood et al. 1958). Birds are generally not a major item in the diet; Errington (1933) found that bobwhite survived the winter equally well in areas occupied by foxes as in areas without foxes.

Gestation has been reported to be 53 days (as in the red fox), 63 days (as in the domestic dog), and 51–63 days, but no record based on conclusive observation exists (Trapp and Hallberg 1975). Breeding in northern latitudes probably occurs primarily in late February or March but may last into May. Probably both parents remain with the young. There is one litter per year, averaging about four young. Family ties weaken in late summer, but the pups may not disperse until January or February. Gray foxes may be territorial, but the related behavioral mechanisms under natural conditions are unknown. Their behavioral interactions with the more widespread and generally more common red foxes are likewise unknown. One surprising characteristic that is widely known is that *Urocyon* climbs trees, not only to seek refuge but apparently also to obtain food (Trapp and Hallberg 1975; Terres 1939; Yeager 1938).

Relationship to People: Gray fox pelts are of little value, and gray foxes provide no sport for fox hunters because they climb the nearest tree rather than provide a long chase. Their preference for woods rather than open country means they are relatively unlikely to encounter people or their hen houses and are also unlikely to have much effect on the rodents that eat our grain. Gray foxes add variety to the wooded areas of Minnesota but probably have little influence on our pocketbooks.

Additional References: Burgdorfer et al. 1974; Chesness et al. 1968; Fox 1970; Fuller 1978; Goldman 1950; Hatfield 1939b; Kavanau and Ramos 1975; Layne 1958a; Layne and McKeon 1956a, b; Lord 1961; Pryor 1956; Richards and Hine 1953; Richmond 1952; Schnurrenberger, Martin, and Lantis 1970; J. E. Wood 1958, 1959.

Family Ursidae
Bears

Bears are moderate- to large-sized mammals, plantigrade, mostly terrestrial, and mostly omnivorous (polar bears are marine animals and strictly carnivorous). Although the canines are well developed, the cheek teeth generally lack the shearing cusps typical of carnivores. The tooth formula is 3/3, 1/1, 4/4, 2/3 × 2 = 42, but the first three adult premolars are small and are often lost in mature individuals. Bears are (or were) widely distributed in North America and Eurasia, and also occur in the Andes. The bears of the Atlas Mountains of North Africa are probably extinct (Stains 1967).

Additional References: Ewer 1973; Vaughan 1978.

Genus *Ursus*
Bears

Ursus americanus
Black Bear

Measurements: There are few museum specimens of black bears, and these are seldom accompanied by standard measurements. Adult weight is generally attained by the age of four or five years. Schwartz and Schwartz (1959) give the following measurements for Missouri specimens: total length 127–198 cm; tail 10–13 cm; hind foot 18–36 cm; ear 14.6 cm; weight 91–272 kg. Poelker and Hartwell (1973: table 37) give the average *body* length and weights of males as 155 cm and 98 kg; of females as 135 cm and 61 kg. In Quebec, Juniper (1978) reports males four years and older ranging from 53.0 to 154.0 kg, females four years and older from 43.5 to 91.0 kg. Judd et al. (1971) give the "average" adult weight in Minnesota as 113–136 kg. Kinsey (1967) indicates that black bears in Minnesota may reach 227–272 kg and lists the heaviest-known Minnesota specimen (I presume a male) as 265 kg.

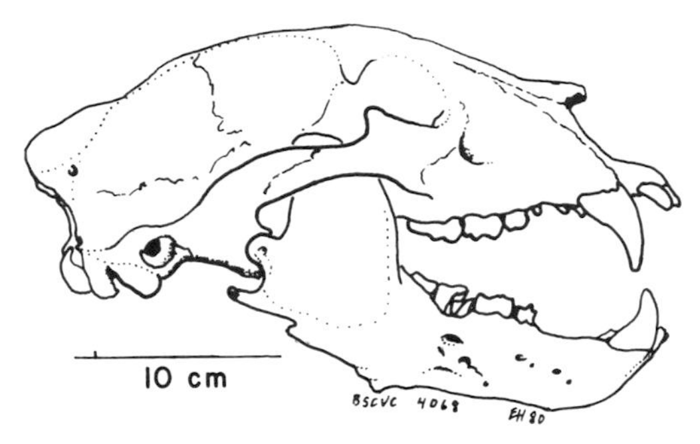

Figure 58. Skull of *Ursus americanus*, black bear, BSCVC 4068. The scarred and perforated region of the mandible and the broken and decayed m2 seem to be the result of a healed fracture, perhaps accompanied by an abcess.

Description: Black bears are large, stout, almost tailless, and generally totally black except for a brown snout. A few individuals are brown (Rogers 1980). Although the skull has large canines, it is easily distinguished from the skulls of other large carnivores by the low, flattened crowns of the cheek teeth and by the last molar which, even in young animals without a full tooth complement, is always larger than the one in front of it.

Range and Habitat: Ursus americanus once occurred throughout most of North America, from northern Mexico to Newfoundland and Alaska. It is now absent from much of the heavily settled East Coast and the intensely farmed interior states, and from parts of the southern Prairie Provinces. In Minnesota, bears are generally restricted to the northeastern third of the state,

Figure 59. *Ursus americanus*, black bear.

though individuals wander great distances and show up in unexpected places. For the most part, black bears reside in well-wooded areas, including wooded swamps. They often forage for berries in cutover areas.

Natural History: Black bears are basically solitary, except for mothers with cubs or yearlings. Adult males and females may associate briefly during the mating season in midsummer, but no permanent bond is formed. Although mating occurs in summer, the fertilized egg does not begin active growth and development for some months. Total gestation lasts seven months or more, and the young (commonly two or three) are born in late January or early February, while the mother is denned up for the winter. They are tiny in

Map 55. Distribution of *Ursus americanus*. ● = township specimens. ○ = other township records, selected locations. ▲ = county specimen. (Map produced by the Department of Biology, Bemidji State University.)

comparison to adults—a newborn bear weighs between 200 and 350 g, about 1/1,000 of the weight of an adult. By the time the young emerge from the overwintering den with the mother, they weigh 2.25 to 2.75 kg. The cubs generally den up with their mother after their first summer and do not disperse until they are over a year old. Some may remain over an additional winter. Sows usually breed every two or three years. Bears are largely diurnal and crepuscular, especially in summer (Amstrup and Beecham 1976).

Bears are omnivores. They eat whatever animal matter is available, but meat is generally a secondary food source. They are capable of killing deer and livestock, but the hoofed mammals that they eat are usually carrion. Bears eat a great variety of berries in season, acorns in the fall, and some soft leafy vegetation. They also have a well-publicized taste for wild honey. The most significant food that people provide for bears is not livestock but garbage.

Both male and female black bears hibernate during the colder months, generally under shelter such as a brush pile, windblown tree, or rock ledge. During the winter sleep, a bear's body temperature drops only a few degrees, but the metabolic rate drops considerably. A heavy layer of fat and a thick fur coat provide enough insulation so that this massive beast can sleep in a den all winter and lose little weight (Arimond 1980). Bears lose appreciable weight, however, after they emerge in the spring, when food is scarce.

Relationship to People: People have curiously ambivalent attitudes toward bears. Like other carnivores, bears have often been despised and persecuted as harmful predators. For many years the Minnesota Legislature authorized bounty payments for bears (and other predators) until it recognized that predators were not necessarily harmful and that the bounty system was a particularly ineffective but expensive means of control. Today, black bears are protected because their value as game animals exceeds any economic damage done by a few individuals.

On the other hand, people often have too little respect for the damage a bear can do—to people. Black bears may not be as aggressive as grizzly bears (Herrero 1972), but they are strong and often unpredictable, and sows are very protective of their cubs. Bears regard campgrounds and garbage dumps as sources of free food, and many vacationing Americans are injured each year because they behave foolishly in the presence of bears (Burghardt et al. 1972; Merrill 1978).

Additional References: Gordon 1977; Jonkel and Cowan 1971; Jordan 1974; Kemp 1972, 1976; Kinsey 1965; Landers et al. 1979; Lindzey and Meslow 1976, 1977a, b; Lindzey et al. 1977; Marks and Erickson 1966; Matson 1946, 1954; McMillin et al. 1976; Maxwell et al. 1972; Northcott and Elsey 1971; Pelton and Burghardt 1976; Rogers 1972, 1974, 1975, 1976; Sauer 1975; Schorger 1946a.

Ursus arctos
Grizzly Bear

Grizzly bears do not occur in Minnesota today, but they may have at one time. When bison roamed the prairie, grizzlies followed the herds, and they may well have ranged as far east as the prairies of western Minnesota. The occipital portion of a skull at Southwest State University (SMSC 169), unearthed on a nearby farm and tentatively identified as a grizzly bear, is actually a Recent or Pleistocene horse, *Equus* sp. (discussion with Robert E. Sloan, 13 July 1978).

The grizzly was formerly designated a distinct species, *Ursus horribilis*, but today it, and all the other large terrestrial brown bears of northern Eurasia and North America, including the Alaska brown bear, are considered one species, *Ursus arctos*.

References: Cauble 1977; Cole 1972; Craighead and Craighead 1972; Gordon 1977; Guilday 1968; Herrero and Hamer 1977; Merrill 1978; Peterson 1965; Spiess 1976; Youngman 1975.

Family Procyonidae
Raccoons and Allies

This family is restricted to North and South America except for the lesser panda of southeast Asia. The classification of the giant panda of Tibet is in dispute. Ewer (1973), following Morris and Morris (1966) and others, classifies it as a procyonid. Stains (1967) and Chorn and Hoffman (1978), following Davis (1964), classify it as an ursid.

Additional References: Goldman 1950; Vaughan 1978.

Genus *Procyon*

Procyon lotor
Raccoon

Measurements: Total length 833 (753–940); tail 265 (227–305); hind foot 118 (108–132); ear from notch 63 (54–68); weight 6.81 kg (4.99–14.52). Marshall (1956) found the average weight of 27 adult males to be 7.9 kg (4.7–11.8) and that of 32 adult females to be 6.9 kg (5.1–9.9). These were summer weights; raccoons put on weight before denning up in the fall.

Description: The black mask across the eyes and the bushy ringed tail make the raccoon unmistakable. The coat is usually a grizzled gray, but blacker as well as relatively brown individuals occur. The tooth formula is 3/3, 1/1, 4/4, 2/2 × 2 = 40. Raccoons are short-eared, stout, and plantigrade, and generally walk with a waddling gait. I have seen one almost pure white individual (in southern Michigan), but it had faint brownish rings on its tail; the daylight was fading, and through my binoculars I could not tell whether the eyes were pink or pigmented.

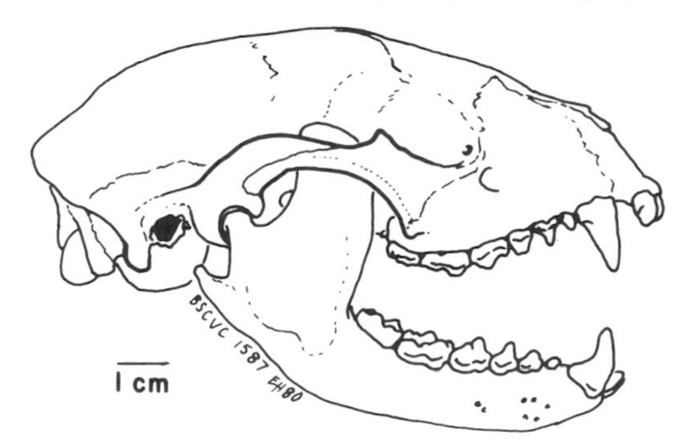

Figure 60. Skull of *Procyon lotor*, raccoon, BSCVC 1587.

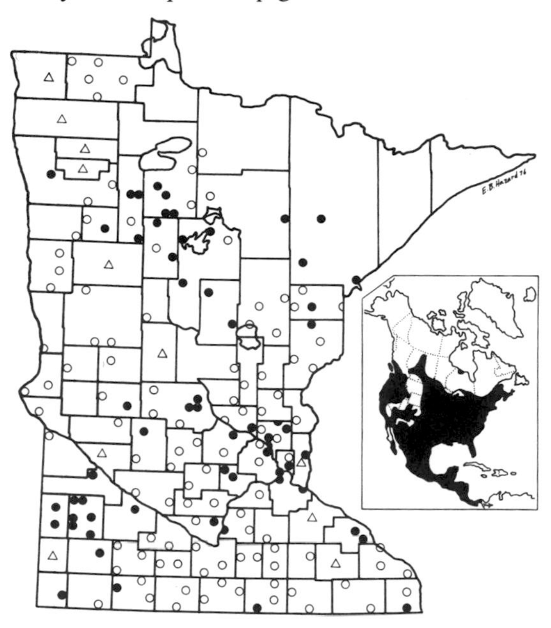
Map 56. Distribution of *Procyon lotor*. ● = township specimens. ○ = other township records, selected locations. △ = other county records. (Map produced by the Department of Biology, Bemidji State University.)

Range and Habitat: In North America, *Procyon lotor* ranges from southern Panama through Central America and Mexico into all the 48 contiguous states and southern Canada. It is absent from parts of the southwestern deserts and the Rockies. Forty or fifty years ago, raccoons were uncommon in Minnesota except in the southeastern counties, but they have since become common except in the coniferous forests of the northeast (Timm 1975). Although we most often associate raccoons with deciduous forests, they

Figure 61. *Procyon lotor*, raccoon.

can also make a home in the prairie, where they typically den underground. In open farming areas, they also den in abandoned buildings, outbuildings, haystacks, and even snowdrifts (Lynch 1974). In the prairie region, raccoons are partial to marshy lowlands (Dorney 1954; Fritzell 1978a).

Natural History: Raccoons are familiar to most of us because they often visit garbage dumps, campsites, and other places where people provide food. If they are not persecuted, raccoons adapt quickly to the presence of people and will live quite happily on handouts (Hoffman and Gottschang 1977). Their natural foods include a variety of animal and vegetable matter, both aquatic and terrestrial. In Minnesota their diet includes crayfish, clams, grasshoppers and other insects, rough fish, frogs, turtle eggs, birds and bird eggs, voles, mice, an occasional muskrat or squirrel, and juneberries, other fruits, acorns, and grain (Schoonover and Marshall 1951). Raccoons obtain most food on the ground or in the water, but they climb well and often have their dens in trees (Berner and Gysel 1967). They are primarily nocturnal (Schneider et al. 1971).

Adults are more or less solitary, but the young generally stay with their mother until they are almost a year old. Mating occurs in early spring, and gestation has been reported to last from 54 to 63 days. Three to six young are born near the end of May, in a hollow tree if one is available. The female and young may abandon the den about two months later (Schneider et al. 1971), and people frequently encounter mothers with young in late summer. Males play no role in child rearing.

Raccoons are among the carnivores that usually den up and remain inactive for the winter, but they do not enter into the kind of low temperature torpor that is characteristic of ground squirrels and jumping mice. In Minnesota, the winter sleep lasts from about the end of October to the end of February. Males generally overwinter alone, females with their young. Communal dens are known, the largest on record being a cellar of an abandoned house in Swift County, which housed 23 overwintering raccoons (Mech and Turkowski 1966).

Relationship to People: Raccoons are valued as a source of fur, sport, and food. Pelts bring a good price when long fur is in style and during periods of nostalgia for the 1920s and 1930s. Raccoon flesh is good but often gamey; the fat is strong and should be removed before cooking. Raccoons carry a variety of parasites, and, as with all game, raccoon flesh should be cooked thoroughly. They also are potential carriers of Q fever, rabies, toxoplasmosis, tularemia, and other diseases (Burgdorfer et al. 1974; Menges et al. 1955; Riemann et al. 1978).

Because raccoons are easy to observe, they provide an aesthetic and educational resource that is hard to measure in dollars. They can, however, be pests and do damage in campgrounds and around lake cottages. Since they like to eat bird eggs and young, and are more numerous than most other predators, raccoons often cause serious losses to waterfowl production in intensively managed refuges (Dorney 1954; Urban 1970).

Additional References: Barash 1974c; Bider et al. 1968; Chesness et al. 1968; Cumbie 1975; Fritzell 1977, 1978b; Longley 1972; Lotze and Anderson 1979; Mech, Barnes, et al. 1968; Mech, Heezen, et al. 1966; Priewert 1961; Sanderson and Nalbandov 1973; Schneider et al. 1971; Schnell 1969; Shirer and Fitch 1970; Stains 1956; Stuewer 1943; Tester 1953; Turkowski and Mech 1968; Twitchell and Dill 1949.

Family Mustelidae
Weasels and Allies

More than half of the native Minnesota carnivores are mustelids. The family occurs on all major land masses except Australia and Antarctica, and occupies a wide range of ecologic niches, especially in the Northern Hemisphere. Mustelids have only two lower molars and one upper one, and the carnassials usually are well developed. Generally, the auditory bullae are flattened and the cranium is elongate. Mustelids typically are five-toed and relatively short-legged. The feet are plantigrade or digitigrade. In most species males are markedly larger than females. Most mustelids are active hunters and more or less specialized for a meat diet.

References: Erickson 1946; Ewer 1973; Heidt et al. 1976; Iversen 1972; Long 1969; Rosenzweig 1966; Stains 1967, 1976a, b; Vaughan 1978; Wright 1963.

Genus *Martes*
Martens and Fisher

The genus *Martes* ranges through most of northern and temperate Eurasia and boreal North America. Old World martens are generally considered separate species from the New World pine marten. All members of the genus inhabit forests or rocky areas. The tooth formula is 3/3, 1/1, 4/4, 1/2 $\times$ 2 = 38.

Martes americana
Marten, Pine Marten

Measurements (one St. Louis Co. male): Total length 615; tail 195; hind foot 80; ear from notch 48; weight 740. Some measurements from the literature: total length, males 500–760, females 460–560; tail, males 180–250; weight, males 0.7–1.8 kg, females about 1 kg (Anon. 1965; Northcott 1977).

Description: The marten is a large, arboreal weasel, about the size of a mink. (Spelling is critical here; the marten is a mammal in the weasel family, the martin a bird in the swallow family.) The marten's thick, luxurious fur is most often a reddish to yellowish brown, somewhat darker on the back and becoming blackish on the bushy tail. There is a distinct pale to orangish yellow bib on the throat. The skull of a marten differs from that of similar-sized mustelids in its more elongate, somewhat doglike rostrum, its less flattened bullae, and the presence of four premolars.

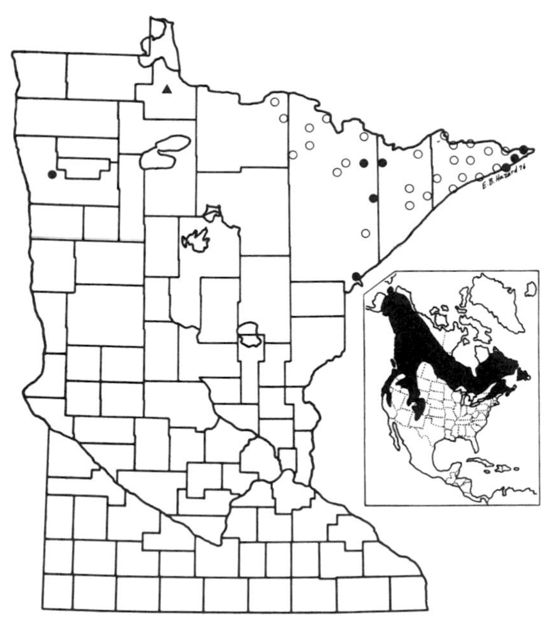

Map 57. Distribution of *Martes americana*. ● = township specimens. Crookston and Duluth specimens are nineteenth-century records. ○ = other township records, selected locations, 1977–79 only. ▲ = county specimen. (Map produced by the Department of Biology, Bemidji State University.)

Range and Habitat: Martes americana is found only in North America, but the closely related *M. martes* occurs in comparable habitat across northern Eurasia. Marten do best in boreal coniferous forest and the forest-tundra ecotone. They reside in a belt from the Maritime Provinces to Alaska, occurring in the contiguous United States only in northern New England, northern Minnesota (and perhaps Michigan and Wisconsin), and the northern Rockies, Cascades, and Coast Ranges. They once occurred as far south as Polk and Crow Wing counties in Minnesota but were eliminated or nearly so from even the northernmost counties during the first decades of this century (Hagmeier 1956). Marten are now completely protected by law and seem to be returning to the northernmost parts of the state (Gunderson 1965; Hazard 1978a; Maxham 1970; Stenlund 1955a). They have apparently been successfully reintroduced into Wisconsin (Anderson and Davis 1978). Marten occurred at least as far south as West Virginia during the Pleistocene (Guilday and Hamilton 1973).

Natural History: In marten, as in many mustelids, delayed implantation occurs. Marten mate in midsummer, but the zygotes do not implant until midwinter (Jonkel and Weckwerth 1963; Wright 1942). Females mate for the first time in their second summer and bear a litter of two to six young (average three to four) in March to May of the following year. The young leave the nest when about two months old (Kelly 1978; Northcott 1977). Adult marten are essentially solitary. Adults of the same sex are more or less evenly spaced in their habitat and thus presumably somewhat territorial. Male home ranges are larger than those of females and overlap more. Female home ranges average as little as 0.7 km^2, those of males as little as 2.4 km^2, when food is abundant. When food is scarce, home ranges may reach 26–39 km^2 (Hawley and Newby 1957; Kelly 1978; Mech and Rogers 1977; Northcott 1977).

Marten are largely but not entirely nocturnal and are active both in trees and on the ground. They prefer tree dens but will utilize rock crevices and other terrestrial shelters (Herman and Fuller 1974; Northcott 1977). They are opportunistic feeders, taking a variety of small and medium-sized mammals (microtines, red squirrels, least chipmunks, northern flying squirrels, shrews, and snowshoe hares), some birds and insects, and various fruits and berries (Cowan

Figure 62. *Martes americana*, marten, and *Clethrionomys gapperi*, red-backed vole.

and Mackay 1950; Koehler and Hornocker 1977; Marshall 1946; Murie 1961; Northcott 1977; Quick 1955; Weckwerth and Hawley 1962). Although their diets vary geographically and seasonally, marten apparently feed most heavily on voles, especially *Clethrionomys*. Hare and grouse do not figure prominently in their diet, and their population fluctuations do not coincide with hare or grouse cycles (Cowan and Mackay 1950; Hawley and Newby 1957; Quick 1955).

Relationship to People: Martes americana has probably never been common enough near agricultural areas to be a threat to the poultry yard. People have valued marten primarily for their superb fur. Unlike many furbearers, marten are easy to trap. A combination of overtrapping and habitat destruction by logging and burning led to the elimination of the marten from much of its range in the Appalachians, the Great Lakes States, and North Dakota. As parts of northern Minnesota revert to a coniferous forest climax, and with the complete protection that the species has enjoyed since 1933, the marten will doubtless become more common. Since marten have a low reproductive potential and are extremely easy to trap, it may never become advisable to legalize trapping (Mech and Rogers 1977).

Additional References: Belan et al. 1978; Cahn 1921; Clem 1977; Cowan and Guiget 1975; de Vos 1951; Dodds and Martell 1971a; Hagmeier 1958, 1961; Leach 1977a, b; Leach and Dagg 1976; Leach and de Kleer 1978; Marshall 1951c; Soutiere 1979; A. Todd 1978; Wright 1953.

Martes pennanti
Fisher

Measurements (nine *males*): Total length 987 (905–1,055); tail 372 (341–397); hind foot 116 (103–123); ear from notch 50 (45–55); weight 3.94 kg (2.86–5.67); (three *females*): total length 884–921; tail 355–359; hind foot 112–115; ear from notch 45–49; weight 2.13–2.15 kg. Blanchard (1964) reported a 9.1 kg live male from

Maine; Wright and Coulter (1967) report that males average 4.9 kg, females 2.5 kg.

Description: Like its cousin the marten, the fisher is an arboreal weasel, but it is bigger, about the size of a fox. Its dense pelage is dark brown, almost black, with a grizzled appearance caused by white bands on the guard hairs. There may be small white spots and streaks on the underside, but there is no distinct light bib. The somewhat doglike skull resembles that of the marten, but it is larger and proportionately wider at the rear of the zygomatic arch. Males are about 15 percent larger, in linear measurements, than females, and weigh about twice as much.

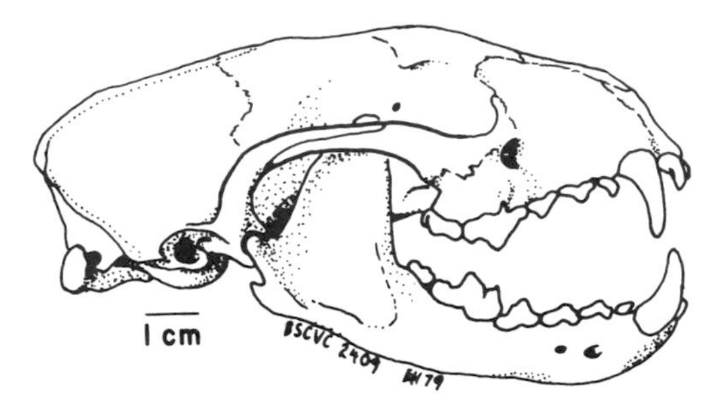

Figure 63. Skull of *Martes pennanti*, fisher, BSCVC 2409.

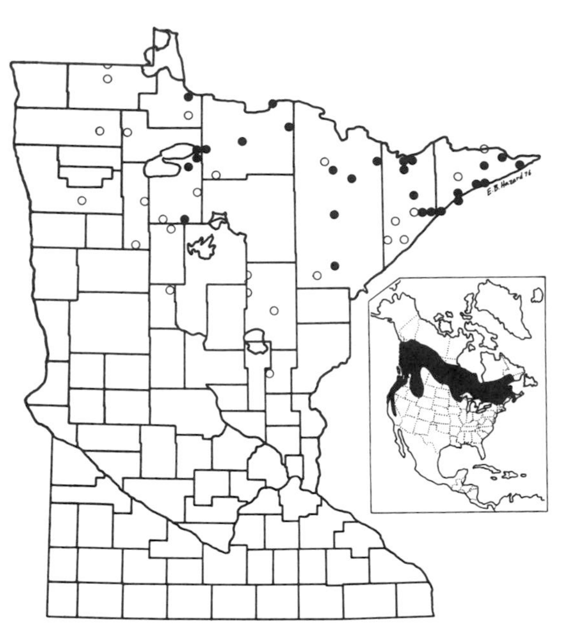

Map 58. Distribution of *Martes pennanti*. ● = township specimens. ○ = other township records, selected locations. (Map produced by the Department of Biology, Bemidji State University.)

Range and Habitat: Martes pennanti is strictly North American. At one time it ranged from the Atlantic to the Pacific coasts of Canada (but not to Alaska) and south into the mountains of the western United States and through New England and the Great Lakes States south into the Middle Atlantic States (Bailey 1926; Hagmeier 1956; Lyon 1936). Unlike the marten, the fisher occurred in deciduous as well as coniferous forests. However, it has been extirpated from the region south of the Great Lakes. In Minnesota, it was once common through most forested regions of the state but was almost extirpated by trapping. Under complete protection until the late 1970s, it has become relatively common again in northeastern Minnesota (Balser and Longley 1966; de Vos 1964; Heaney and Birney 1975).

Natural History: The fisher eats a variety of animal and vegetable matter but concentrates on flesh. Being larger than the marten, the fisher can feed on larger food, including porcupines (de Vos 1951), occasional young deer, and probably marten. It also takes hares, red squirrels, mice, and birds. (*Martes pennanti* apparently acquired its common name because it supposedly fed mainly on fish, but it does not.) Fisher are solitary except during mating and when rearing their young. Males may not breed until two years of age, although females generally breed when one year old (Wright and Coulter 1967). Gestation lasts about 11 months, but delayed implantation accounts for most of this period. In late March or April, the female bears a litter, most often of two or three young. Fisher generally den in tree holes or hollow logs and may den up during severe weather.

Relationship to People: People have substantially restricted the range of the fisher—directly by trapping for the fisher's valuable fur and indirectly by eliminating much of the forest.

Martes pennanti, however, is more flexible than *M. americana*, being at home in both deciduous and coniferous forest. The fisher is also known to do well in second-growth forests. It was probably never completely eliminated in Minnesota and is now common enough that complete protection is no longer necessary or even desirable (Balser and Longley 1966). The fisher population is closely monitored by the Department of Natural Resources, and there is a limited harvest.

Additional References: Boise 1975; Cahn 1921; Clem 1977; Coulter 1960; Davison et al. 1978; Dodds and Martell 1971b; Hamilton and Cook 1955; Leach 1977a, b; Leach and Dagg 1976; Leach and de Kleer 1978; Parmalee 1971; Powell 1977, 1978, 1979a, b; Rand 1944; Weckwerth and Wright 1968.

Genus *Mustela*
Weasels and Minks

Mustela is a widespread and diverse genus in both North America and Eurasia. Two (perhaps three) Minnesota species also occur in Eurasia. The tooth formula is 3/3, 1/1, 3/3, 1/2 × 2 = 34.

References: Ewer 1973; Hall 1951b; Stains 1967; van den Brink 1968.

Mustela erminea
Ermine, Short-tailed Weasel

Measurements (*males*): Total length 307 (289–330); tail 85 (75–94); hind foot 40 (35–43); ear from notch 19 (16–22); weight 112.7 (90.6–142.5); (*females*): total length 251 (234–285); tail 65 (53–76); hind foot 32 (29–37); ear from notch 16 (12–19); weight 75.4 (43.0–125.0).

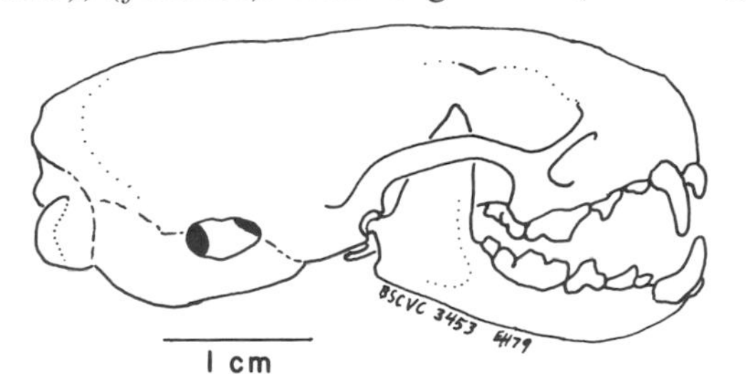

Figure 64. Skull of *Mustela erminea*, ermine, BSCVC 3453.

Description: Ermine are short-legged, long-bodied carnivores, with small, elongate heads. Females weigh about as much as a chipmunk, and males are up to twice as heavy as females. The tail is usually less than one-third of the total length. In winter, ermine are white (sometimes stained with yellow), except for the tip of the tail, which is black all year. In summer, they are dark brown above and white or yellowish white below, with white fur on the inner sides of the hind legs. *M. erminea* is distinctly smaller than *M. frenata*, the long-tailed weasel, and skins or skulls of the two species can usually be distinguished on the basis of size if the sex is known. But note that male *M. erminea* are often about the size of female *M. frenata*. The rostrum of the skull of the ermine is tapered, the postorbital processes are relatively rounded, postorbital constriction is slight, and the bullae are flatter than those of the long-tailed weasel.

Range and Habitat: The ermine is a boreal, Holarctic species. In Great Britain, *M. erminea* is called the stoat, and the word "ermine" is reserved for the valued winter pelage. Although the same species occurs in both northern North America and northern Eurasia, European *M. erminea* are larger on the average than American specimens (van den Brink 1968). This size difference may relate to the presence in North America, but not in Eurasia, of a larger weasel, *M. frenata*. The ermine is primarily a denizen of northern forests, ranging only as far south as northern California. It also occupies open country at the prairie edge and lives on the tundra to the Arctic coast (Youngman 1975). *M. erminea* probably occurs throughout Minnesota, but it is uncommon in the southwestern counties.

Natural History: Ermine are generally solitary, terrestrial hunters, but they sometimes hunt in pairs. Their major food is voles and mice, but they also take chipmunks, shrews, rabbits, small birds, snakes, lizards, and frogs, as well as in-

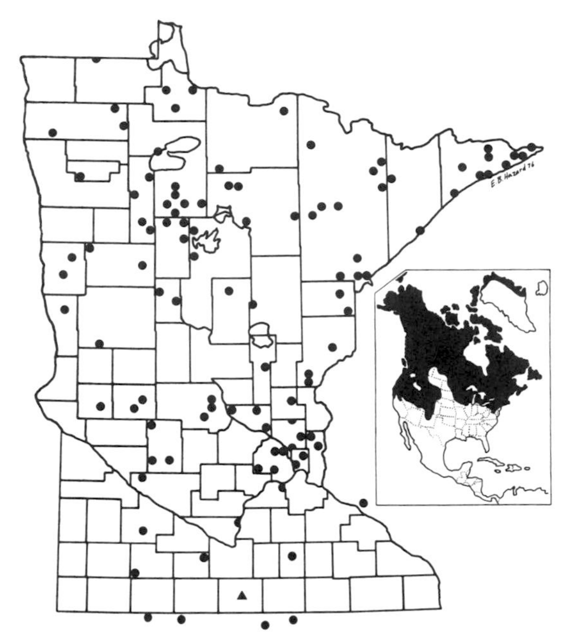

Map 59. Distribution of *Mustela erminea*. ● = township specimens. ▲ = county specimen. Iowa specimens reported in Bowles 1975 (Map produced by the Department of Biology, Bemidji State University.)

sects, earthworms, and other invertebrates (Aldous and Manweiler 1942; Day 1968). Polderboer (1942) found least weasel (*M. nivalis*) remains in an ermine food cache. Ermine occasionally wreak havoc in henhouses. Because males are much larger than females they can take bigger prey. Ermine may make their dens in a vacant chipmunk hole or in such shelter as a pile of rocks or brush. Nesting material often consists of the fur or feathers of prey. Ermine breed in midsummer; the total gestation, including a long period of delayed implantation, is eight or nine months (Shelden 1972; Wright 1942). A litter of four to nine young is born about April. The young disperse by the end of summer. Males sometimes help in rearing the young (Jackson 1961).

Throughout its range in North America, *M. erminea* turns white in winter. This obviously adapts it better to hunting and escaping enemies in the snow (Powell 1973; Rosenzweig 1966). However, snow cover is probably not the stimulus for the change. This is not surprising, because the onset and the disappearance of snow are highly variable, and the molt takes some weeks. The apparent stimulus for the molt is the highly dependable change in photoperiod (day length) that occurs in the fall and spring. Temperature may have some effect on the spring molt (Rust 1962).

Relationship to People: People value ermine (and other weasels) for their pelts, but otherwise commonly think of them as chicken thieves and wasteful killers of wildlife. Actually, domestic fowl constitute an insignificant part of the total diet of weasels. Weasels' major impact on people is to exert some measure of control on populations of mice and voles. At the present rate of trapping weasels for fur, we do not endanger their populations.

Additional References: Chesness et al. 1968; Hall 1945; Hamilton 1933b; Ralls 1977; Simms 1978.

Mustela nivalis
Least Weasel

Measurements: Total length 180 (157–195); tail 33 (27–37); hind foot 21 (18–25); ear from notch 8–13; weight 39.8–56.0.

Description: The least weasel is the smallest member of the order Carnivora. In summer, it is brown above and whitish below, and the underside sometimes has brown flecks. In winter, most least weasels are white. The tail is only about 3 cm long and does not have a black tip in winter or summer, though it may have a few black hairs. Least weasels can also be clearly distinguished from other weasels by a peculiar property; their fur fluoresces under ultraviolet light (Latham 1953). (CAUTION: Do not use this technique without proper eye protection; UV radiation can damage human eyes.) The skull resembles that of other weasels but is smaller and more rounded. Males are somewhat larger than females.

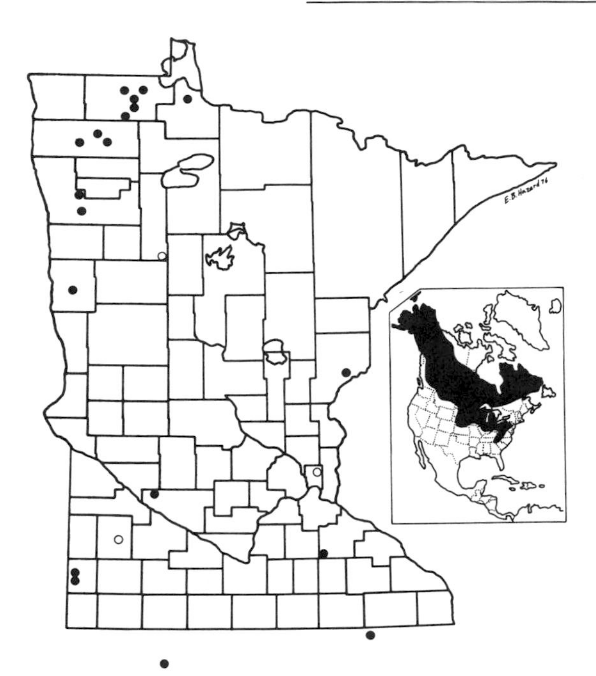

Map 60. Distribution of *Mustela nivalis*. ● = township specimens, selected locations. ○ = other township records. Iowa specimens reported in Bowles 1975. (Map produced by the Department of Biology, Bemidji State University.)

Range and Habitat: The least weasel was once thought to be restricted to North America and was named *Mustela rixosa*; it was considered a distinct species from the Old World *M. nivalis* (which is called simply the weasel in Britain). Allen (1933) showed "*M. rixosa*" to be conspecific with some Asiatic races of *M. nivalis*. The consensus among American mammalogists is that only one Holarctic species is involved which would, therefore, bear the older name, *M. nivalis*. Ewer (1973), however, maintains it has not been proved that *M. rixosa* and *M. nivalis* intergrade in Europe.

In North America, the least weasel ranges in a broad band from Alaska across central Canada into the north central and northeastern United States and eastern Canada. It apparently does not reach the Atlantic coast. In many areas, its recorded distribution is sporadic. The least weasel probably occurs throughout Minnesota. In some places it seems to prefer areas with water at or near the surface (Beer 1950a), but it is also found in upland woods and fields, and in Iowa seems to be more tolerant of sparse cover than either *M. frenata* or *Spilogale* (Polder 1968).

Natural History: The biology of the least weasel is poorly known. Unlike other mustelids, it apparently has more than one litter per year, and young may be born throughout the year. Delayed implantation probably does not occur. Heidt et al. (1968) showed the gestation period to be 35 days. Litters number from one to 10 young, but four to six is probably most common. The least weasel generally nests below ground or under some shelter on the surface. Mole runs and mouse or vole dens are often used, and the fur of such small mammals may be used, along with grass, to make nests (Polderboer 1942).

The least weasel probably feeds primarily on mice and voles, as well as on insects, other invertebrates, ground-nesting birds, and carrion (Mullen and Pitelka 1972). The larger weasels sometimes eat *M. nivalis* (Polderboer 1942; Rosenzweig 1966). The least weasel is about intermediate in size between the ermine and the short-tailed shrew. It is thus small enough that it must consume a prodigious amount of flesh, probably about 30 percent of its body weight per day, in order to prevent excess heat loss. It presumably feeds at all hours, though in captivity it is most active in the dark (Kavanau 1969).

Relationship to People: The least weasel is a mouser and is too small to prey on most species that people value, so it may be of economic benefit when abundant. It is probably more common than specimens in collections would indicate. Because there are apparently several litters per year, the least weasel can respond to favorable conditions by rapidly increasing its numbers.

Additional References: Cahn 1937; Elwell et al. 1973; Hamilton 1933b; Heidt 1970, 1972; Heidt and Huff 1970; King 1975; Nagel 1972; Polderboer 1948b; Polderboer et al. 1941; Swanson and Fryklund 1935; Swenk 1926; van den Brink 1968; Wurster-Hill 1973.

Mustela frenata
Long-tailed Weasel

Measurements (males): Total length 425 (398–459); tail 152 (131–165); hind foot 48 (44–57); ear from notch 23 (22–25); weight 316.0 (243.0–419.0); *(females)*: total length 356 (302–412); tail 107 (96–139); hind foot 42 (35–45); ear from notch 20 (17–23); weight 198.5 (103–276.1).

Description: The long-tailed weasel is larger than the ermine, though female *Mustela frenata* are about the size of male *M. erminea*. The black-tipped tail of *M. frenata* is over one-third of the total length. In summer, long-tailed weasels are brown above and white below, but the white does not extend onto the insides of the hind legs. The feet are brown. In the North, but not in the South, long-tailed weasels turn white in winter. At middle latitudes, some individuals turn white, some do not, and others develop an intermediate winter pelage. The sides of the rostrum of the skull of *M. frenata* are more parallel than those of *M. erminea*, the postorbital processes more pointed, the postorbital constriction more marked, and the bullae less flattened. If the sex is known, skulls can be distinguished on the basis of size alone.

Figure 65. *Mustela frenata*, long-tailed weasel, and *Zapus hudsonius*, meadow jumping mouse.

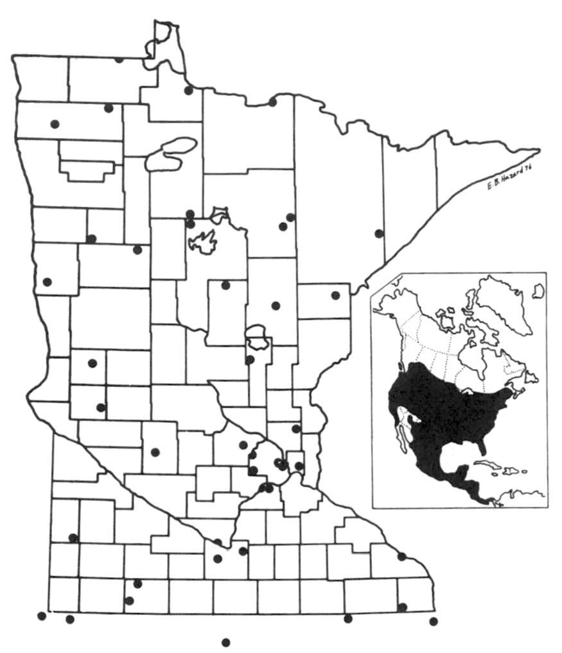

Map 61. Distribution of *Mustela frenata*. ● = township specimens. Iowa specimens reported in Bowles 1975. (Map produced by the Department of Biology, Bemidji State University.)

Range and Habitat: The long-tailed weasel ranges from southern Canada through almost all of the contiguous United States, Mexico, and Central America to northern South America. It may occur throughout Minnesota, but it is not common in the northern and northeastern counties. It resides in open country, brush, and woodland, and is said to prefer areas near water. Wetland drainage and stream channelization may have diminished suitable habitat in Minnesota (Choate 1972).

Natural History: Like other members of its genus, *M. frenata* is a strict carnivore. It hunts both night and day. More than 90 percent of its diet typically consists of small mammals. Male long-tailed weasels are large enough to prey on cottontail rabbits, but the bulk of the prey is mice and voles (Jackson 1961; Quick 1951). Presumably the elongate bodies, short fur, and short legs of long-tailed and other weasels make them particularly adapted to pursuing small mammals in burrows and runways. The predatory effectiveness of this shape must help them compensate for what Brown and Lasiewski (1972) call "the cost of being long and thin," namely metabolic inefficiency. Weasels have a greater surface area than "normally" shaped mammals of like weight and are unable to conserve heat by rolling into a ball. As a consequence, when subjected to cold temperatures, the metabolism of *M. frenata* (and probably other weasels) is 1.5 to 2 times as high as that of other mammals of comparable weight.

Although mainly a mouser, *M. frenata* probably also preys to some extent on *M. erminea*. Ermine may be a more effective size than long-tailed weasels for predation on mice, and Rosenzweig (1966) suggests that *M. frenata* may be able to compete successfully with *M. erminea* for the same food resource only by also preying on *M. erminea*. The three North American weasels are sympatric in broad areas in the northern United States. Their ecologic interactions merit further study.

Delayed implantation occurs in long-tailed weasels, and the gestation period varies from seven to 11 months. Ovulation, fertilization, and early cleavage occur after breeding, but the blastocyst does not implant in the uterine wall until 27 days or less before birth the following spring (Wright 1942, 1948). Four to eight young are born in April or May. Older females as well as newly mature females (three or four months old) breed in the summer. The den is usually in a hole in the ground and is likely to be in a bank or under a log. Apparently male *M. frenata* sometimes bring food to females and their young in the den.

Relationship to People: Same as ermine's relationship to us (see p. 134). The "ermine" of the fur trade includes both species, as well as some *M. nivalis*, which are often misidentified as small or young ermine.

Additional References: Basrur 1968; Boxall 1979; Chesness et al. 1968; Hamilton 1933b; Kavanau and Ramos 1975; Latham 1952; Leopold 1937; Polder 1968; Polderboer 1948a; Polderboer et al. 1941; Svendsen 1976; Wright 1947.

Mustela vison
Mink

Measurements (*males*): Total length 618 (559–705); tail 196 (168–228); hind foot 67 (61–73); ear from notch 25 (24–27); weight 1,007 (964–1,361); (seven *females*): total length 513 (460–546); tail 176 (160–187); hind foot 58 (54–64); ear from notch 18.5–24; weight 598.3–794.

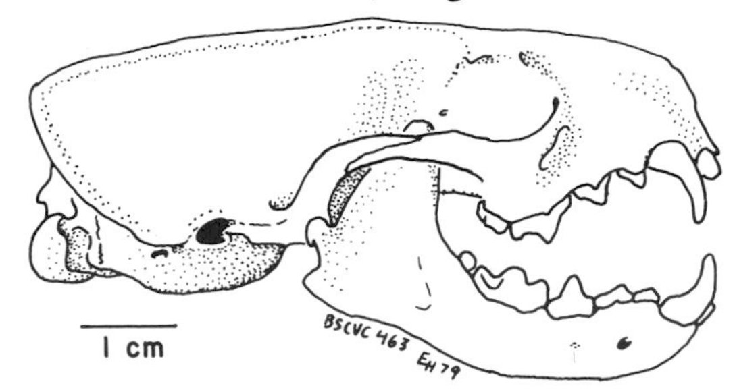

Figure 66. Skull of *Mustela vison*, mink, BSCVC 463.

Description: Mink have the elongate bodies and short legs characteristic of weasels, but they are larger and stouter. The tail is about one-third of the total length, somewhat bushy, and blackish toward the tip. The dense pelage is a uniform, sometimes blackish brown, except for a patch of white on the chin and some white flecks on the chest. The coat color is the same in winter and summer. Rare genetic variants occur (as they do in most species), and some of these "mutations" are prized by mink ranchers. Males are appreciably larger than females. The skull of a mink is about the size of that of a marten or a striped skunk, but a mink can be distinguished from either of these by a combination of characters: the number of upper cheek teeth (four), the weasellike form of the skull, the peculiar hourglass shape of M1, and the relative prominence of the bullae. The anal glands of mink produce a strong secretion, particularly evident during breeding.

Range and Habitat: As presently classified, *Mustela vison* is restricted to North America. It is

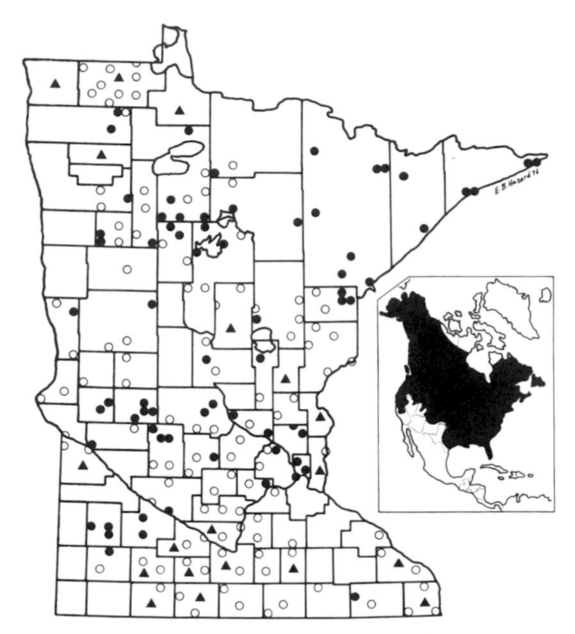

Map 62. Distribution of *Mustela vison*. ● = township specimens. ○ = selected other township records. ▲ = county specimens. (Map produced by the Department of Biology, Bemidji State University.)

found throughout most of the United States but does not reach the Rio Grande and is absent from most of the Southwest. It ranges into Canada and Alaska north of tree line yet is generally absent from the High Arctic. Van den Brink (1968) regards the European and American forms as conspecific. American mink have repeatedly escaped from captivity in Great Britain and continental Europe, and have become feral there (Clark 1970; Cuthbert 1973; van den Brink 1968). Mink occur in various habitats, but always near more or less open water (Marshall 1936). They are most common where trees or rocks provide shelter along lakeshores, marshes, and the banks of streams.

Natural History: Mink are successful on land and in water, moving and hunting effectively in both habitats without extreme specialization for either. They occupy dens (their own or someone else's) in holes in the ground (often these are protected by tree roots), in rock piles, in the banks of streams, in hollow logs or stumps, and in old muskrat houses (which they have often helped vacate). Females may range over as much as 8 ha, and males have much larger ranges (Marshall 1936). One April morning by the Mississippi near Lake Bemidji, I watched two mink wage a vigorous and noisy battle on the rocks under a railroad bridge. This suggested territoriality, but I was unable to determine the sex of the two mink, which were of similar size.

Mink are largely nocturnal but will also hunt in the daytime. They generally hunt alone. The diet is both aquatic and terrestrial and includes muskrats, rabbits, red squirrels, waterfowl and other birds, snakes, frogs, fish, and crayfish and other invertebrates. Mink sometimes store food in their dens in winter (Yeager 1943), and there is a record of mink feeding on bats (Goodpaster and Hoffmeister 1950). Because part of their diet is aquatic and because aquatic food chains often concentrate pollutants such as DDT derivatives and polychlorinated biphenyls, mink may be more vulnerable than most mammals to poisoning by such chemicals (Auerlich et al. 1972; Sherburne and Dimond 1969).

Breeding takes place in late winter, and mink may first breed as yearlings. Delayed implantation occurs in mink, but the delay is not generally as long as in most mustelids. Gestation lasts from 40 to 75 days (Enders 1952), of which the embryo is implanted in the uterine wall for 30–32 days. A litter of three to ten (most often four to six) young is born in April or May. The young remain with the female until the end of summer. Some authors report that mink are promiscuous, but others hold that both parents bring food to the young, which suggests a monogamous relationship.

Relationship to People: People and mink affect each other in several ways. Mink are trapped for their valuable fur and also bred for it. Mink ranching today accounts for the bulk of the furs used commercially. Mink are also the major predators on Minnesota's leading furbearer, the muskrat. Under most conditions, the interaction

Figure 67. *Mustela vison*, mink.

of the two species is well balanced and both can be harvested on a sustained-yield basis (Errington 1938, 1943, 1954b). Mink occasionally raid henhouses and lower the production in fish hatcheries, but these are not their major sources of food. People profoundly affect mink (and all other aquatic species) by their land and water management practices. Appropriate habitat is crucial to any species, and the most serious and permanent harm to mink, muskrat, and waterfowl has been wrought not by hunting or trapping by people, and not by predation by other animals, but by stream channelization, wetland drainage, and dam construction (Choate 1972).

Additional References: Birney and Fleharty 1968; Chesness et al. 1968; Duby and Travis 1972; Dunstone and Sinclair 1978a, b; Gilbert 1969; Guilday and Hamilton 1978; Hibbard 1957; Long and Howard 1976; Northcott 1974; Poole and Dunstone 1976; Rust et al. 1965; Sargeant et al. 1973; Schladweiler and Storm 1969; Schladweiler and Tester 1972; Sinclair et al. 1974.

Genus *Gulo*

Gulo gulo
Wolverine

Measurements (from Jackson 1961 and Peterson 1966): *Males*—total length 960–1,075; tail 200–250; hind foot 173–205; ear from notch up to 55; weight 13.6–19.1 kg; *females*—total length 725–947; tail 170–200; hind foot 165–180; ear from notch 45 or more; weight 10.0–12.7 kg. Rausch and Pearson (1972) record average weights of males as 14.2 kg, of females as 9.4 kg.

Description: Wolverine are Minnesota's largest mustelids, about as heavy as a bobcat or a medium-sized dog. Males are about one-third heavier than females. The fur is primarily dark brown, both above and below. The forehead is

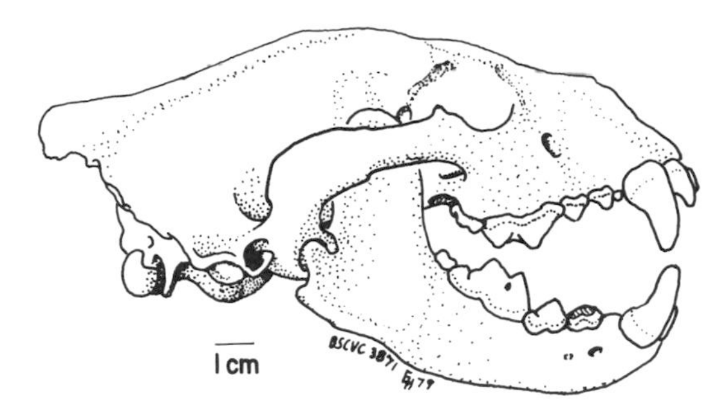

Figure 68. Skull of *Gulo gulo*, wolverine, BSCVC 3871, p1 and p2 absent.

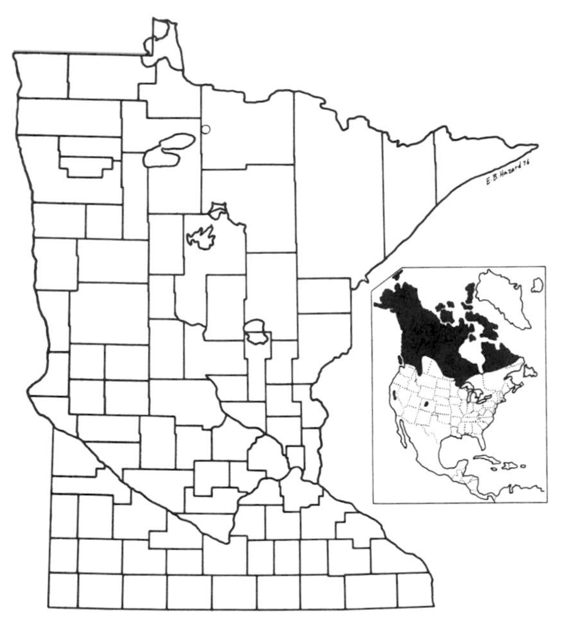

Map 63. May 1982 sighting of *Gulo gulo* by David Bosanko and David Parmelee. (Map produced by the Department of Biology, Bemidji State University.)

lighter, a grayish or yellowish brown, and two broad yellowish bands extend back from the shoulders along the sides of the body and converge on the rump and base of the bushy tail. Irregular creamy spots usually mark the throat. The broad skull is over 85 mm wide, and the braincase is no more elongate than it is in nonmustelid carnivores. The skull is somewhat diamond-shaped in dorsal view, owing to the short braincase and great breadth at the zygomata. The jaw cannot be removed from the skull because the mandibular fossa forms more than half a cylinder around the mandibular condyle. The tooth formula is 3/3, 1/1, 4/4, 1/2 × 2 = 38. The carnassials are particularly massive.

Range and Habitat: Gulo gulo is a boreal, Holarctic species (Rausch 1953). In North America, it once ranged from the Arctic Coast south through the tundra and taiga into the northern part of the eastern deciduous forest and into the mountains of the West (Barker and Best 1976; Cory 1912). In Eurasia, it had a comparable range. (The American wolverine was once considered a distinct species, *G. luscus*, and a few authors still use this name. A variant spelling of the common name, "wolverene," seems to have disappeared from the literature.) The wolverine is at home in both open and wooded country.

Wolverine are now restricted, in both the New and the Old Worlds, to the more inaccessible, northern parts of their range. They used to be relatively common in northern Minnesota, but trappers almost extirpated them by the 1920s. Occasional specimens have been taken since then, and they are probably now resident in small numbers in the northeastern counties (Franey 1953; Jackson 1922a; C. E. Johnson 1922, 1923a, 1930; Nowak 1973). In recent decades, the wolverine has also extended its range southward and eastward in Montana (Newby and McDougal 1964; Newby and Wright 1955). Haugen (1961) suggests that a live wolverine taken in Iowa in 1960 somehow arrived there with human aid, but Bowles (1975) believes that it got there "under natural circumstances."

Natural History: The wolverine is solitary, wide-ranging, omnivorous, and primarily a scavenger. Most of its diet is meat; although a wolverine can kill a deer, most of the big game it eats is carrion. The large carnassials, sturdy skull, and massive jaw muscles are well suited to cracking the bones of deer carcasses left by wolves, lynx, and cougars. Wolverine also eat porcupines, beaver, ground squirrels, voles, fish, berries, roots, and any food cached by people that they can get at. They are primarily terrestrial, but they climb well. They cover great distances with a peculiar loping gallop; individuals may have a home range over 100 km in diameter. Adults probably do not tolerate other adults of the same sex in the same area. Such behavior insures that wolverine populations will always be sparse. In the period 1835–39, the American Fur Company shipped 1.5 million muskrat pelts but only 15 wolverine pelts (D. R. Johnson 1969). Sparse populations are to be expected in scavengers, because they are necessarily less abundant than the predators that are their providers (van Zyll de Jong 1975b).

Reports on breeding in wolverine vary, and dates probably differ from region to region. Mating occurs in late spring or early summer and the father plays no other familial role. The blastocysts implant the following winter, and two to five young are born in late winter or early spring (Rausch and Pearson 1972; Wright and Rausch 1955). The young disperse in the fall or remain with their mother over the winter.

Relationship to People: Gulo has a fantastic reputation among trappers for ferocity, malice, cunning, and gluttony. Even where wolverine occur, they are so sparse that they can have little effect on other wildlife that is of value to people. However, they do remove furbearers and bait from traplines and destroy food stores of trappers and woodsmen. They also often mark what they do not use with their powerful musk. Wolverine are trapped for their fur, but primarily for local use. Frost from a person's breath does not cling tightly to wolverine fur, so it can be easily brushed off with a gloved hand. The fur is therefore prized as trim for parka hoods (Quick 1952).

Additional References: Birney 1974; Boles 1977; Burkholder 1962; Kurtén and Rausch 1959; Moyle 1975; Myrhe and Myrberget 1975; Schorger 1946b.

Genus *Taxidea*

Taxidea taxus
Badger

Measurements (provided by Richard Lampe, personal communication): *Males*—total length

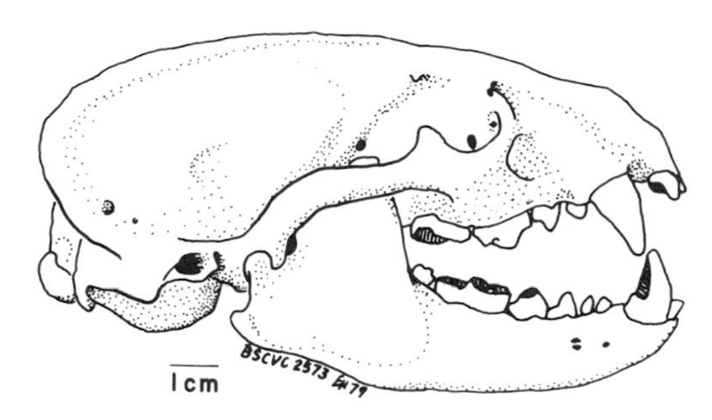

Figure 69. Skull of *Taxidea taxus*, badger, BSCVC 2573.

725–860; tail 130–190; hind foot 106–122; ear from notch 45–55; weight 6.25–11.25 kg; *females*—total length 710–780; tail 110–157; hind foot 100–120; ear from notch 47–55; weight 5.29–9.09 kg.

Description: Badgers are broad, low-slung, medium-sized mammals with short legs and long hair. Their front legs are particularly powerful, their long, heavy claws well suited to digging. The pelage on the back and sides is yellowish gray, and distinctly grizzled because of the banding pattern on the long guard hairs. The feet, neck, crown, and nose are dark brown or blackish, the cheeks are white, and a white stripe runs from the nose over the top of the ear. The skull is broad, especially at the back where a broad nuchal crest serves as the site of attachment of massive neck muscles. The auditory bullae are more rounded than those of most mustelids. As in wolverine, the condyloid processes of adults are typically locked in place by the mandibular condyle. This plus fusion of the dentaries results in a rigid jaw articulation. Consequently, the upper and lower canines wear against each other continually and are sharpened in the process (Long 1975a). The tooth formula is 3/3, 1/1, 3/3, 1/2 × 2 = 34.

Range and Habitat: Taxidea taxus occurs only in the New World, and the genus is monotypic. There are other genera of badgers in the Old World (Neal 1948). *Taxidea* ranges from northern Mexico throughout the western United States into southwestern Canada, and east through the north central states into northern Ohio and the

Ontario shore of Lake Erie. It is known from Connecticut and New York, but these eastern individuals may be descendents of badgers that escaped or were released from New York fur farms (Hamilton 1943b; Hamilton and Whitaker 1979; Nugent and Choate 1970). Badgers are open-country animals. They have spread to the north and east as forest has given way to meadow, pasture, and cropland, and as ground squirrels and pocket gophers, the major prey of badgers, have moved into these new habitats (Leedy 1947; McGee 1965; Nugent and Choate 1970; Schantz 1953; Snyder 1935).

Natural History: Badgers are highly specialized for preying on fossorial rodents by digging them out or by ambushing them from underground positions (Balph 1961; Knopf and Balph 1969). In addition to ground squirrels and pocket gophers, the diet of badgers includes voles, mice, occasional small birds, reptiles, amphibians, and insects and other invertebrates (Errington 1937a; Snead and Hendrickson 1942). I examined the

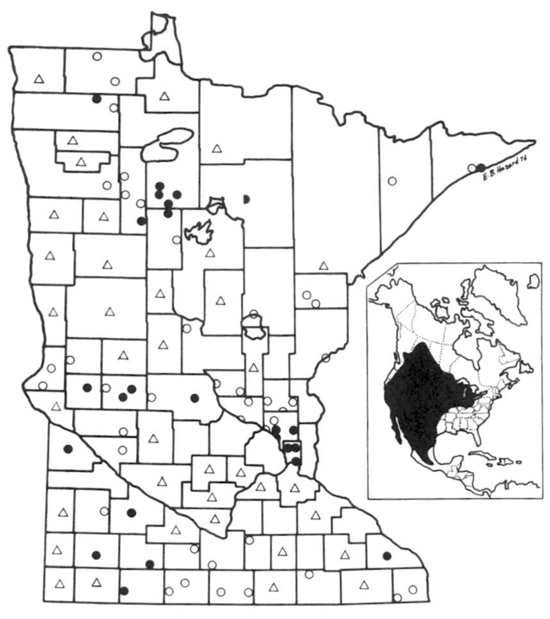

Map 64. Distribution of *Taxidea taxus*. ● = township specimens. ○ = other township records, selected locations. △ = other county records. (Map produced by the Department of Biology, Bemidji State University.)

Figure 70. *Taxidea taxus*, badger.

stomach contents of two road-kill badgers in southern Michigan. One, an adult, had recently eaten four *Spermophilus tridecemlineatus*, one adult *Peromyscus* and four nest young, one *Microtus pennsylvanicus*, two *Zapus hudsonius*, and one toad (*Bufo*). The stomach of the other, a subadult, contained only two dozen bumble bee pupae.

Badgers commonly dig their own extensive burrows, but they also enlarge those of other mammals. Their burrows may be over 8 m long and 2 m deep. Here badgers seek refuge, raise their young, and, in the northern part of their range, den up for the winter. Except when denned up or when a female has young in the den, badgers move frequently from den to den. Sargeant and Warner (1972) radio-tracked a badger at the Cedar Creek Natural History Reservation, on the Anoka Sand Plain, from late July until early January. It was strictly nocturnal and fed primarily on *Geomys bursarius*, the major prey of *Taxidea* in east central Minnesota (Lampe 1976). From 29 July to 27 September it utilized 46 dens in a home range of about 761 ha. From 30 September to 30 November it occupied a nearby smaller area of 53 ha, utilizing 23 dens. From 2 December to 9 January it holed up in a single den but ventured out on nights as cold as −29° C.

Badgers mate in late summer, but the blastocysts do not implant until February or March. The litter, typically of three or four young, is born about a month after implantation. The young generally remain with their mother until late summer. Both males and females mature sexually at about 14 months of age (Wright 1966, 1969). As adults, American badgers are solitary, in contrast to the gregarious European badger, *Meles meles*, an omnivorous forest dweller (Neal 1948).

Relationship to People: American badgers are not economically important furbearers. They raid hen houses on rare occasions and are said to molest beehives. However, none of the three professors at Bemidji State University who keeps bees as an avocation has ever had trouble with the local badgers. The major economic significance of badgers in Minnesota is their role in controlling populations of ground squirrels, pocket gophers, and voles. They have also served as subjects for extensive behavioral research at the Cedar Creek Natural History Reservation in Anoka and Isanti counties (Lampe 1976).

Additional References: Chesness et al. 1968; Crowe and Strickland 1975; Lindzey 1978; Long 1964, 1972c, 1975b; Wurster-Hill 1973.

Genus *Spilogale*

Spilogale putorius
Eastern Spotted Skunk

Measurements (*males*): Total length 470–550; tail 178–220; hind foot 48–52; weight 0.4–1.3 kg; (*females*): total length 445–482; tail 165–200; hind foot 43–47; weight 0.3–0.6 kg (based primarily on Iowa specimens, as reported by Jackson 1961).

Description: Both this species and the striped skunk typically have conspicuous black and white markings. This is warning coloration: most predatory mammals presumably associate a skunk's markings with its powerful, fetid secretion after a single encounter and thereafter leave skunks alone. Unfortunately, some farm dogs are slow learners. The white markings of spotted skunks generally comprise four stripes that begin on the forequarters but curve downward and break into patches on the hindquarters. A white patch marks the forehead, and the long, bushy tail is tipped with white. In some specimens from the Mankato area, the white markings are nearly absent. *Spilogale* is the size of a small housecat, and males average somewhat larger than females. The skull is seldom longer than 50 mm. The tooth formula is 3/3, 1/1, 3/3, 1/2 × 2 = 34.

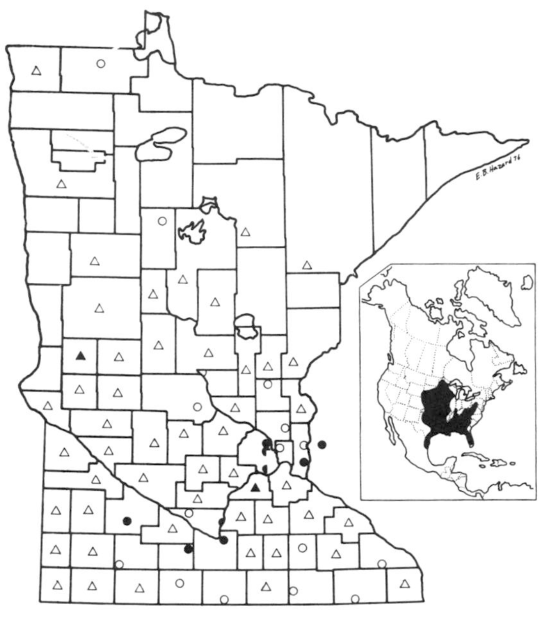

Map 65. Distribution of *Spilogale putorius*. ● = township specimens. ○ = selected other township records. ▲ = county specimen. △ = other county records (mostly from Swanson et al. 1945). (Map produced by the Department of Biology, Bemidji State University.)

Range and Habitat: East of the Mississippi, *S. putorius* ranges from the Gulf Coast north to approximately the Ohio Valley. West of the Mississippi, it ranges north to northern Minnesota and west to western Nebraska. A western species, *S. gracilis*, ranges from central Mexico to southern Montana and southwestern British Columbia. Burt (Burt and Grossenheider 1976) and Jackson (1961) treat *S. putorius* and *S. gracilis* as a single species. Most of the scattered records for the spotted skunk in Minnesota are from the southern half of the state, but one is from Roseau County (Swanson 1934) and one from northern Hubbard County (Hazard 1978b). *Spilogale* inhabits open and brushy areas and is often found in farmyards. Farmland populations in Iowa run as high as five per km^2 (Crabb 1948).

Natural History: Spotted skunks in the northern states den up for the winter, but they are not true hibernators and may emerge on mild winter days.

Figure 71. *Spilogale putorius*, eastern spotted skunk, and *Picoides villosus*, hairy woodpecker.

In winter,they feed mostly on small rodents. In summer they eat many insects but also a little of everything else—small mammals, birds and their eggs, lizards, snakes, frogs, carrion, garbage, fruits, other soft wild vegetation, and corn. They are almost exclusively nocturnal and largely terrestrial, but they will climb trees to escape potential enemies. A spotted skunk may, however, confront an intruder by standing on its hands with its hind feet in the air, its tail arched forward, and its face toward its adversary (C. E. Johnson 1921). In this position, the anal glands also face forward, and if the intruder does not retreat, it may be sprayed with a liquid reputedly more acrid than that of a striped skunk.

Spilogale is somewhat communal. Adults are not territorial, and different spotted skunks often use the same den on successive days. Most dens are on the surface, under woodpiles or other shelter. Underground dens are usually renovated ground squirrel burrows (Crabb 1948; Polder 1968).

In *Spilogale putorius*, the preimplantation period lasts only about two weeks. Mating usually takes place in April, and the total gestation lasts 50 to 65 days (Mead 1968a). In Minnesota, the litter of four to six young is probably most often born in June. In the south, there may be two litters per year. In the western form, *S. gracilis*, the preimplantation period lasts 200–220 days, and the total gestation lasts 230–250 days (Foresman and Mead 1973, 1974; Foresman et al. 1974; Mead 1968b).

Relationship to People: Until recently, farming in the Midwest benefited spotted skunks in many ways. Farmers drained marshy areas and provided shelter such as outbuildings, woodpiles, and haystacks. Farms also provided food in the form of house mice, rats, and stored food, and farmers eliminated some potential predators (Choate et al. 1973). Present trends toward fewer farmsteads, removal of permanent fences, continuous cultivation of corn with no crop rotation, less hay, and elimination of brush with herbicides destroy the cover that *Spilogale* needs (Polder 1968). People are of two minds concerning spotted skunks. The "hydrophobia cat" can transmit rabies, and it sometimes kills chickens. On the other hand, *Spilogale* is an effective rat killer, and rats are generally a more serious threat to chickens and their eggs than are skunks. The spotted skunk is overrated as a rabies carrier. In the period 1951–72, the Minnesota Department of Health reported fewer total cases of rabies in spotted skunks ("civet cats") than in domestic cats, cattle, dogs, or striped skunks (Hazard 1973). The fur of spotted skunks is currently of little value.

Additional References: Chesness et al. 1968; Crabb 1941; Greensides and Mead 1973; Hall and Kelson 1959; Heidt and Hargraves 1974; H. N. Johnson 1959; Mead 1972, 1975; Selko 1937; Van Gelder 1953, 1959.

Genus *Mephitis*

Mephitis mephitis
Striped Skunk

Measurements: Most specimens in Minnesota collections are subadult or young. The following measurements are from Jackson 1961. *Males*: total length 630–760; tail 220–280; hind foot 72–82; weight 2.5–4.5 kg (weights of fat males just before denning up may reach 6 kg); *females*: total length 540–650; tail 200–265; hind foot 69–76; weight 1.8–4.1 kg, up to 5.4 kg in late fall.

Description: This animal is larger, stouter, and less weasellike than the spotted skunk. It is black

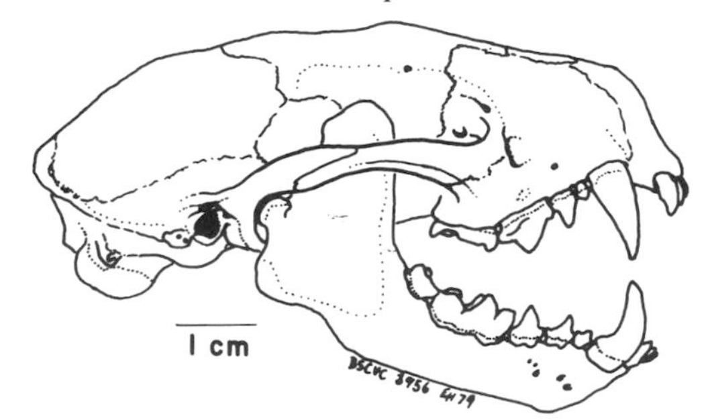

Figure 72. *Skull of Mephitis mephitis*, striped skunk, BSCVC 3956.

except for a thin white stripe on the forehead and two longitudinal white stripes that fork from a white patch on the crown of the head and nape of the neck and continue along the upper sides to the rump. The extent of the white areas varies, and a few individuals are nearly all black. There is usually some white hair in the black bushy tail. The tooth formula is 3/3, 1/1, 3/3, 1/2 × 2 = 34, and the skull is generally over 55 mm long.

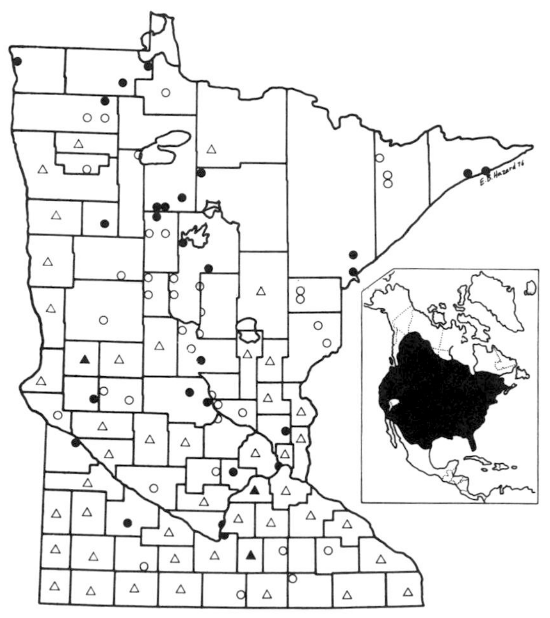

Map 66. Distribution of *Mephitis mephitis*. ● = township specimens. ○ = other township records, selected locations. ▲ = county specimens. △ = other county records. (Map produced by the Department of Biology, Bemidji State University.)

Range and Habitat: Mephitis mephitis ranges throughout the 48 contiguous states and into northern Mexico and southern Canada. It occupies a variety of habitats but is generally not common in continuous closed forest or in extensive lowlands where the water table is close to the surface. In cultivated areas, it often favors fencerows which may support a high concentration of mice. Striped skunks occur throughout Minnesota but are not common in the northeastern coniferous forests.

Natural History: Striped skunks are nocturnal, terrestrial, and largely solitary. They live primarily in underground dens, which they often dig for themselves, preferably on slopes with good drainage (Allen and Shapton 1942; Scott and Selko 1939; Selko 1938b; Shirer and Fitch 1970; Verts 1967). Mutch (1977), however, found that in Delta Marsh, Manitoba, winter dens are in wet, low-lying habitat rather than sandy ridges or dikes. The only times a skunk is likely to associate with other skunks is when the female is raising young and when two or more den up for the winter, usually about mid-November (Sunquist 1974). Often, a male and one or more females will occupy a single winter den (Gunson and Bjorge 1979). During mild periods in winter, striped skunks sometimes emerge from their winter dens.

Striped skunks usually breed in late February or in March. In years when spring comes late, females in dens with a male are likely to mate earlier than other females. Gestation is about 63 days (Houseknecht 1969), so the litter, typically of four to six young, is usually born in May. Pregnant females and mothers are particularly intolerant of males. The young are weaned in about two months and probably disperse when about three months old.

Striped skunks are omnivores; their diet varies greatly depending on the season and local availability of suitable food. When insects are abundant, they generally are the most common food. During the colder months, various fruits are the major item in the diet. *Mephitis* also eats small mammals, earthworms, crustaceans, mollusks, carrion, birds, nuts, and garbage (Hamilton 1936). In fall 1976, many adult skunks invaded Bemidji, perhaps in response to a dearth of wild fruits resulting from severe summer drought. Apparently, many overwintered in town, and the next summer they produced a crop of young skunks who interacted interestingly with the townspeople and their pet dogs. In marshes, striped skunks are often important predators on duck eggs (Bailey 1971; Lynch 1972).

The striped skunk is most noted for its powerful scent. This affords a high degree of protection against mammalian predators, who generally learn to associate the smell and the conspicuous warning coloration of the skunk after one unpleasant encounter. Predatory birds have a poor sense of smell and are not so easily put off. One of the more serious predators on skunks is the great horned owl.

Relationship to People: Skunks are significant predators on crop-destroying rodents and insects. They occasionally raid hen houses and are often serious predators on eggs in areas of concentrated waterfowl breeding. On the other hand, they are also important furbearers. Those with the least white fur are most valuable, because the coarser white fur must be discarded.

In many areas, skunks have become the major wildlife reservoir of rabies (Kaplan and Koprowski 1980; Robinson 1980). The incidence of skunk-carried rabies in Minnesota, mostly in corn-producing areas, is relatively high. Cases of rabies in striped skunks ranged from 50 to 291 per year over the period 1951–72 (Hazard 1973). Schnurrenberger, Martin, and Koch (1970) urge that skunks not be kept as pets because they may have been exposed to rabies before being captured.

Additional References: Allen 1939; Burgdorfer et al. 1974; Casey and Webster 1975; Hamilton 1937, 1963; Houseknecht and Huempfner 1970; Houseknecht and Tester 1978; H. N. Johnson 1959; Mead 1963; Mutch and Aleksiuk 1977; Selko 1938a; Shadle 1956; Storm 1972; Storm and Verts 1966; Verts 1963b; Verts and Storm 1966.

Genus *Lutra*

Lutra canadensis
River Otter

Measurements: Total length 900–1,220; tail 300–475; hind foot 112–133; weight 6.8–11.3 kg, occasional males up to 13.6 kg (Jackson 1961).

Description: The river otter is large and weasel-like, with short, powerful legs, a relatively small head, and a long, thick, and tapering tail. Its nose is broad and blunt, its ears and eyes are small, and its eyes face distinctly forward. The otter's dense fur is dark brown, a bit lighter on the underside, and often grayish or silvery on the throat. The toes are fully webbed. The braincase is broad, the rostrum short, and the auditory bullae are flattened. Male otter are slightly larger than females. The otter has five cheek teeth in both the upper and lower jaws, and is the only mammal in Minnesota with the tooth formula 3/3, 1/1, 4/3, 1/2 × 2 = 36.

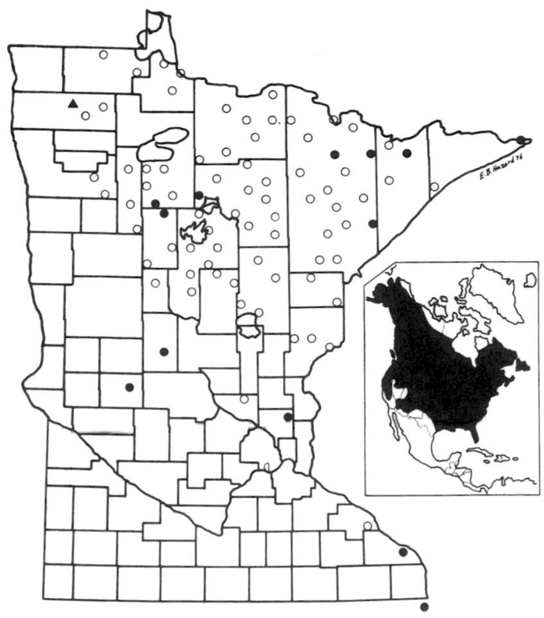

Map 67. Distribution of *Lutra canadensis*. ● = township specimens. ○ = other township records, selected locations. ▲ = county specimen. Iowa specimen reported in Bowles 1975. Inset = presettlement continental distribution. (Map produced by the Department of Biology, Bemidji State University).

Range and Habitat: Lutra canadensis ranges throughout the 48 contiguous states except for arid parts of the southwest, and north into Canada and Alaska except for the northernmost parts. Van Zyll de Jong (1972) designated American otters as a separate genus, *Lontra*, retaining *Lutra* for Old World otters. However, Sokolov

(1973) considers this distinction invalid. Within its wide range, the river otter is found only near lakes and streams, and near salt water in some coastal areas. It is generally not common in heavily settled areas.

Natural History: Otter are Minnesota's most aquatic carnivores. They feed primarily on fish and aquatic invertebrates, particularly crayfish. They also take frogs, tadpoles, salamanders, and an occasional muskrat. The percentages of forage fish, panfish, game fish, crayfish, and other items in the diet vary according to season and locality (Greer 1955; Knudsen and Hale 1968; Lagler and Ostenson 1942; Ryder 1955; Sheldon and Toll 1964; Toweill 1974).

Otter are sociable during much of the year and are often playful. An otter may play with a fish it catches much as a cat plays with a mouse. Otter also like to slide and often form chutes in the mud or snow by repeatedly sliding down

Figure 73. *Lutra canadensis*, river otter.

banks into the water. Adults often tolerate one another and may be seen in pairs. After mating, delayed implantation occurs, followed by the birth of usually two to four young in late winter or early spring. The male breeds the female again shortly after the birth of this litter, after which he is driven away from the family. A male, presumably the father, may join the family group when the young are a few months old (Jackson 1961). The total gestation varies from 10 months to just over a year (Liers 1951). The young stay with the mother for the better part of their first year. Females can breed for the first time when they are two years old, before they reach their full size. Males do not usually breed successfully until age six or seven, so there is often a marked size difference within a mated pair (Hamilton and Eadie 1964).

Otter swim gracefully and rapidly, most of the power being supplied by sinusoid movements of the body and tail. They must swim actively to stay afloat (Tarasoff et al. 1972). On land they move with an exaggerated weasellike lope, alternately arching and straightening the back (Severinghaus and Tanck 1948). Although they may travel long distances over land in dispersing or seeking a mate, they generally make their homes near water. Otter often take over the bank burrows of beaver or muskrats, or an old beaver lodge. They may also hole up under a log, an overhanging tree, or the exposed roots of a tree near water. They sometimes use open nests in brushy areas in marshes or swamps.

Relationship to People: Otter generally avoid people, but the young are tractable and can be domesticated. Old World otter have been trained to catch fish for people. Otter fur is particularly luxuriant. Fortunately for this species, it is harder to trap than are marten and fisher. Although otter eat game fish, they are seldom abundant enough to affect fish populations in natural waters. They can, however, be a problem in a fish hatchery. From an aesthetic standpoint, otter are among the most pleasing mammals to watch.

Additional References: Friley 1949; Henderson 1978c; Hooper and Ostenson 1949; R. P. Johnson 1973; Liers 1953, 1958; Stephenson 1977; Tabor and Wight 1977; Tarasoff and Kooyman 1973a, b.

Family Felidae
Cats

Cats are medium- to large-sized carnivores. The rostrum is shorter than that of most other carnivores, and the cranium is rounded. The teeth are highly specialized for shearing flesh, and most species are primarily meat eaters. Cats are digitigrade, usually have strongly curved, retractile claws (Gonyea and Ashworth 1975), and often are arboreal. Most cats are not good long-distance runners, but rely on stealth and ambush to obtain prey. The Old World cheetah is an exception, as was *Felis trumani* (or *Acinonyx trumani*) of the North American Pleistocene (Adams 1979; Martin et al. 1977). In contrast to many canids, most felids do not form lasting social groups (Kleiman and Eisenberg 1973). Cats are widespread, occupying all the continents except Antarctica and Australia (Gonyea and Ashworth 1975; Kleiman and Eisenberg 1973; Stains 1967).

Additional References: Ewer 1973; Guggisberg 1975; Hemmer 1978; Jorgensen and Mech 1971; Leyhausen 1978; Vaughan 1978.

Genus *Felis*
True Cats

Adams (1979) and other authors place the cougar in a separate genus, *Puma*.

Felis concolor
Cougar, Mountain Lion

Measurements: Total length 1,800–2,590; tail 700–900; hind foot 254–295; average weight, male, about 50 kg, female, about 40 kg (Jackson 1961).

Description: Adult cougars are typically a uniform buff above, somewhat lighter below. Their throats may be nearly white. There is generally a dark stripe down the midline of the back. The fur is thick but short, the triangular ears are untufted, and the tail is long, slender, and cylindrical, not bushy. Males average slightly larger than females. Unlike adults, young cougars have dark spots. The tooth formula is 3/3, 1/1, 3/2, 1/1 × 2 = 30.

Range and Habitat: When Europeans first came to the Americas, *Felis concolor* was probably the most widespread New World mammal. It occurred from the forests of western and eastern Canada, throughout the area of the 48 contiguous states, and south throughout Latin America to southern Argentina. By and large, this range coincides with the composite range of cervids other than caribou and moose. North of Mexico, cougars now reside in the mountains of the western states and Canada, and range east into the Prairie Provinces and along the Gulf Coast into Louisiana, Arkansas, and coastal Mississippi. There is also a population in peninsular Florida, and there are persistent reports of sightings of cougars in the southern Appalachians and in Maine and adjacent Canada. There are also frequent enough reports of cougar sightings or tracks in North Dakota, Minnesota, Manitoba, and western Ontario to suggest that individuals occasionally travel east into our area (Buchwald 1978; Bue and Stenlund 1953; Dear 1955; Magnus 1956; Nero and Wrigley 1977; Thomson 1974). Individual cougars may range over hundreds of kilometers, and there is no firm evidence that northern Minnesota and adjacent areas harbor a resident breeding population.

Natural History: Cougars take a variety of prey but feed primarily on deer (Hornocker 1970a). Adults are solitary and hunt both at night and in the daytime. In mountainous habitats, cougars migrate between high summer ranges and low winter ranges along with the deer. During the rut, male mule deer and wapiti are less cautious than females and are likely to become exhausted from breeding and fighting. Owing to the behavior of adult male deer, and to the relative ease of killing young animals, cougars take a disproportionate percentage of adult male and young deer and wapiti, as opposed to adult females. They also take a disproportionate number of old adults, again probably as a result of opportunity rather than deliberate choice.

The female cougar raises the litter of two or three kittens, and they remain with her for 18 to 20 months. Females generally breed only in alternate years (Hornocker 1970a; Robinette et al. 1961). In the northern Rockies, adults have home ranges conservatively estimated at 13 to 78 km^2. Males generally range more widely than females (Hornocker 1969, 1970a).

Relationship to People: Cougars have seldom been able to survive in heavily settled areas, at least north of Mexico. In Minnesota, the occurrence of cougars is too sporadic for them to be of economic significance. Cougars avoid people and will rarely attack even when cornered. They thus present little direct threat to people. In some areas, human hunting is inadequate to regulate populations of wapiti and mule deer, and predation by cougars is necessary to prevent overpopulation and consequent habitat deterioration (Hornocker 1970b). *F. concolor couguar*, the subspecies native to Minnesota and most of the eastern states, is on the Endangered Species list.

Additional References: Jackson 1955; Norris-Elye 1951; Robinette et al. 1959; Seidensticker et al. 1973; Seton 1909; Toweill and Meslow 1977; Wright 1959, 1972; Young and Goldman 1946.

Genus *Lynx*
Lynxes and Bobcats

The skulls of both the lynx and the bobcat may be told from normal skulls of the genus *Felis* by the absence of P2; the tooth formula is 3/3, 1/1, 2/2, 1/1 × 2 = 28. Some authors treat *Lynx* as a subgenus of *Felis*. Within the group, some regard the New World lynx (here distinguished as *L. canadensis*) as conspecific with *L. lynx* of the Old World.

Lynx canadensis
Lynx, Canada Lynx

Measurements: Total length 875–1,000; tail 100–120; hind foot 215–250; weight 7.3–15.9 kg, rarely heavier (Jackson 1961).

Description: The lynx is about twice as large as a good-sized housecat and has a short, stubby tail and relatively long legs. Its paws are large, its ears have pronounced tufts, and there are large ruffs on either side of the face below the ears. The long fur is generally a pale gray mixed with pale brown, and without black spots on the belly. The measurements of male lynx average about 5 percent greater than those of females. There is considerable variation in pelage and size of lynx, but several characters distinguish the lynx from the bobcat. The tail of the lynx has a totally black tip, as though it had been dipped in ink. The feet of the lynx are larger than those of the bobcat. On the underside of the skull two foramina at the medial and rear edge of the bulla, the posterior lacerate foramen and the anterior condyloid foramen, open separately at the surface. In the bobcat, they open into a common depression. The interorbital breadth of adult lynx always exceeds 30 mm, and the presphenoid is wide at the rear, exceeding 6 mm.

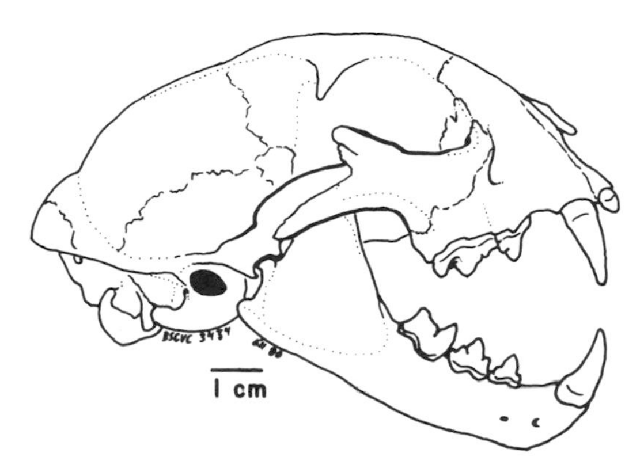

Figure 74. Skull of *Lynx canadensis*, Canada lynx, BSCVC 3434.

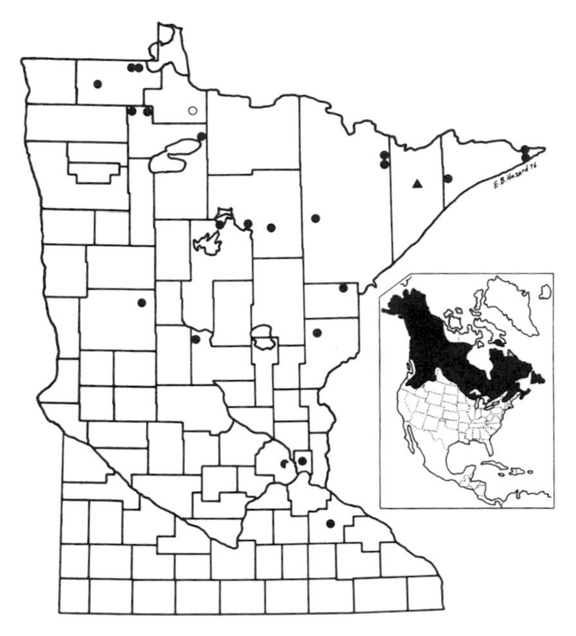

Map 68. Distribution of *Lynx canadensis*. ● = township specimens. ○ = furbuyer record. Hennepin, Ramsey, and Goodhue specimens are from the 1972–73 irruption. (Map produced by the Department of Biology, Bemidji State University.)

Range and Habitat: Lynx canadensis originally ranged throughout the boreal forests of North America and the mixed coniferous-deciduous forests of the northeastern and Great Lakes States, and past the forest edge into the Canadian and Alaskan tundra. Lynx are now generally absent from more settled areas. In Minnesota, they seldom range outside the northeastern counties. However, individuals may move great distances when food runs out. In the winters of 1962–63 and 1972–73, presumably as a result of a drop in the supply of snowshoe hares, many lynx wandered south, and individuals were taken in such unlikley spots as Morrison, Ottertail, Hennepin, and Ramsey counties (Mech 1973b). In July 1963, one was shot in Shelby Co., Iowa, some 200 km south of the Minnesota border (Rasmussen 1969). As with snowy owls in years when the lemming supply drops, probably few immigrants make it back to the North Woods.

Natural History: The lynx is a solitary hunter that relies on stealth and ambush. It is at home both in the trees and on the ground. In a winter tracking study in Alberta, the length of the average pounce when hunting was 1.7 m (Nellis and Keith 1968). Lynx averaged 0.42 kills per night. The average cruising distance and cruising radius varied inversely with the degree of hunting success, which was dependent at least as much on snow conditions as on abundance of prey. Overall, cruising distances (km traveled per night) ranged from 0.8 to 19.2, averaging 8.9 km when hunting was difficult to 4.8 km when it was easier. The respective cruising radii averaged 2.9 and 1.6 km. In most areas, the principal food of lynx is snowshoe hares. Lynx also eat considerable carrion (especially during and after deer season), grouse of various species, and red squirrels. They can bring down a deer, especially when snow makes escape difficult, but lynx are generally not major predators on deer. Lynx will also take foxes and porcupines (Brand et al. 1976; Hancock et al. 1976; Nellis and Keith 1968; Nellis et al. 1972; Saunders 1963a; van Zyll de Jong 1966).

Mating occurs in January or February, and the one to four young are born 62 days later in March or April (Jackson 1961; Peterson 1966). The den is made under shelter at ground level. There is only one litter per year, and both sexes are sexually mature at one year of age. Yearlings sometimes remain with the mother (Saunders 1963b), which suggests that some females may not breed every year.

Relationship to People: The lynx is rare in

Figure 75. *Lynx canadensis*, Canada lynx, and *Lepus americanus*, snowshoe hare.

densely populated areas. A major ecologic effect of the lynx may be to act as a check on the density of snowshoe hare populations (Gilpin 1973; Nellis et al. 1972; Weinstein 1977). It does not pose a significant threat to livestock or deer. Its fur has some value.

Additional References: Brand and Keith 1979; Buchwald 1978; Finerty 1979; J. F. Fox 1978; Gunderson 1978; Henderson 1978d, f; Koehler et al. 1979; Mech 1977g, 1980; More 1976; Saunders 1964; van Zyll de Jong 1971, 1975a.

Lynx rufus
Bobcat

Measurements: Total length 800–1,015; tail 130–180; hind foot 155–197; weight 6.8–15.9 kg, rarely to 18 kg; average weight, male, about 11 kg, female, about 7.3 kg (Jackson 1961).

Description: The overall build and size of the bobcat are much like those of the lynx. The bobcat generally is redder, especially in summer, its feet are smaller, the ear tufts and sideburns are not as pronounced, and the coat, including the belly fur, is definitely spotted. The tail is one-third to one-half again as long as that of the lynx, and an irregular succession of black and whitish markings extends to its tip. The posterior lacerate foramen and the anterior condyloid foramen open into a common depression, the interorbital breadth is less than 30 mm, and the presphenoid is narrow for its full length, never as wide as 6 mm. As in the lynx, male bobcats are generally larger than females.

Range and Habitat: Lynx rufus ranges from central Mexico north through the contiguous United States into southernmost Canada. It occupies a wide variety of habitats but is generally not common in the mature boreal forest occupied by the lynx. The bobcat has retreated less than the lynx in the face of human settlement and is still present in many of the less densely populated and "developed" areas of its pre-Columbian range.

Natural History: Occupying a diversity of habitats, the bobcat hunts a variety of food. It feeds commonly on hares and rabbits, and ground and tree squirrels. It will also take various mice, pocket gophers, young deer, birds, young livestock, muskrats, porcupines, and even such carnivores as foxes, housecats, and various mustelids (Gashwiler et al. 1960; Marston 1942; Nussbaum and Maser 1975; Pollack 1951; Westfall 1956). The main prey in northern Minnesota, where bobcats are common, is *Lepus americanus* (Petraborg and Gunvalson 1962; Rollings 1945).

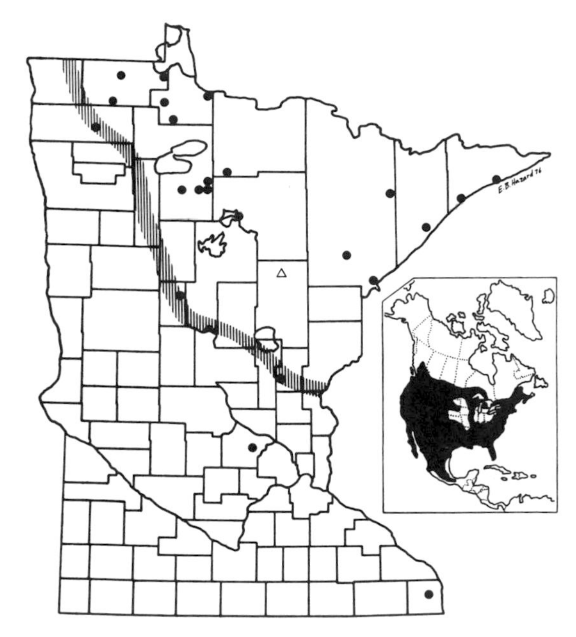

Map 69. Distribution of *Lynx rufus*. ● = township specimens. △ = other county record. Area northeast of hatching = primary bobcat range (after Berg 1980b). (Map produced by the Department of Biology, Bemidji State University.)

Like most cats, bobcats are generally solitary. The home ranges of females are usually mutually exclusive and smaller than those of males. The home ranges of males overlap those of other males and of females (Bailey 1974). Home range boundaries are advertised by scent marks. The female raises the young alone, although the male may rejoin the family when the kittens are partly grown. The litter, typically of two to four young, is generally born in April, May, or June, after a gestation of 50 to 62 days. The young remain with the female until about January (Jackson 1961). If a female is not impregnated in late winter, she may come into heat again and bear a litter in late summer or early fall (Gashwiler et al. 1961). Sometimes there are two litters a year in the South, but not in Minnesota.

Relationship to People: The bobcat is not abundant enough in most stock-raising areas to be a serious threat to livestock, though it may be a problem in some localities. It may at times contribute significantly to the control of populations of rodents and leporids. The fur is of some value.

Additional References: Beasom and Moore 1977; Berg 1980; Crowe 1975a, b; Cumbie 1975; Fritts and Sealander 1978a, b; Hamilton and Hunter 1939; Henderson 1978e; Laycock 1979; McCord 1974; Robinson and Grand 1958; Young 1958.

ORDER ARTIODACTYLA
Even-toed Ungulates

Artiodactyls include pigs, hippopotami, camels, giraffes, deer, pronghorned antelope, Old World antelope, cattle, sheep, and goats. A diagnostic feature of the order is the astragulus (ankle bone). Both its upper and lower articulating surfaces are movable, an arrangement unique to the Artiodactyla. The weight-bearing axis of the foot passes between the third and fourth digits. Camels are digitigrade, walking on the toes, but most other artiodactyls are unguligrade, walking on hoofs that are modified claws (Koopman 1967). Because each foot has two hoofs, one on the third toe and the other on the fourth, artiodactyls are often referred to as even-toed ungulates and are commonly called cloven-hoofed. Many bear horns or hornlike structures on the head (Geist 1966).

The majority of artiodactyls chew their cud, or ruminate. These ruminants are a particularly successful group and have been the dominant large land-dwelling herbivores since the Late Oligocene, about 40 million years ago. All native Minnesota artiodactyls are ruminants. The artiodactyls have a wide natural distribution, including all major land areas except Australia and Antarctica, and they occur almost everywhere as domesticated or otherwise introduced species. Because the remains of domestic artiodactyls may be mistaken for those of wild species, I have included domestic as well as wild artiodactyls in the skull key. See also the section on domestic and feral species.

There are few museum specimens of artiodactyls from Minnesota, and some of those lack measurements. Measurements in the following species accounts are from the literature.

Additional References: Frick 1937; Geist 1974; Greaves 1978, 1980; W. D. Hamilton 1963, 1971; Hudson 1976; Lent 1974; Matthew 1934; Owen-Smith 1977; Radinsky 1978; Todd 1975; Van Valen 1971a; Vaughan 1978.

Family Cervidae
Deer

In most species of deer, males each year grow and subsequently shed a pair of antlers, which are bony and usually branched outgrowths of the frontal bone. Females do not generally bear antlers, except in caribou. There are no upper incisors in deer, or in other ruminants. The premaxilla instead bears a tough, horny pad. The lower canine in deer and other ruminants is incisiform. The family Cervidae occurs throughout North and South America and Eurasia, but is absent from all but the northwestern part of Africa and from Australia, except by introduction.

References: de Vos et al. 1967; Mathisen 1972; Pimlott 1967; Vaughan 1978; Walters and Bandy 1972; Walters and Gross 1972.

Genus *Cervus*
Holarctic Deer

The American and subsequently the east Asian forms were once classified as a separate species, *C. canadensis*, considered to be distinct from *C. elaphus*, the red deer of Europe. All forms are now generally treated as a single species. "Elk" is an unfortunate appellation for the American form (although it will no doubt continue to be used) because it properly refers to the European forms of *Alces alces*, the same species as our moose. "Wapiti" is a Shawnee term, and perhaps is preferable, though "elk" is far more commonly used.

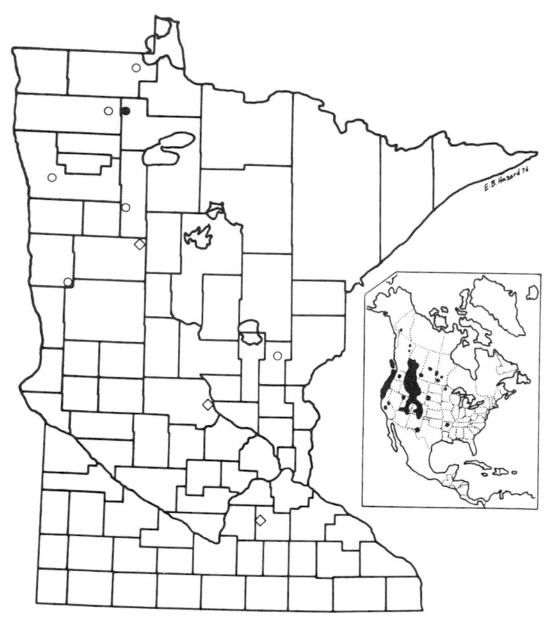

Map 70. Distribution of *Cervus elaphus*. ● = recent township specimen. ○ = other township records. ◇ = old bones. (Map produced by the Department of Biology, Bemidji State University.)

Cervus elaphus
Wapiti, American Elk

Measurements (males): Total length 2,285–2,745; tail 140–157; hind foot 610–710; weight 295–376 kg; *(females)*: total length 2,135–2,440; tail 114–125; hind foot 580–655; weight 227–295 kg (Jackson 1961). Fashingbauer (1965b) gives the maximum and average weights of bulls as about 450 and 320 kg, with cows weighing about 45 kg less than bulls.

Description: The wapiti, or American elk, is a large cervid which generally weighs more than twice as much as a white-tailed deer. The winter coat is brownish gray on the back and sides, becoming blackish brown on the neck, head, and feet. There is a buff rump patch. The coats of cows and young bulls show less contrast. The summer pelage is shorter and generally redder or yellower than that of winter.

The antlers of a bull are characterized by long beams that curve outward and backward from the skull; a series of tines projects upward from each beam. The tooth formula is 0/3, 1/1, 3/3, 3/3 × 2 = 34. No other extant Minnesota deer normally has an upper canine.

Range and Habitat: Cervus elaphus is Holarctic, originally ranging throughout most of Eurasia and North America. In the New World, this once widespread mammal is now confined largely to the Rocky Mountains of Canada and the United States and to parts of the Pacific states. Various populations, including a herd of a few thousand in northern Lower Michigan, have been reestablished by introductions, but other introductions have failed. Wapiti were once common in Minnesota, especially in the prairie and the prairie-hardwood forest transition, but they last occurred as natives in the state early in the present century. A small herd was later established by introduction in semi-confinement in Itasca State Park, Clearwater County. In 1935, 27 elk from this herd were released in northwestern Beltrami County. The number of animals in this herd has fluctuated, but the herd is still there, and individuals, presumably wanderers from this herd, occasionally show up elsewhere in the state (Fashingbauer 1965b; Swanson 1940).

Natural History: Cow elk are sexually mature at three years of age. Bulls may be physiologically capable of breeding at two years of age but probably seldom father offspring before the age of four. Mature bulls gather a harem of cows, breeding them as they come into heat. Harem bulls generally do not eat regularly, and as one mature bull becomes exhausted, another displaces him. Very old and very young bulls are normally prevented from breeding by mature bulls (Murie 1951). Gestation reportedly lasts 249–262 days, and the calf (seldom more than one) is born in May or early June (Fashingbauer 1965b). Cows isolate themselves at calving time, but soon come together to form a band of cows and calves. Bulls remain apart in small groups during the summer. They divide the cows into harems during the fall rut, after which all ages and sexes come together in large winter herds.

Wapiti eat a variety of grasses, sedges, forbs, and deciduous and coniferous woody plants. They also like domestic forage such as grasses and alfalfa, either fresh or as hay, and various grains.

Relationship to People: Minnesota has a herd of wapiti today because the species was reintroduced into the state some years ago and later transplanted to its present location. Not all wildlife biologists agree that the introduction was wise. Whether it was or not, the herd provides a good example of the complexities involved in game management, particularly in introductions or reintroductions.

It might at first seem desirable to have this native, majestic species back in the state. The habitat in northwest Minnesota is suitable, if not optimal, and is relatively wild. But it is not entirely wild, and wapiti are not sedentary. Farmers in northwest Minnesota grow, cut, and stack hay, and a few wapiti can eat a great deal of hay. Timber wolves, cougars, grizzly bears, and black bears prey on wapiti (Cole 1972; Hornocker 1970a, b). Today no cougars or grizzlies reside in Minnesota. Black bears occur in the area occupied by the herd, and there are wolves nearby, but probably neither species is able to keep the herd in check. The herd, however, seems to be stable, neither increasing in numbers nor expanding its range. Hunting cannot possibly be a factor, because wapiti are completely protected by Minnesota law. Perhaps it is something in the hay.

Additional References: Cahn 1921; Church and Hines 1978; Craighead et al. 1972, 1973; Haag 1962; Hansen and Clark 1977; Hansen and Reed 1975; Knight 1970; Kufeld 1973; Pickford and Reid 1943; Ricklefs 1977; Rounds 1977; Schultz and Bailey 1978; Severinghaus and Darrow 1976; Telfer 1978.

Genus *Odocoileus*
North American Deer

This genus comprises two species, the mule deer and the white-tailed deer, both of which occur in Minnesota. The tooth formula is 0/3, 0/1, 3/3, 3/3 × 2 = 32.

Odocoileus hemionus
Mule Deer

Measurements (*males*): Total length 142–148 cm; tail 15–20 cm; hind foot 41–59 cm; ear 12–25 cm; weight 50–215 kg; (*females*): total length 132–152 cm; tail 11–20 cm; hind foot 41–51 cm; ear 12–24 cm; weight 31.5–72 kg (Banfield 1974).

Description: In the field, it is often difficult to distinguish the rare mule deer from the much commoner white-tail (see the next account). The adults are the same size and have similar coloring; the fawns of both species are spotted. The antlers of a mule deer buck, however, are strikingly different from those of a normal white-tail buck. The main beam of a mule deer's antler rises vertically and somewhat rearward and outward, and then forks. In mature bucks, each branch forks again. This pattern does not normally occur in white-tails. The two species can also be told apart in other ways. The mule deer's tail is small and black-tipped, not nearly so conspicuous as the flag of a white-tail. A mule deer runs with the tail held down, a white-tail with the tail erect. Although the gray-brown winter pelage of a mule deer is much like a white-tail's, in summer a mule deer is somewhat yellower than a white-tail.

It is widely acknowledged that mule deer have larger ears than do white-tails, but this is hard to document. Banfield (1974) writes that the white-tail has "smaller ears" (p. 391) but gives the ear measurements of male white-tails as 14–24 cm and those of male mule deer as 12–25 cm. Fashingbauer (1965a) indicates that mule deer have larger ears but provides no figures. Cowan (1956) and Kellogg (1956) list several distinguishing characters but do not include ear length. However, fig. 14 in Cowan (1956, p. 337) shows the ears of a mule deer as about one-third longer than those of a white-tail. Seton (1929) reports a mule deer buck with ears 10″ long, perhaps the basis for the maximum length of 25 cm reported by Banfield (1974). I have found no records of ear length for museum specimens of mule deer, but have seen both species in the wild and believe that mule deer have distinctly larger ears. For ear measurements of white-tail, see the data supplied by P. D. Karns and W. Wanek in the next account.

Range and Habitat: Mule deer inhabit western North America from southeast Alaska and northwest Saskatchewan to the Dakotas, west Texas, northwest Mexico, and Baja California. They favor brushy and wooded habitats, often living in higher and drier areas than do white-tails. In Minnesota, the mule deer at present occurs sporadically, and there may be no resident population (Erickson and Bue 1954; Fashingbauer 1965a; Gunderson 1948). It also occurs irregularly in western Iowa (Bowles 1975; Kline 1959; Sanderson 1956). Apparently mule deer were native to northwestern Minnesota until the late nineteenth century (Swanson et al. 1945). Cowan (1956: fig. 13, p. 334) shows a population in a north-south band from Kittson and Roseau to Wilkin counties, isolated from the mule deer in the western Dakotas. The widespread distribution of records in map 71 is probably testimony to the great distances an individual may travel.

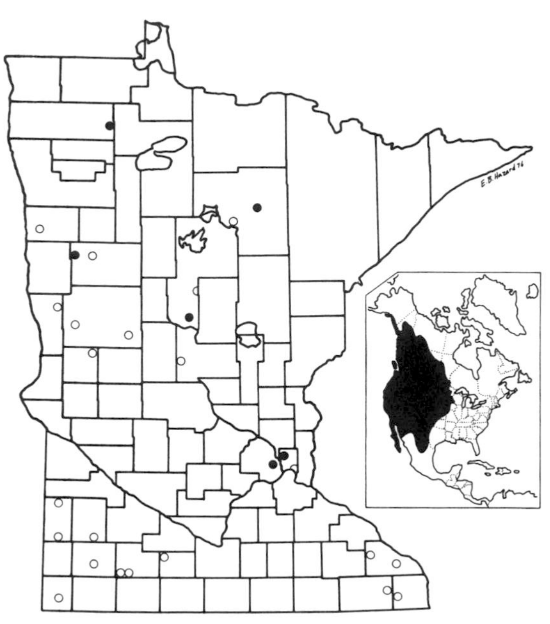

Map 71. Distribution of *Odocoileus hemionus*. ● = township specimens. ○ = other township records. (Map produced by the Department of Biology, Bemidji State University.)

Natural History: The life history of the mule deer closely parallels that of the white-tail except that, in much of its range, the mule deer migrates substantial distances between high altitude summer feeding areas and low-altitude wintering areas. Rut (mating season) occurs from late October well into December, and the fawns (typically twins for mature does) are born about seven months later. Mule deer feed on a diversity of grasses, forbs, and brush. In southern Alberta, at least, dead vegetation is a major winter food of mule deer (Geist 1981). Mule deer probably do not compete significantly with cattle on jointly used rangeland (Dusek 1975). The major natural predator of the mule deer in much of the West is the cougar. In some areas, predation by cougars is necessary to keep mule deer population in check; human hunting itself cannot keep mule deer from exceeding the carrying capacity of their range (Hornocker 1970a).

Relationship to People: Mule deer are too uncommon in Minnesota to be of economic significance. They are, consequently, rare enough to be a source of excitement when seen.

Additional References: Anderson et al. 1974; D. L. Baker et al. 1979; Einarsen 1956; Goldberg and Haas 1978; Julander et al. 1961; Kline 1959; Krausman 1978; MacConnell-Yount and Smith 1978; Medin and Anderson 1979; Quay and Müller-Schwarze 1971; Rees 1971; Seton 1929; Telfer 1978.

Odocoileus virginianus
White-tailed Deer

Measurements (*males*): Total length 1,800–2,150; tail 280–360; hind foot 510–538; ear from notch (two, W. Wanek and P. Karns, personal communication) 14.3 and 16 cm; weight about 110 kg (68–223 kg); (*females*): total length 1,600–2,000; tail 255–325; hind foot 480–520; ear from notch 15.1 cm (13.8–16.5 cm [17 females, P. D. Karns, personal communication]); weight about 73 kg (41–95 kg). (All figures are from Jackson 1961, except for ear measurements.) Petraborg and Burcalow (1965) record a Cook Co. specimen that dressed out at 182 kg, equivalent to a live weight of over 227 kg.

Description: The head, neck, back, and sides of an adult white-tail are reddish brown for three or four months in summer, and brownish gray in fall, winter, and early spring. The coat is longer in winter than in summer, giving the deer a bulkier form, and the individual hairs are thick and hollow, providing effective winter insulation. The throat, belly, rump, and edging and underside of the tail are white. Young fawns are spotted. If you see a moderate-sized wild deer in Minnesota, the chances are overwhelming that it is a white-tail, not a mule deer. The antlers of a typical male white-tail easily distinguish it from a mule deer buck. The main beam of a mature white-tail antler grows upward and slightly backward, then curves forward and outward, and a series of upright tines arise from it. The tail is distinctive for both sexes and all ages. It is brown above with a white fringe, is entirely white below, and is longer and broader than a mule deer's tail. When relaxed, the tail partly obscures the light rump patch. A fleeing white-tail deer clearly identifies itself with its conspicuous, erect white flag (Hirth and McCullough 1977). The ears of a white-tail are smaller than those of a mule deer. In both species, bucks average larger than does.

Range and Habitat: The white-tail ranges from southern Canada, through most of the United States except for parts of the West Coast and southwestern states, and into Mexico, Central America, and northern South America. Obviously, it occupies a variety of habitats, but generally favors relatively open wooded cover,

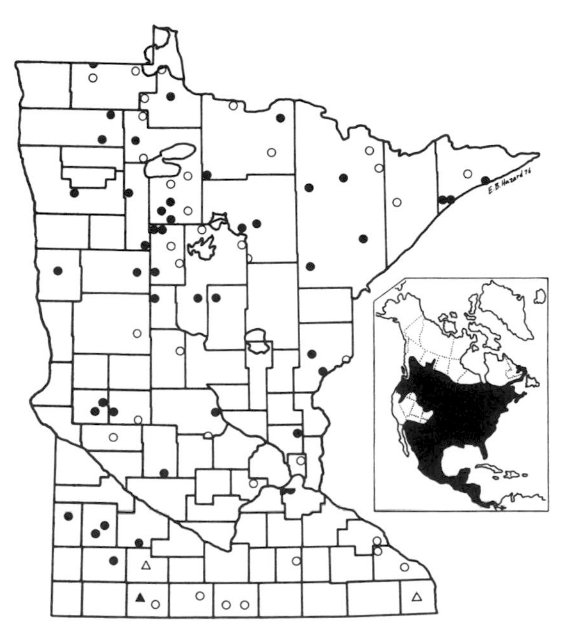

Map 72. Distribution of *Odocoileus virginianus*. ● = township specimens. ○ = other township records, selected locations. ▲ = county specimen. △ = other county records. (Map produced by the Department of Biology, Bemidji State University.)

avoiding both dense woods and expanses of open prairie. In Minnesota and adjacent states, white-tails prefer deciduous and mixed wooded uplands in summer and, particularly in the north, coniferous lowlands in winter (Kohn and Mooty 1971; McCaffery 1976; Ozoga 1968; Petraborg and Burcalow 1965; Stenlund and Gunvalson 1957).

Natural History: White-tail bucks typically drop their antlers in December, when the mating season has passed. Antler regrowth begins in

March. During the spring and summer, the growing antlers are covered with furry skin, or "velvet." By the end of August, growth generally has ceased, and, during September, the bucks rub off the dead velvet, often conspicuously marking and injuring the trunks of small trees in the process (Kile and Marchinton 1977). In summer bucks travel singly or in small groups and generally ignore does and fawns. By mid-October the bucks begin to take a sexual interest in does and become aggressive toward one another. They often battle each other with their antlers for pos-

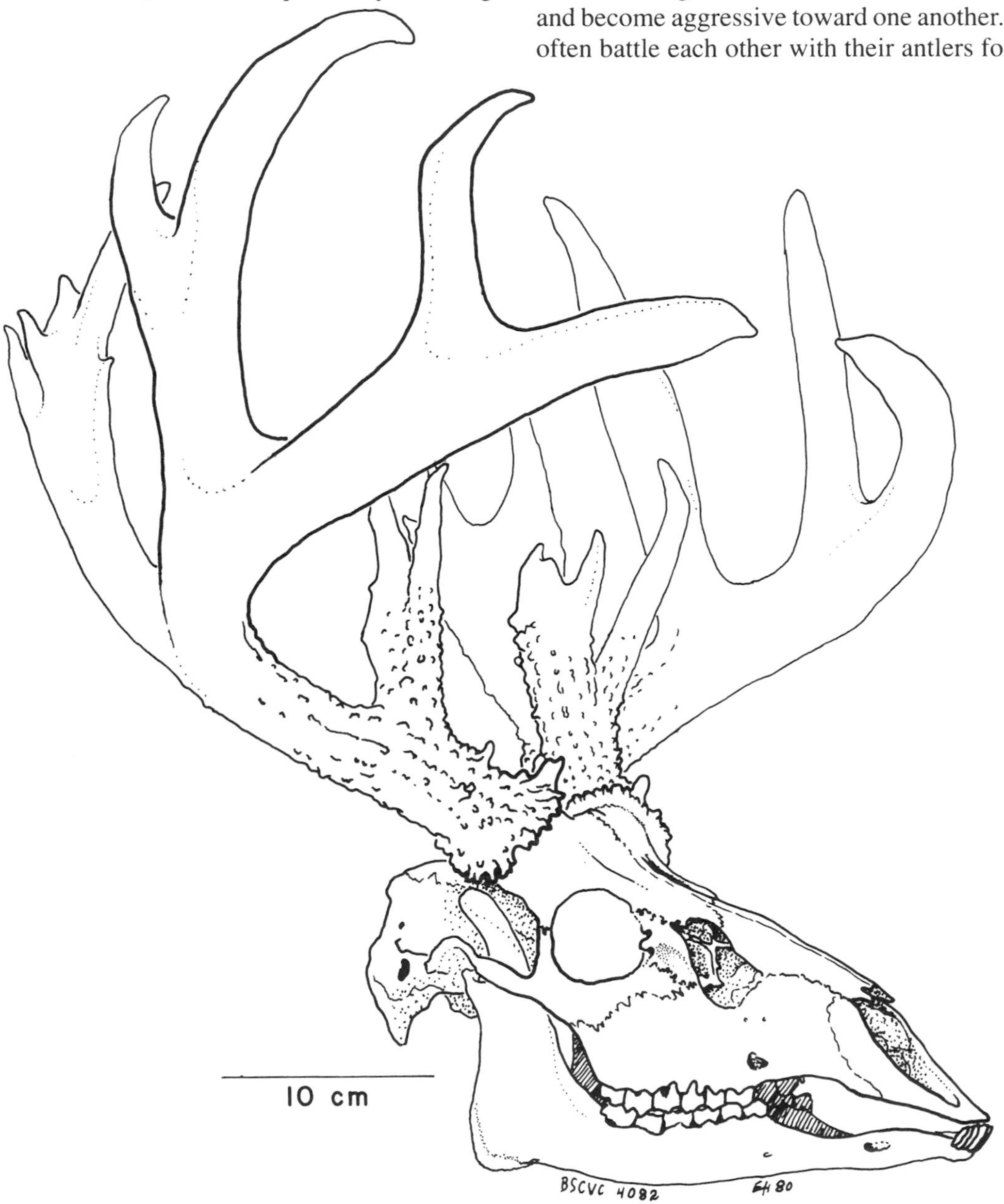

Figure 76. Skull of *Odocoileus virginianus*, white-tailed deer, BSCVC 4082. This old buck, hit by a car east of Bemidji in the fall of 1977, had well-worn teeth and was probably at or past the peak of his antler-growing ability.

session of a doe, occasionally with lethal results. White-tail males do not long remain interested in one doe, nor do they maintain harems. By the end of November, most does have been bred. In southern Minnesota, about 20 percent of the fawn does are bred; most breeding of fawn does occurs in late November and early December (Haugen 1975).

Gestation in white-tail does lasts 196–213 days (Verme 1969), and most fawns are born in late May and early June. Mature does most often bear two young, but yearlings generally bear only one. The young are spotted, but molt to the solid brownish gray winter coat in September.

Mothers vigorously lick their newborn young. This process may imprint the doe with the particular odor of her own young, enabling her to distinguish them from other fawns (Townsend and Bailey 1975). Fawns usually remain with their mothers through their first winter, and sometimes a doe in winter is accompanied by both first- and second-year fawns.

White-tails eat a variety of woody vegetation: leaves and twigs in summer, and buds, twigs, young bark, and conifer foliage in winter. In season they also eat various forbs, mushrooms, and fallen acorns (Erickson et al. 1961; Harlow et al. 1975; Kohn and Mooty 1971; McCaffery et al. 1974). Acorns and waste corn are major winter foods in parts of southern Minnesota (Dorn 1971). Particularly favored browse includes white cedar, mountain maple, red osier and other dogwoods, and the hazels (Coblentz 1970; Fashingbauer and Moyle 1963). Deer can live on white cedar alone for months without losing weight, but white cedar is easily overbrowsed. White-tails can and do eat aspen, but by itself it is a poor diet and excessive amounts may decrease a deer's ability to extract nutrients from other food (Ullrey et al. 1971). Except for white cedar, most of the preferred browse species in northern Minnesota, where most Minnesota deer live, are shade-intolerant upland species. They grow best in early successional stages of brush or open deciduous woodland, not under the closed canopy of mature climax forest. Deer in northern Minnesota spend the summer primarily in such upland areas. In the winter, however, most northern deer seek shelter under the protective evergreens of coniferous swamps. John Mathisen, Wildlife Biologist on the Chippewa National Forest, told me that 80 percent of the wintering areas on the Chippewa are in cedar or black spruce swamps.

Relationship to People: For better or worse, there is probably more human concern for *Odocoileus virginianus* than for any other wild Minnesota mammal. There is as much controversy surrounding the white-tailed deer as the timber wolf; indeed, much of the controversy involves the interaction of these two species. The deer is of interest as an object of aesthetic beauty, a potential danger on some highways (Allen and McCullough 1976), a pest to some farmers and orchardists, an economic asset to tourist and sporting goods businesses, a problem herbivore in some reforestation projects, and, perhaps primarily, as Minnesota's major big game species (Erickson et al. 1961; Petraborg and Burcalow 1965). These interests are interrelated and often conflicting. It is perhaps impossible to manage any one area in such a way that all reasonable interests in the deer population are met. Some values must take precedence in some areas, different values in others.

The management of the white-tail is complicated by one overriding fact: Minnesota's major deer herd lives on borrowed time in marginal habitat. We have grown so accustomed to a substantial deer herd in the north woods that we have come to think of its presence there as normal. Actually, in pioneer times the white-tail was most common in the deciduous woods of southeastern Minnesota, and few lived north of Morrison and Chisago counties. Agriculture and urbanization made southern Minnesota less suitable for deer, and lumbering made the north country more suitable. Dense, relatively uniform forests of mature conifers with little undergrowth were replaced by a more varied habitat consisting largely of early successional stages: brushlands and relatively open deciduous, mixed, or coniferous woodlands with a substantial brush understory. Deer do well in such habitats. Except in the extreme northeast, lumbering and agriculture were accompanied by the elimination of the deer's major predator, the gray wolf, and by reduction or elimination of other large carnivores.

The white-tail population has increased under the conditions that people have created in Minnesota, but the continued prosperity of the northern deer herd depends on continued manipulation of the environment. Most of the early successional habitats, if left to themselves, will slowly revert to mature stands of spruce, fir, and pine, vegetation that will support few if any deer (Stenlund 1963). Various management tools are available for the maintenance of a suitable habitat mix, and some of these are by-products of economical forestry. For example, the clear-cutting of small blocks of mature timber permits new browse to flourish adjacent to good cover (Fashingbauer 1961; Halls and Alcaniz 1968; Ozoga 1968; Wetzel et al. 1975).

Although people have modified northern biotic communities in the white-tail's favor, they have not changed the weather (Crête 1976b; Ozoga and Gysel 1972; Ransom 1967; Rongstad and Tester 1969; Verme 1968). Deer yard up in cedar and spruce swamps during winters with heavy snow. If population density is too high, the deer will overbrowse the yard, many (especially fawns) will starve, and some pregnant does will resorb fetuses, thus decreasing the coming fawn crop. During such winters, deer on a tight energy budget stand a greater chance of becoming exhausted when frightened or chased, as they may be by people on snowmobiles (Dorrance et al. 1975; Richens and Lavigne 1978; Severinghaus and Tullar 1975; Wanek 1971). Deer yards are not as inaccessible as they once were.

In order to prevent habitat deterioration and mass winter starvation, predation on the deer herd is necessary (Hansen and Brown 1950; Hoskinson and Mech 1976; Kolenosky 1972). The timber wolf can do part of this in the northeasternmost counties, but wolves are not compatible with human activities in much of the rest of Minnesota. Here, if properly regulated according to sound game management principles, hunting is not only a source of sport and meat for the rifleman and archer but also a necessary conservation measure.

Additional References: Bahnak et al. 1979; Carlsen and Farmes 1957; Crête 1976a; Dahlberg and Guettinger 1956; Drolet 1978; Floyd et al. 1979; Frenzel 1965; Hirth 1977; Hunt and Mangus 1954; Irwin 1975; Jacobsen 1979; Karns 1979b; Kellogg 1956; Larson et al. 1978; Mautz 1978; McCullough 1979; McMillan et al. 1980; Mech and Frenzel 1971; Mech, Frenzel, and Karns 1971; Mech and Karns 1977; Moen 1968; Nelson 1971; Petraborg and Gunvalson 1962; Saunders 1973; Seal et al. 1978; Severinghaus and Cheatum 1956; Stenlund 1958; Tester et al. 1964; Verme 1977; Wanek 1972–75; Woolf and Harder 1979; Zagata and Moen 1974.

Genus *Alces*

Alces alces
Moose

Measurements (*males*): Total length 2,440–2,800; tail 76–110; hind foot 760–840; weight 385–544 kg; (*females*): total length 2,040–2,590; tail 90–122; hind foot 725–810; weight 329–385 kg (Jackson 1961).

Description: The moose is a large blackish brown deer about the size of a horse. Its coat fades to a lighter brown during the winter. Its legs are long, with large hoofs and well-developed dew claws, which ensure good footing in soggy habitats, and its tail is quite short. The nose is large and pendulous, and a prominent dewlap hangs from the throat of adults of both

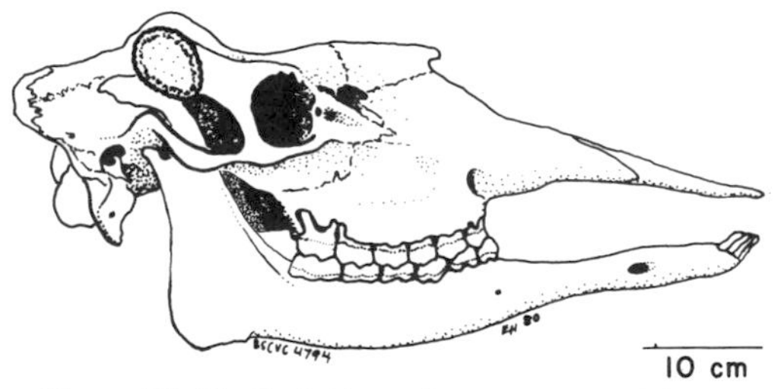

Figure 77. Skull of *Alces alces*, moose, adult male, BSCVC 4794. Antler bosses restored, based on another specimen.

sexes. Bull moose have large, palmate antlers. The tooth formula is 0/3, 0/1, 3/3, 3/3 × 2 = 32.

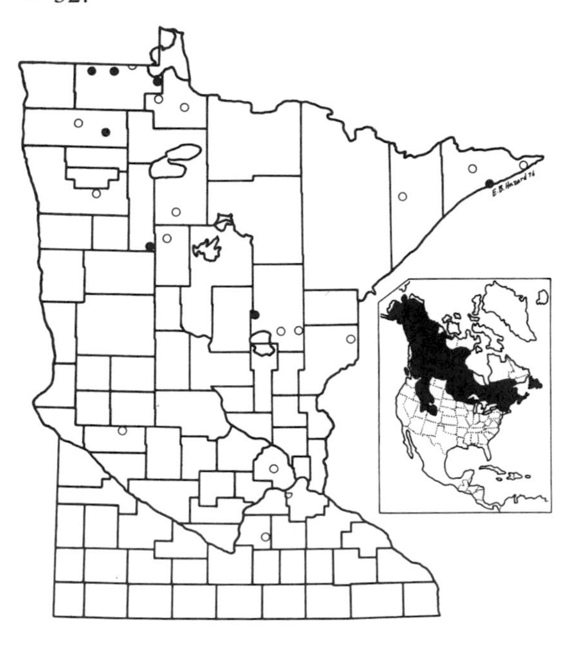

Map 73. Distribution of *Alces alces*. ● = township specimens. ○ = other township records, selected locations. (Map produced by the Department of Biology, Bemidji State University.)

Range and Habitat: Alces alces is Holarctic, inhabiting the boreal forest biome of Eurasia and North America. At one time New World moose were recognized as a distinct species, *Alces americana*, but mammalogists now consider our moose and the Old World elk to be conspecific. Its range in North America extends south from Alaska and Canada into northern New England, northern Minnesota, and the mountains of the western states. Moose have been reintroduced into Upper Michigan. In Minnesota the areas of greatest abundance of moose presently include the rugged terrain of the northeastern counties, and the boggy lowland north and west of the Red Lakes (Anon. 1965; Berg 1971; Idstrom 1965; Karns 1967b; Ledin and Karns 1963). Before the evergreen forests of northern Minnesota were first logged, moose were abundant as far south and west as Chisago, Wadena, and Roseau counties, but they were uncommon in northeasternmost Minnesota, where caribou were the dominant deer. Throughout its range, *Alces alces* favors early successional stages with abundant deciduous browse, although it may seek cover in mature, dense conifer stands in winter (Berg 1971; Irwin 1975; Karns et al. 1974; Peek et al. 1976; Phillips and Siniff 1973).

Natural History: Moose are relatively solitary, although they may aggregate in small groups, sometimes in response to a concentrated food source (Houston 1974; Peek et al. 1974). Bulls begin to grow new antlers in April. The velvet is lost in August, and the antlers are shed in December, January, or later. The rut lasts from sometime in September through October. After a gestation of 240 to 246 days, one or two calves are born in May or June (Peterson 1955). During the calving period, the cows are solitary and aggressive.

Moose eat a variety of browse, including willow, beaked hazel, red osier dogwood, aspen, mountain maple, white birch, and balsam fir. During the summer they also feed extensively on aquatic plants; moose swim well and feed on submerged as well as surface aquatics (Idstrom 1965; Peek et al. 1976). Forbs and grasses appear to be of little importance in their diet. Most of the terrestrial plants in the diet are typical of early successional stages; thus moose are most abundant in areas that have been burned or cut over recently.

The diet of moose broadly overlaps that of white-tailed deer, so the two species are potential competitors. Moose succeed best in areas where cold weather or snow depth put deer at a disadvantage. The bulkier moose conserves heat more efficiently than a deer does, and, whereas half a meter of snow may force deer to yard up, moose can roam freely until the snow is twice as deep (Kelsall 1969). Why does the moose not move south, expanding its range into good white-tail habitat in north-central Minnesota? Occasionally, moose are seen outside their two major areas

Figure 78. *Alces alces*, moose.

of concentration, but they generally do not establish resident populations in these new locations. In areas where winters are not too harsh, white-tails have a competitive advantage over moose because of their greater resistance to a parasitic roundworm, *Parelaphostrongylus tenuis*. The larvae of *P. tenuis* pass out of the deer in the feces, and are ingested by a snail or slug. There they undergo three molts. Deer or moose become infected by accidentally eating the gastropod while grazing. The worms move from the gut via the bloodstream to the nervous system (Harold Borchers, personal communication). Where white-tails are abundant, 60 percent or more are infected (Karns 1967a, 1980); the adult worms reside on the surface of the deer's brain but apparently do little harm. When *P. tenuis* infects moose, however, it burrows into the brain, killing the moose.

White-tailed deer reached peak numbers in cutover habitats in northern Minnesota in the 1920s and 1930s, and the moose population declined markedly at the same time. Most of the decline in moose apparently resulted from infection by *P. tenuis*, introduced by the white-tail (Karns et al. 1974; Saunders 1973). A subsequent decline in white-tail populations caused mainly by severe winters and succession leading to less suitable habitat has allowed moose populations to recover to the point where carefully controlled hunting can be allowed and is in fact desirable to prevent moose from exceeding the carrying capacity of their habitat.

Relationship to People: The human activity that has had the most substantial effect on moose has been the modification of northern Minnesota habitats; people have made the Quetico-Superior area less suitable for caribou and more suitable for moose but have encouraged the white-tail with its resident parasite *P. tenuis* to invade the former moose range. Perhaps some northern Minnesota areas cannot successfully maintain abundant populations of both white-tails and moose (Karns 1979a). However, various historical records, including the diaries of E. L. Brown, a trapper and market hunter in the border counties at the turn of the century, mention the frequent occurrence in the same area of white-tails and moose, as well as caribou (Cole 1979 and personal communication). Further research will be needed to determine the compatibility of these cervids in nature. It seems reasonable to me that some of the state's suitable habitat should be managed primarily for moose.

Additional References: Allen 1974; Allen and Mech 1963; Breckenridge 1946; de Vos 1958; Franzmann and Arneson 1976; Franzmann and Le Resche 1978; Frenzel 1974; Hansen et al. 1973; Joyal 1976; Joyal and Scherrer 1978; Karns 1973; Krefting 1974; Ledin 1966; Peek 1974; R. O. Peterson and Allen 1974; Rounds 1977, 1978; Telfer and Cairns 1978; Van Ballenberghe and Peek 1971; M. L. Wolfe 1977.

Genus *Rangifer*

Rangifer tarandus
Caribou, Woodland Caribou

Measurements (*males*): Total length 1,840–2,260; tail 110–150; hind foot 550–670; weight about 160–180 kg; (*females*): total length 1,660–2,030; tail 100–140; hind foot 500–600; weight about 88–120 kg (Jackson 1961). Fashingbauer (1965c) indicates that bulls in good condition weigh 170–193 kg, cows under 136 kg.

Description: The caribou is a good-sized, relatively short-legged deer that averages larger than a white-tail but smaller than a wapiti. The pelage of bulls in winter is grayish brown on the back and lighter on the belly. The neck and often the shoulder and flanks are an ivory color, and a whitish beard appears on the throat. There is generally less of a contrast in the coloring of cows. Both sexes have a prominent whitish rump patch, and both have antlers, although those of a bull are larger. The main beams of a bull's antlers sweep up, back, and outward from the head, and there are three tines, each of which is actually a palmate structure with several points. Usually the brow tine of one antler is larger than the other and lies forward over the midline of the head. The tooth formula is the same as the wapiti's (0/3, 1/1, 3/3, 3/3 × 2 = 34), but the upper canine may remain below the gum line.

Range and Habitat: Rangifer tarandus has a circumpolar distribution in the tundra and taiga biomes. Several species were formerly recognized in the genus, some of them primarily tundra or "barren ground" inhabitants (called reindeer in the Old World), the others primarily woodland forms (Banfield 1961; Harper 1955; Macpherson 1965; Youngman 1975). In North America, woodland caribou once ranged throughout Canada and Alaska south of tree line and north of the prairie, and into the contiguous United States in the northern Rockies, the upper Great Lakes States, and northern New England. In general, the caribou is now gone from those southern parts of its range which people have substantially altered. In Minnesota, caribou once ranged as far south and west as Carlton, Mille Lacs, and Kittson counties, and were common in the deep woods of the north-central and northeastern counties (Fashingbauer 1965c). They were last recorded in Minnesota in 1935, in the area between Upper Red Lake and Lake of the Woods (Breckenridge 1935; Nelson 1947). An

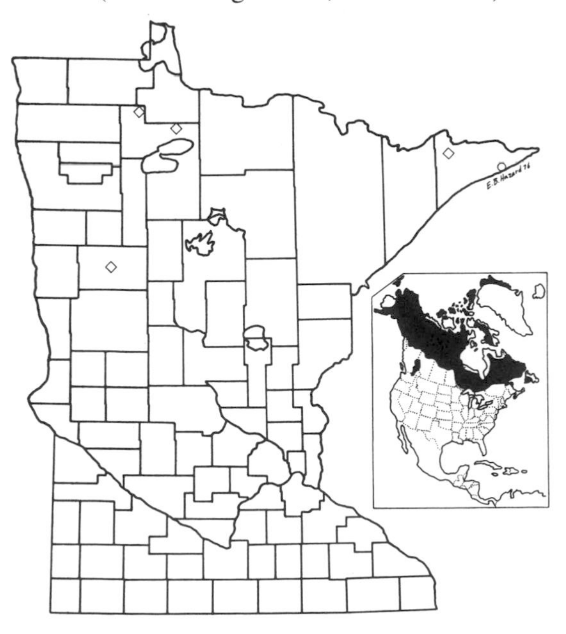

Map 74. Distribution of *Rangifer tarandus*. ○ = 1981 sighting. ◇ = approximate sites of old bones. (Map produced by the Department of Biology, Bemidji State University.)

attempt to reintroduce caribou in 1938–42 failed (Moyle 1975). (At least one, perhaps two, caribou were sighted by Bill Peterson and others near Grand Marais in the winter of 1980-81. See Peterson 1981.) Unlike the two species of deer that now inhabit northern Minnesota, the woodland caribou favors mature evergreen forest rather than open deciduous brush or woodland.

Natural History: Woodland caribou travel in small groups, rather than in the large herds typical of barren ground caribou (P. Karns, personal communication). Rut occurs in mid-October, and the polygamous mature bulls maintain harems (Bergerud 1973). Calves (only one in about 85 percent of the births) are born near the end of May after a gestation of about 228 days (Bergerud 1971b; Shoesmith 1976). The young are precocial and soon able to roam with the adults. Caribou move more or less continually in search of food. The diet varies seasonally (Bergerud 1971a, 1972, 1974b; Miller 1974). Woody browse is eaten during most of the year, but green sedges are important in spring, fungi figure prominently in the summer diet, and lichens are eaten throughout the year, especially in summer, fall, and winter. When the ground is bare, lichens of the genus *Cladonia* (which includes "reindeer moss") predominate, whereas in the winter a staple food item is the various lichens that characteristically grow on the trunks and branches of mature conifers. Lichens grow slowly; thus it is a long time before a browsed patch recovers its former luxuriant growth. Caribou herds therefore have to forage over large areas to find adequate food (Bergerud 1974b). The caribou is more than the equal of the moose in its resistance to the cold, and its ability to utilize lichens as a primary food ensures its dominance in mature coniferous forest, such as existed in northeastern Minnesota before the advent of the white man. Caribou, however, are apparently nearly as vulnerable to the trematode *Parelaphostrongylus tenuis* as moose are (P. Karns, personal communication) (see p. 164). Moyle (1975) and Marshall and Miquelle (1978) suggest that the reintroduction of caribou into the present white-tailed deer range may be unsuccessful. It may, however, be possible for these two cervids to coexist if there are places where they could occupy different habitats in the same general area (Cole 1979).

Relationship to People: The caribou was a major source of meat, hides, and implements for the Inuit and Indians as well as a major resource for the early white trappers and loggers. Overhunting is a serious problem in some present-day caribou habitats (Banfield and Tener 1958; Bergerud 1967) and no doubt played a major role in the disappearance of caribou from Minnesota (Bergerud 1974c). But, as is apparent from this account and those on white-tailed deer and moose, manipulation of the habitat for forestry, agriculture, recreation, or other human activities has been the primary agent in determining which deer will be dominant in a given area. The drama involves the interplay of many factors; the three deer, human land use practices, hunting (legal and illegal), predators such as the lynx and wolf, the weather, other browsers such as the snowshoe hare, perhaps other lichen feeders like the northern flying squirrel, at least one internal parasite, and, to some extent, perhaps every species that lives in the north woods. Such interweaving of ecologic relationships is characteristic of any biotic community, but it seems particularly well illustrated by the natural history of the white-tail, the moose, and the caribou in northern Minnesota. The complexity of such relationships belies any simple solution, even the deceptively innocent suggestion that if we kept hands off, everything would be all right.

Additional References: Bergerud 1974a, 1975; Cahn 1921; Calef and Lortie 1975; Euler et al. 1976; Holleman et al. 1979; Karns and Crichton 1978; Kelsall 1968; Miller and Robertson 1967; Timm 1975.

Family Antilocapridae
Pronghorns

This family has been restricted to North America since its first appearance in the Middle Miocene, some 20 million years ago. It was more diverse in the past, with several genera coexisting during some geologic epochs (Frick 1937; Furlong 1935; Stirton 1938; Webb 1973). Today there is only one species in the family. Antilocaprids resemble bovids in many ways; O'Gara and Matson (1975) suggest that pronghorns be classified as a subfamily within the family Bovidae.

Additional References: Koopman 1967; Vaughan 1978.

Genus *Antilocapra*

Antilocapra americana
Pronghorn

Measurements (*males*): Total length 125–145 cm; tail 6–17 cm; hind foot 39–43 cm; ear 15 cm; weight (Oregon) 50.5 kg (45.4–62.8); (*females*): total length 135 cm; tail 6 cm; hind foot 39 cm; ear 14 cm; weight (Oregon) 41.5 kg (32.2–47.5) (Banfield 1974).

Description: The pronghorn is a delicate, graceful ungulate with unique blackish horns arising just above the eye sockets, which are set high on the head. The top of the horn curves backward, and there is a short prong on the front. The keratinaceous horn (which is fibrous but apparently does not consist of fused hairs as was once thought) is supported by a core of bone. The core is permanent, but the sheath is shed each winter after the rut, being forced off by the new horn growing from the skin beneath it (O'Gara 1969b; O'Gara and Matson 1975; O'Gara et al. 1971). The dorsal pelage of the pronghorn is tan; the flanks, belly, lower jaws, and two transverse bars on the throat are white. The black markings on the head are more extensive in bucks than in does, and bucks average somewhat larger than does. The tooth formula is 0/3, 0/1, 3/3, 3/3 × 2 = 32.

Range and Habitat: Antilocapra is most characteristic of the shortgrass prairie. It occurs from northern Mexico through the prairies and deserts of the western states into the southern Prairie Provinces. Pronghorns are rare in the eastern

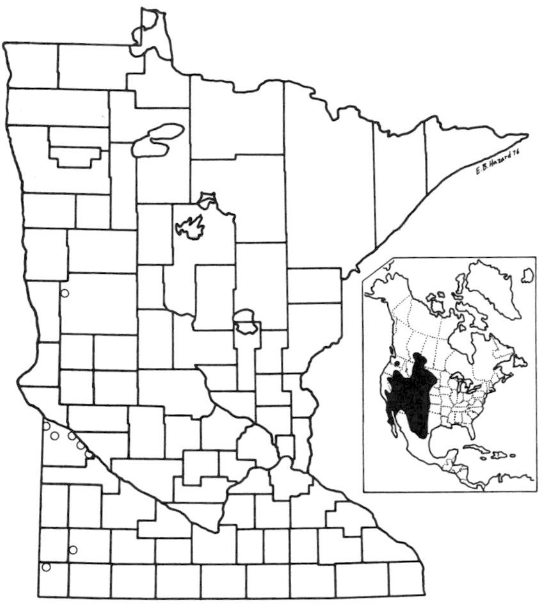

Map 75. Sightings of *Antilocapra americana* by the Minnesota Department of Natural Resources. (Map produced by the Department of Biology, Bemidji State University.)

Dakotas, but some occasionally wander east into western Minnesota (Fashingbauer 1965e). A herd of 11 pronghorns (see photograph in Henderson 1978g) overwintered in Lac Qui Parle County in 1977–78. They were last seen in April 1978, after which they apparently returned to South Dakota. I know of no specimens from Minnesota that are preserved in collections.

Natural History: Pronghorns are sociable and live in small herds (Kitchen 1974). Bucks are territorial during the fall rut (Bromley 1969; Gilbert 1973; Kitchen and Bromley 1974), and does separate themselves from the herd at fawning time. Does usually bear twins (Howard 1966). The pronghorn has unusually good eyesight for a ruminant. In unfenced high country pronghorns migrate some distance from summer to winter feeding grounds. They are unusually fast, achieving speeds up to 96 kph, and can sustain a speed of about 60 kph for several kilometers. They feed mostly on forbs and prairie shrubs.

Relationship to People: Pronghorns presently have no economic significance in Minnesota, though many Minnesotans are willing to travel west to hunt them. In states where they are common, they are of interest to hunters and to those who simply enjoy watching a graceful creature run free. They are not generally considered significant food competitors of cattle (Schwartz and Nagy 1976).

Additional References: Autenrieth and Fichter 1975; Bayless 1969; Bruns 1977; Buechner 1950; Dow and Wright 1962; Einarsen 1948; Ellis and Travis 1975; Jones 1960; Martinka 1967; O'Gara 1969a, 1978; Seal and Hoskinson 1978; Wesley et al. 1970, 1973; Wright and Dow 1962.

Family Bovidae
Cattle and Allies

This is the largest family of artiodactyls, and it has undergone an extensive adaptive radiation in the Old World, especially in the tropics. A few species reached North America during the Pleistocene. No native bovids occur in South America, Madagascar, or in the Australian region. In the form of introduced livestock and several game species, the family now has a worldwide distribution. Only one species of bovid is native to Minnesota, and it is no longer found here in the wild.

References: Koopman 1967; Vaughan 1978.

Genus *Bison*

Bison bison
Bison

Measurements (*males*): Total length 3,050–3,800; tail 550–815; hind foot 585–660; weight about 725–910 kg; (*females*): total length 1,980–2,280; tail 430–535; hind foot 460–555; weight about 410–500 kg (Jackson 1961).

Description: Everyone used to know what a bison looked like, when a nickel was trolley or subway fare and an Indian-head nickel had a "buffalo" on the reverse. Actually, "buffalo" is a term better reserved for various bovids of the Old World tropics. The New World bison is a large cattlelike mammal, with horns in both sexes, a dark brown coat, a relatively short tail, and a hump on the shoulders where the neural processes of the vertebrae are unusually long. There is marked sexual dimorphism, and the resemblance to domestic cattle is more apparent in a bison cow than in a bull. The bison bull has a higher shoulder hump than does the female, larger horns, a beard, and a heavier cape of shaggy hair over the shoulders, forequarters, and roof of the skull (Guthrie 1966; Lott 1974). The tooth formula is 0/3, 0/1, 3/3, 3/3 × 2 = 32.

Range and Habitat: Bison once roamed the open country of North America from the Appalachians to the Rockies and from the Gulf of Mexico to the Northwest Territories. Large herds numbered in the hundreds of thousands, and the total population may have been as high as 70 million. The species was almost brought to extinction in the

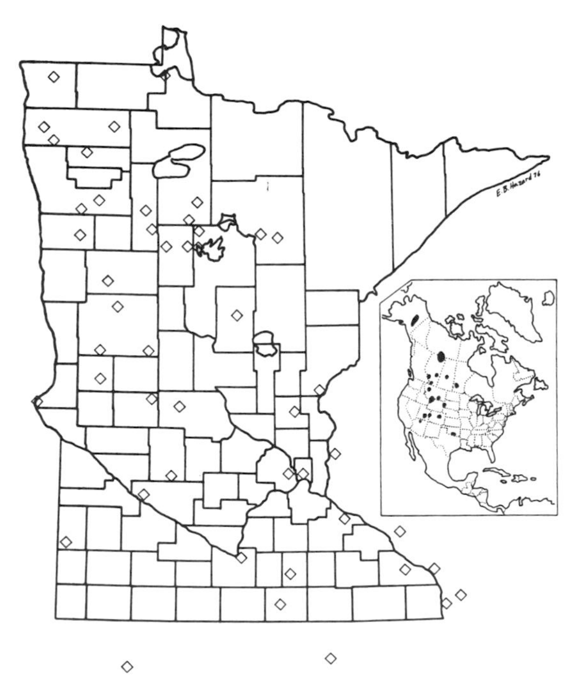

Map 76. Selected sites of old *Bison* bones, both *B. bison* and the extinct *B. occidentalis* (all but two locations provided by Robert Bright). (Map produced by the Department of Biology, Bemidji State University.)

nineteenth century. Fortunately, a few semicaptive herds were preserved, and bison may be seen today in Canada and the United States in a relatively wild state in a few national parks and similar reserves (Choquette et al. 1972; Meagher 1974). A related species, the wisent, exists in even smaller numbers in Europe.

Until two centuries ago, bison were characteristic of western Minnesota prairies and also occurred in semi-open country along the Mississippi. They had become rare in Minnesota by 1830 and were gone by 1880 (Fashingbauer 1965d). Roughly 4,000–6,000 years ago, during the hypsithermal interval, the prairie extended much farther east, but some low spots in northern Minnesota retained their characteristic bogs. Bison took refuge in these bogs during fires, and often perished there. In May 1980, T. E. Jessen, R. L. Jessen, R. C. Melchior, and I collected bones of several bison in a marl deposit in southern Beltrami County, and in October 1967, R. C. Melchior and his students collected the remains of two dozen or more bison in peat in Clearwater County. The measurements of the Clearwater specimens suggest that they are intermediate between *B. bison* and the extinct *B. occidentalis*. The localities in map 76 all refer to finds of bones, not to living animals.

Natural History: Bison are markedly gregarious animals. During most of the year bulls congregate separately from cows and calves. During the rut, mature bulls attend individual cows that are in heat, rather than maintain harems. Rut extends over several weeks from midsummer to early fall. Gestation lasts nine to ten months, and a female usually bears only one calf. Bison feed on a variety of grasses and some forbs. Before the prairie was fenced, bison herds migrated over areas a few hundred kilometers in diameter (Fashingbauer 1965d; Haugen 1974).

Relationship to People: When they were numerous, bison were a source of meat, leather, and other products (Agenbroad 1973; Reeves 1973; Wheat 1967). The remnant herds are now mainly a source of aesthetic, historic, and scientific interest, and a reminder of how wasteful and heedless people can be. Bison have been crossbred with beef cattle with some success.

Additional References: Dagg and Taub 1970; Garretson 1938; Guthrie 1970; S. A. Hall 1972; Mahan et al. 1978; McHugh 1958; Meagher 1973; Reynolds et al. 1978; Roe 1970; Rorabacher 1970; Shakleton et al. 1975; Soper 1941.

Figure 79. *Bison bison*, bison.

Domestic and Feral Mammals

Domestic mammals are those that people keep and breed. Examples include racehorses, dairy cows, retrievers, and white rats. Most cannot survive for long in the wild, but some can. People have selectively bred most of these species for so long that they differ substantially from their wild ancestors. Remains of all but the smallest domestic mammals are likely to be found outdoors, most often in modified habitats. The following is, to my knowledge, a complete list of species that people keep in Minnesota.

Oryctolagus cuniculus, domestic rabbit
Mesocricetus auratus, golden hamster
Meriones and other genera, gerbils
Rattus norvegicus, laboratory rat
Mus musculus, laboratory mouse
Mycastor coypus, nutria or coypu
Chinchilla sp., chinchilla

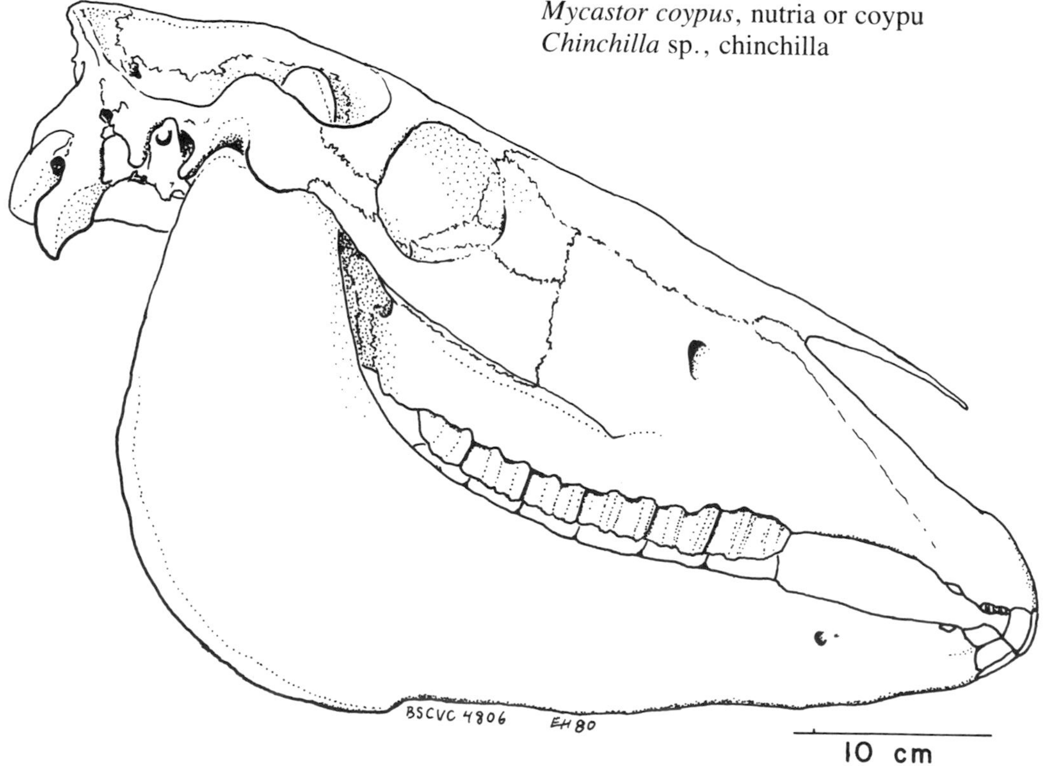

Figure 80. Skull of *Equus caballus*, domestic horse, five-year-old male, BSCVC 4806.

Cavia porcellus, guinea pig
Canis familiaris, domestic dog
Vulpes vulpes, fur farm fox
Mustela putorius, domestic ferret
Mustela vison, ranch mink
Felis catus, house cat
Equus caballus, horse
E. asinus, donkey
E. caballus × *asinus*, mule
Sus scrofa, domestic pig and wild boar
Cervus dama, fallow deer
(Various other deer are kept in private "deer parks" in Minnesota)
Bos taurus, cow
Ovis aries, sheep
Capra hircus, goat

Feral mammals are domesticated mammals that, without human assistance, succeed in "staying alive" (as Errington often put it) in the wild. This may be detrimental to native plants, animals, or habitats. Fortunately, most species in the above list cannot do this in Minnesota. Our wild Norway rats and house mice are exotic, but not feral, because they are derived from stock that has never been domesticated. Nutria have been kept on Minnesota fur farms, but escapees apparently cannot survive our winters. Feral populations of nutria exist on the Middle Atlantic Coastal Plain, and in the Deep South, southwestern Michigan, and northern Ohio (Hamilton and Whitaker 1979; Willner et al. 1979). Feral dogs occasionally survive in Minnesota (Kreeger 1979), and may interbreed with wolves or coyotes. I doubt that this has been a significant factor in the population genetics of coyotes or wolves in Minnesota. Feral populations of house cats often exist in modified habitats. Wild boars have, unfortunately, been introduced as game animals in the southeastern states and in California, where they have crossed with escaped domestic stock (Wood and Barrett 1979). To my knowledge, there are no such populations in Minnesota. Exotic deer sometimes escape from privately operated deer farms, but I know of no established feral populations in Minnesota. The skin and skull of one unidentified deer from Mahnomen Co. are in the Bell Museum collection (MMNH 13291).

Additional References: Beach and Gilmore 1949; Beck 1973, 1975; Bekoff 1979; Berger 1977; Bourdelle 1939; Christian 1975; Clark 1975; Coman and Brunner 1972; Denny 1974; Doucet 1973; Elton 1958; Fox 1978; George 1974; Hafez 1969; Jope 1976; Kreeger 1977; Leyhausen 1965, 1978; Lowry and McArthur 1978; Nesbitt 1975; Olsen and Olsen 1977; Pearson 1964; Progulske and Baskett 1958; Riemann et al. 1978; Sadleir 1969; J. P. Scott 1967; M. D. Scott and Causey 1973; Sweeney et al. 1971; N. B. Todd 1977, 1978.

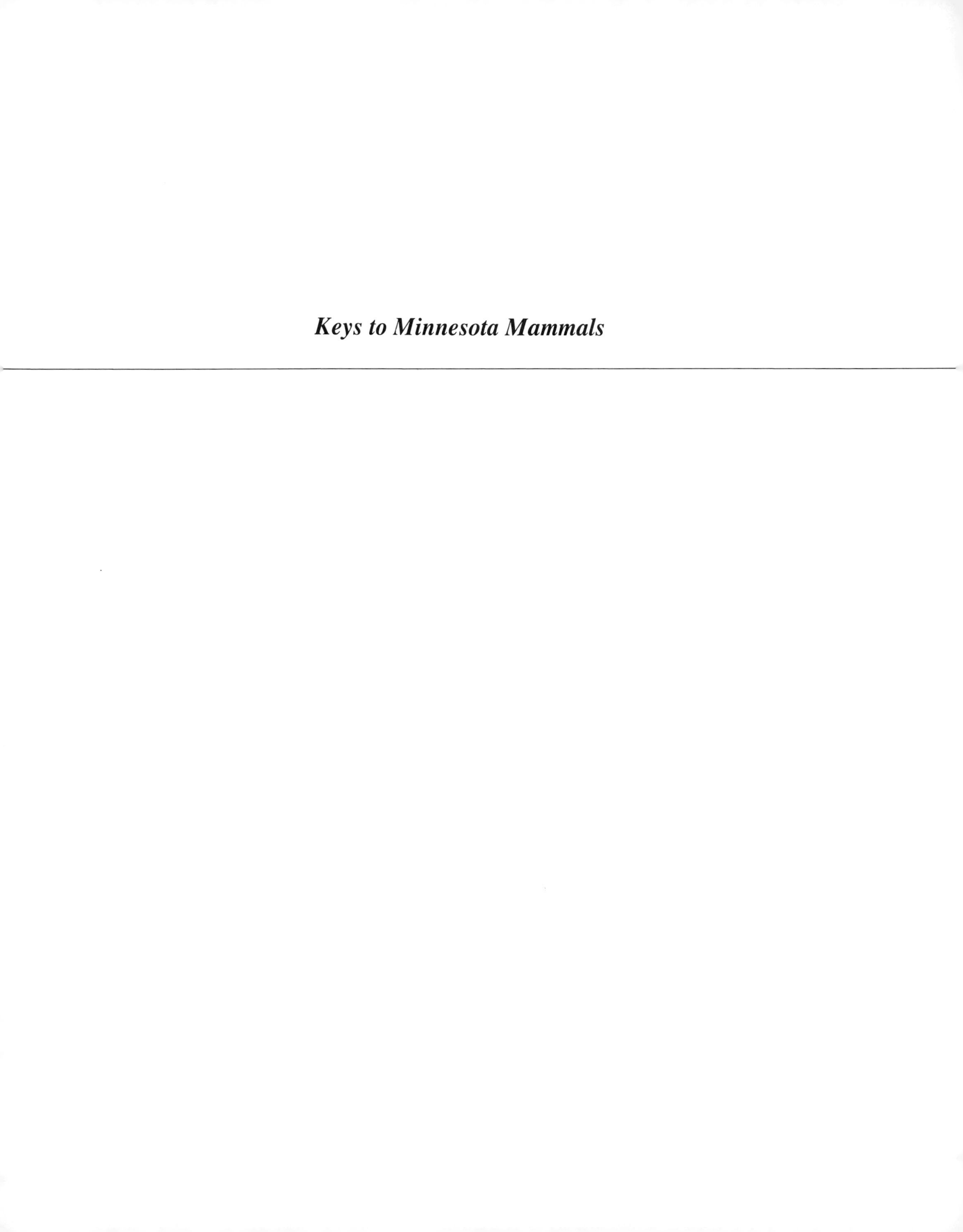

Keys to Minnesota Mammals

Keys to Minnesota Mammals

A key is a written (and sometimes illustrated) identification device. Ideally, with the key and appropriate measuring aids, you should be able to identify a single specimen collected in the geographic area covered by the key, without having to compare it with other specimens. For the two keys that follow, you will need a millimeter rule and perhaps metric calipers, and a hand lens or low-power microscope. If you are trying to identify a study skin, it should be one with standard measurements already on the label. Standard measurements can only be made from intact specimens, not from prepared skins. You can, on the other hand, make your own measurements on skulls.

To use the key, read the first couplet (alternatives *1a* and *1b*) at the beginning of the key, examine the specimen, and decide which alternative best fits the specimen. Then go to the couplet to which that alternative directs you, and repeat the process (always reading both alternatives) until, instead of directing you to another couplet, your chosen alternative indicates a particular species. Normally, this should give you the correct identification. For example, suppose you have an unknown mammal skull. It has 32 teeth, seven in each side of the lower jaw and nine in each side of the upper jaw, so it cannot be choice *1a* in the first couplet in the skull key on page 184. Alternative *1b* leads to couplet *2*. There is no postorbital bar (alternative *2b*), so you go to couplet *5*. There is a large diastema between the two single upper incisors, and the next tooth, the canine, is prominent; alternative *5a* leads to *6*. In addition to the aforementioned median diastema, there is a sharp posterior edge on the canine, and the skull is under 25 mm long. It is therefore a bat, order III, Chiroptera. The incisors are 1/3 (choice *1a* in the Chiroptera key on page 187), leading to couplet *2*, genus *Lasiurus*. Since the skull is 18.5 mm long, it keys out in alternative *2b* as *Lasiurus cinereus*. As you become more familiar with the keys and with the mammals of the state, you can skip unnecessary couplets and go directly to the part of the key that refers to the group in question. Even the uninitiated will not need to start with the first couplet if a skin has wings, or long ears and a cottony tail, or if a skull has horns or antlers.

Most of the couplets are redundant, giving two or more distinguishing characters. This serves as a check; it is essential when there is overlap between some of the characters; and it may allow you to proceed even though a specimen is incomplete or abnormal. For instance, many of the skulls found outdoors are missing some bones. "Upper incisors absent," by itself, is not much help in keying out a skull that lacks premaxillaries. Likewise, "slender zygomata present" may be untrue of damaged mole skulls. One cannot cover all possible difficulties, but it would be an interesting and useful endeavor for knowledgeable mammalogists to construct and publish keys to partial specimens, particular bones (e.g., mandibles, bacula), juvenile specimens, and the like.

Many authors combine skin and skull characters in a single key, but I have constructed two keys, for various reasons. It is common to have only a skin or a skull—a skin for which the skull is missing, a fresh or perhaps live specimen with most skull characters inaccessible, a skull that someone found in a woods. Also, some species or groups separate readily on skin characters but not on skull characters, and vice versa. The two keys follow different routes, depending on what characters, for skins or skulls independently, seemed to best separate groups at a particular juncture. If you do have both a skin and a skull, use the two keys independently to serve as a check on your identification.

The present keys are limited to Recent species that occur in Minnesota or that may once have. They will not work well outside Minnesota. The skin key includes wild species only, with few exceptions, because most people easily recognize dogs, guinea pigs, goats, and such. The skull key, on the other hand, includes a number of domesticated species of moderate to large size, some of which occasionally go feral, and all of which may show up occasionally as skulls in the fields or woods. I have not included small pets such as hamsters and gerbils.

An important reminder about keys: a key is an identification tool, not a phylogeny. Key characters are those that best enable people to distinguish species of mammals from one another, not necessarily those that best inform us about the evolutionary relationships among species of mammals. Most important, a key character should not be regarded as "the difference" between two species, particularly in an adaptive or ecologic sense. An extra loop on M2 nicely separates the skull of *Microtus pennsylvanicus* from that of *M. ochrogaster*, but the adaptive differences between the two species involve such things as reproductive potential, ability to conserve water, and adaptibility to modified environments, none of which preserve well in museum specimens. Much fruitless argument in the older literature has resulted from regarding key characters as the sole or primary differences among species.

Rare Species. Throughout both keys, asterisks (*) mark species that should be considered unusual finds. These are rare species known to occur in the state, or species not known to occur but that will possibly show up, or species that once occurred here and that might be found as skulls or skeletons. If a specimen keys out as rare, or as a common species outside its normal range, it might be well to run it through the key(s) again to be sure. If the original identification still seems likely, please bring it to the attention of a professional biologist with some competence in mammalogy. The Minnesota colleges and universities listed below maintain collections of mammals:

Bell Museum of Natural History 10 Church Street, S.E. University of Minnesota	Minneapolis	55455
Bemidji State University	Bemidji	56601
Concordia College	Moorhead	56560
Gustavus Adolphus College	St. Peter	56082
Mankato State University	Mankato	56001
Moorhead State University	Moorhead	56560
Saint Cloud State University	St. Cloud	56301
Saint John's University	Collegeville	56321
Southwest State University	Marshall	56258
University of Minnesota, Duluth	Duluth	55812
University of Minnesota, Morris	Morris	56267
Winona State University	Winona	55987

Of the institutions listed, only the Bell Museum currently (1981) meets minimum standards for research collections of Recent mammals as determined by the American Society of Mammalogists. It is best to donate rare or otherwise significant specimens to such collections.

You can also bring specimens to the regional offices of the Minnesota Department of Natural Resources, listed below. (Note that some of the species marked with an asterisk are protected and can only be taken in season by a licensed hunter or trapper, or by a professional biologist under the provisions of a special permit.)

Regional Offices of the Minnesota Department of Natural Resources

Region	*Location*
1	Bemidji
2	Grand Rapids
3	Brainerd
4	New Ulm
5	Rochester
6	St. Paul

Keys should use reliable and obvious characters, not necessarily original ones. Few of the characters used in my keys are new. Most of them appear in one or more of the following references, and probably elsewhere: Banfield 1974; Burt 1957; DeBlase and Martin 1974; Eddy and Hodson 1961; Fisler 1970; Glass 1951; Gunderson and Beer 1953; Hamilton and Whitaker 1979; Jackson 1961; Lawlor 1979; Peterson 1966.

Key to Adult Skins and Whole Specimens

An asterisk before a species denotes an unusual find.

Key to the Orders

1a. Each foot terminating in one or two hooves, lateral digits reduced or absent . . . . . *2*.

1b. Feet not terminating in hooves, at least four digits on hind feet bearing claws . . . . *3*.

2a(1a). Mesaxonic, each foot terminating in a single hoof . . . PERISSODACTYLA, VII.

2b. Paraxonic, each foot terminating in two hooves (cloven-hoofed), most with small, reduced hooves on side toes . . . . .ARTIODACTYLA, VIII.

3a(1b). Forelimbs developed as wings, flight membrane supported by highly elongated fingers and attached to sides of body, similar membrane extending from insides of hind legs to tail . . . . CHIROPTERA, III.

3b. Forelimbs not developed as wings . . . *4*.

4a(3b). First hind toe opposed, thumb clawless; tail long, naked, round, and tapering, blackish at base but whitish for most of its length; female with marsupium; male with testes anterior to penis . . POLYPROTODONTA, I.

4b. All functional toes clawed, none opposable; tail, if naked and tapering, not marked as above; no marsupium in females; testes, if apparent, posterior to penis in males . . *5*.

5a(4b). Front feet with five claws (first toe may be rudimentary and high on foot); tail not flattened laterally or dorsoventrally, and not a cottony tuft . . . . *6*.

5b. Front feet with four claws, or, if five, the tail flattened or a cottony tuft shorter than the ear . . . . *7*.

6a(5a). Length of head and body less than 115 mm, or, if longer, no pinna present and no white on belly; eyes reduced; pelage without distinct guard hairs . . INSECTIVORA, II.

6b. Length of head and body more than 115 mm; pinna present; eyes not reduced; pelage with distinct guard hairs and underfur . . . . CARNIVORA, VI.

7a(5b). Tail a cottony tuft, shorter than ear; hind foot with four well-clawed toes; soles of feet densely and fully furred . . . . .LAGOMORPHA, IV.

7b. Tail longer than ear; hind foot with five well-clawed toes; soles of feet not densely and fully furred . . . . RODENTIA, V.

Key to Genera and Species by Orders

I. Order POLYPROTODONTA . . . . Didelphidae, *Didelphis virginiana*, p. 17.

II. Order INSECTIVORA

1a. Pinna absent; forefeet wider than 10 mm and much larger than hind feet; head plus body length greater than 100 mm; shoulder girdle and forelimbs placed forward, neck not apparent . . . . Talpidae, *2*.

1b. Pinna present but reduced; forefeet narrower than 6 mm and not larger than hind feet; head plus body length less than 100 mm; shoulder girdle and forelimbs well behind head, not obscuring neck region . . . . . Soricidae, *3*.

2a(1a). Tip of snout bearing 22 fleshy papillae; pelage black, without sheen; tail over 45 mm long, constricted at base, spindle shaped; eyelids not fused . . *Condylura cristata*, p. 30.

2b. Tip of snout without papillae; pelage grayish, with a silvery sheen; tail shorter than 45 mm, cylindrical, not swollen; thin fused eyelids covering eyes . . *Scalopus aquaticus*, p. 28.

3a(1b). Pinna present but merely a rim around ear opening; tail shorter than 30 mm and less than one-half of body length . . . . . . *4*.

3b. Pinna distinct; tail usually longer than 30 mm and more than one-half of body length . *5*.

4a(3a). Total length more than 90 mm; tail longer than 17 mm; pelage normally slate black or dark gray . . . . *Blarina brevicauda*, p. 25.

4b. Total length less than 90 mm; tail shorter than 17 mm; pelage brown . . . . *Cryptotis parva*, p. 27.

5a(3b). Total length more than 135 mm; hind feet fringed with medial and lateral rows of short, stiff hairs . . . *Sorex palustris*, p. 22.

5b. Total length less than 135 mm; hind feet shorter than 17 mm and not fringed with stiff hairs . . . . *6*.

6a(5b). Color of sides sharply distinct from color of back; total length more than 110 mm . . . . *Sorex arcticus*, p. 23.

6b. Color of sides not sharply distinct from color of back; total length less than 105 mm . *7*.

7a(6b). Tail longer than 30 mm; CAUTION: See *7b* . . . . *Sorex cinereus*, p. 20.

7b. Tail shorter than 34 mm; CAUTION: positive identification of this species as opposed to *Sorex cinereus* requires examination of the skull . . . . *Microsorex hoyi*, p. 24.

III. Order CHIROPTERA, Family Vespertilionidae

1a. Pelage frosted (tips of many hairs white); half or more of dorsal surface of interfemoral membrane well furred . . . . *2*.

1b. Pelage not frosted; interfemoral membrane naked, or less than basal quarter well furred . . . . *4*.

2a(1a). Pelage blackish brown and frosted; basal half of dorsal surface of interfemoral membrane well furred . . . . *Lasionycteris noctivagans*, p. 36.

2b. Pelage yellow-gray, yellowish, or reddish, and frosted; dorsal surface of interfemoral membrane fully furred . . . . *3*.

3a(2b). Pelage reddish or yellow-red, and frosted; yellow-white patch on each shoulder; forearm shorter than 44 mm; ear not rimmed with black . . . . *Lasiurus borealis*, p. 39.

3b. Pelage yellow-gray and frosted, throat yellow; forearm longer than 44 mm; rim of ear black . . . . *Lasiurus cinereus*, p. 40.

4a(1b). Total length more than 105 mm; forearm at least 44 mm long; ears, wings, and interfemoral membrane naked and blackish . . . . *Eptesicus fuscus*, p. 38.

4b. Total length less than 100 mm; forearm no longer than 43 mm . . . . *5*.

5a(4b). Ear rounded, tragus less than 7 mm long; fur of back yellowish; wingspread no greater than 240 mm; forearm no longer than 36 mm . . . . . . *Pipistrellus subflavus*, p. 37.

5b. Ear more or less pointed, tragus at least 7 mm long; back fur brownish; wingspread at least 250 mm; forearm at least 34 mm long . . . . *6*.

6a(5b). Ear, when folded forward in a fresh

specimen, not extending beyond snout; tragus 7–8 mm long, less than one-half of length of ear from notch, tip of tragus somewhat rounded; pelage with a coppery sheen . . . . *Myotis lucifugus*, p. 33.

6b. Ear, when folded forward, extending at least 4 mm beyond snout; tragus usually 10–12 mm long, more than one-half of length of ear from notch, and pointed; pelage with less sheen and brassy rather than coppery . . . . *Myotis keenii*, p. 35.

IV. Order LAGOMORPHA, Family Leporidae

1a. Pelage not white in winter, nape of neck rusty in all seasons; hind foot shorter than 115 mm . . . . . . *Sylvilagus floridanus*, p. 44.

1b. Pelage white in winter, nape never rusty; hind foot longer than 115 mm . . . . . . *2*.

2a(1b). Total length less than 550 mm; ear from notch shorter than 75 mm; tail shorter than 55 mm; hind foot shorter than 150 mm; black fur on tip of ear not a distinct patch on outer half of tip of ear . . . . *Lepus americanus*, p. 46.

2b. Total length over 550 mm; ear from notch longer than 85 mm; tail longer than 60 mm; hind foot longer than 150 mm; a distinct patch of black fur on outer half of back of ear tip . . . . *Lepus townsendii*, p. 48.

V. Order RODENTIA

1a. Back, sides, and tail covered with spines Erethizontidae, *Erethizon dorsatum*, p. 109.

1b. Pelage not spiny . . . . *2*.

2a(1b). Tail well haired with longest hairs at least 10 mm long, *or* tail scaly and flattened horizontally . . . . *3*.

2b. Tail with short fur or naked but not flattened horizontally . . . . *14*.

3a(2a). Tail scaly, flattened horizontally; hind feet webbed . . . . Castoridae, *Castor canadensis*, p. 79.

3b. Tail well haired; feet not webbed . . . . Sciuridae, *4*.

4a(3b). Head plus body over 350 mm long; feet black; tail bushy and less than one-fourth of total length . . . . *Marmota monax*, p. 56.

4b. Head plus body shorter than 350 mm; feet seldom black; tail more than one-fourth of total length . . . . *5*.

5a(4b). Five or more longitudinal stripes on back and sides . . . . *6*.

5b. Back unstriped; sides unstriped or with one black stripe . . . . *8*.

6a(5a). Back with alternating narrow, yellowish brown stripes and broader dark brown stripes, each dark stripe enclosing a row of light dots, stripes not continuing onto face; longest front claw longer than 5 mm . . . . *Spermophilus tridecemlineatus*, p. 60.

6b. Black and white stripes on back, stripes continuing onto side of face, no stripes punctuated by a line of dots; longest front claw shorter than 5 mm . . . . *7*.

7a(6b). Rump reddish, black and white stripes not continuing to base of tail . . . . *Tamias striatus*, p. 52.

7b. Rump similar to back, stripes continuing to base of tail . . . . *Eutamias minimus*, p. 54.

8a(5b). Fur very soft; tail fur flattened into a distinct horizontal plane; a loose fold of furred skin extending from each front leg to corresponding hind leg . . . . *9*.

8b. Fur not unusually soft; tail fur not flattened into a distinct horizontal plane; no loose fold of skin on sides. . . . . *10*.

9a(8a). Hairs on belly lead colored at base . . . . . . *Glaucomys sabrinus*, p. 73.

9b. Hairs on belly white to roots . . . . *Glaucomys volans*, p. 71.

10a(8b). Tail distinctly bushy, more than 40% of total length; ear from notch longer than 18 mm . . . . *11*.

10b. Tail not bushy to moderately bushy, less than 40% of total length; ear from notch shorter than 17 mm . . . . *13*.

11a(10a). Total length less than 400 mm; a distinct black stripe along the side in summer, ears distinctly tufted in winter . . . . *Tamiasciurus hudsonicus*, p. 69.

11b. Total length more than 400 mm; no black stripe along the side at any season . . . . *12*.

12a(11b). Back and sides gray to brownish gray, many body and tail hairs tipped with white; belly fur white, grayish white, or brownish (especially in northern Minnesota); occasional individuals black or blackish; pelage never distinctly orange-brown . . . . *Sciurus carolinensis*, p. 63.

12b. Back and sides grayish washed with orange, underparts and feet distinctly orangish, tail hairs tipped with orange . . . . *Sciurus niger*, p. 67.

13a(10b). Tail moderately bushy, at least 100 mm long; back and sides a mixture of brownish gray and blackish, obscurely spotted, the spotting often tending to occur in lines across the back . . *Spermophilus franklinii*, p. 62.

13b. Tail not bushy, shorter than 95 mm; back and sides a uniform grayish tan . . . . *Spermophilus richardsonii*, p. 58.

14a(2b). Eyes and pinnae much reduced; forefeet larger than hind feet, and with longer claws; external (fur-lined) cheek pouches present *and* tail under 35% of total length . . . . Geomyidae, *15*.

14b. Eyes and pinnae moderate to large; forefeet not larger than hind feet and not bearing unusually long claws; external (fur-lined) cheek pouches absent *or* if present, tail over 44% of total length . . . . *16*.

15a(14a). Pelage on back gray-brown with little sheen; white on throat and sometimes on chest; forefeet only moderately enlarged; middle foreclaw about 3 mm deep at base and chord shorter than 9 mm from underside of toe to tip of unworn claw (examination of skull advised) . . . *Thomomys talpoides*, p. 75.

15b. Pelage on back a rich brown and glossy; white fur absent from underside or confined to chin; forefeet much enlarged; middle foreclaw at least 4 mm deep at base and chord usually longer than 9 mm from underside of toe to tip of unworn claw (examination of skull advised) . . . . . *Geomys bursarius*, p. 76.

16a(14b). External cheek pouches present . . . . *Heteromyidae, *Perognathus flavescens*, p. 78.

16b. External cheek pouches absent . . . . *17*.

17a(16b). Tail at least 57% of total length; belly hairs white to bases; hind foot about one-third of length of head and body . . . . Zapodidae, *18*.

17b. Tail no more than 53% of total length; belly hairs not white to bases; hind foot one-fourth or less of length of head and body . . . . *19*.

18a(17a). Tail rarely white-tipped; upper parts olivaceous . . . . *Zapus hudsonius*, p. 106.

18b. Terminal 18 to 22 mm of tail white-tipped; upper parts, especially sides, with distinct orange tinge . . . . *Napaeozapus insignis*, p. 107.

19a(17b). Total length over 300 mm; body length over 190 mm; long tail nearly naked and either cylindrical and tapering or laterally compressed; belly not white . . . . *20*.

19b. Total length under 200 mm; body length under 140; tail naked or furred but not laterally compressed; belly fur always pigmented at base, tips white or pigmented . . . . . . *21*.

20a(19a). Total length over 440 mm; dense brown underfur overlain by glossy brown guard hairs; long blackish tail sparsely haired, laterally compressed; hind toes fringed with stiff hairs . . . . *Ondatra zibethicus*, p. 97.

20b. Total length under 440 mm; fur not luxuriant; pelage brown or grayish brown mixed with black hairs or, sometimes, completely black or blackish; sparsely haired tail tapering; hind toes not fringed with stiff hairs . . . . *Rattus norvegicus*, p. 102.

21a(19b). Underparts distinctly white or light gray; body length more than 80 mm; weight more than 17 g . . . . *22*.

21b. Underparts not distinctly white or light gray; *or*, if belly fur tipped with white or whitish gray, then body length less than 90 mm; weight less than 16g, pelage on back a mixture of black and light ochraceous buff hairs . . . . *25*.

22a(21a). Tail 30% or less of total length; weight usually over 35 g; back and sides gray to brownish gray; belly hairs tipped with light gray . . . *Onychomys leucogaster*, p. 88.

22b. Tail 35% or more of total length; weight usually under 35 g; back and sides distinctly brownish (gray in subadults and juveniles); belly hairs white tipped . . *Peromyscus*, *23*.

NB: The three forms of *Peromyscus* in the Upper Midwest are sometimes difficult to distinguish. Subadults are especially difficult. The following couplets should separate most adults.

23a(22b). Tail sharply bicolor, a distinct edge between dark dorsal hair and white ventral hair . . . . *24*.

23b. Tail dark above and light below, but dark upper pelage grading into lighter lower pelage; total length 180 mm or less; tail 42 to 48% of total length; ear from notch 15–17 mm . . . . . . *Peromyscus leucopus*, p. 86.

24a(23a). Total length less than 165 mm; tail less than 44% of total length; ear from notch 14 to 16 mm . . . . *Peromyscus maniculatus bairdii*, p. 84.

24b. Total length more than 170 mm; tail 46 to 52% of total length; ear from notch 17 mm or more . . . . *Peromyscus maniculatus gracilis* (near Canadian border possibly intergrading with *P. m. maniculatus*, with somewhat shorter ears and tail), p. 85.

25a(21b). Tail 40% or more of total length; ears and eyes prominent . . . . *26*.

25b. Tail 30% or less of total length; ears partly or largely hidden in fur; eyes more or less beady . . . . *27*.

26a(25a) Pelage silky, individual hairs about 7 mm long; pelage on back a mixture of black and light ochraceous buff hairs; tips of belly hairs generally whitish with a distinctive edge between grayish belly and buffy sides; tail well furred, scales or annulations not apparent; hind foot usually less than 18 mm long; weight usually under 15 g . . . . *Reithrodontomys megalotis*, p. 83.

26b. Pelage coarse, individual hairs about 5 mm long; upper parts brownish, or yellowish brown mixed with black; no sharp break between upper parts and somewhat lighter buffy gray underparts; tail sparsely haired, scales or annulations apparent; hind foot 17 mm or longer; weight usually more than 14 g . . . . *Mus musculus*, p. 104.

27a(25b). Tail more than 22 mm long . . . *28*.

27b. Tail less than 22 mm long . . . . . . *32*.

28a(27a). Back reddish (lighter in winter than in summer), distinctly contrasting with grayish sides; tail usually 30 to 45 mm; total length less than 155 mm . . . . *Clethrionomys gapperi*, p. 90.

28b. Back and sides similar, dark brown or grizzled . . . . *29*.

29a(28b). Yellowish on top of nose . . . . *30*.

29b. Not yellowish on top of nose . . . . . *31*.

30a(29a). Tail shorter than 38 mm . . . . *Phenacomys intermedius* (identification should be confirmed by a professional mammalogist) p. 91.

30b. Tail length 38 mm or more . . . . *Microtus chrotorrhinus*, p. 94.

31a(29b). Belly washed with yellowish (examination of skull advised) . . . . *Microtus ochrogaster*, p. 95.

31b. Belly gray, no yellowish cast (examination of skull advised) . . . . *Microtus pennsylvanicus*, p. 92.

32a(27b). Velvety, molelike fur; back and sides auburn or chestnut, belly buffy; ear from notch 11 mm or less . . . . *Microtus pinetorum*, p. 96.

32b. Fur not velvety; back grizzled brown, belly frosted gray . . . . *Synaptomys*, *33*.

33a(32b). Hairs at base of ears not brighter than rest of fur; six teats in females . . . . *Synaptomys cooperi*, p. 99.

33b. Hairs at base of ears brighter than rest of fur; eight teats in females . . . . *Synaptomys borealis*, p. 101.

VI. Order CARNIVORA

1a. Five clawed toes on forefeet and hind feet . . . . *2*.

1b. Four clawed toes on hind feet; first toe on

forefoot rudimentary and high on foot, other four functional . . . . *18*.

2a(1a). Pelage black (occasionally brown or cinnamon) throughout except for a brown muzzle and often a white spot on chest; tail rudimentary, no longer than rump fur; adult weight 90 kg or more . . . . Ursidae, *Ursus americanus*, p. 124.

2b. Pelage not uniformly black with brown muzzle and occasional white chest spot; tail apparent; weight less than 35 kg . . . . . *3*.

3a(2b). Tail with alternating buff and black rings and with black tip; black "mask" across eyes . . . Procyonidae, *Procyon lotor*, p. 127.

3b. Tail without alternating rings; no black mask across eyes . . . . Mustelidae, *4*.

4a(3b). Pelage dark brown, except for irregular creamy markings on throat and dorsolateral bands of pale buff which meet on upper base of tail; weight 6–15 kg . . . . *Gulo gulo*, p. 139.

4b. Pelage and weight various, but not in the above combination . . . . *5*.

5a(4b). Foreclaws partly concealed by fur; no median white stripe or spots on forehead (but winter pelage may be almost entirely white) . . . . *6*.

5b. Foreclaws not partly concealed by fur; a median white mark on forehead; skin never normally almost wholly white . . . . *16*.

6a(5a). Entire underparts white or yellowish white; *or* pelage entirely white, sometimes with black tail tip. . . . . *Mustela*, except for *M. vison*, *7*.

6b. Entire underparts not white or yellowish white; pelage never normally all white, with or without black tail tip . . . . *13*.

7a(6a). Tail without distinct black tip; tail shorter than 40 mm . . . *Mustela nivalis*, p. 134.

7b. Tail with distinct black tip at all seasons; tail longer than 50 mm . . . . *8*.

8a(7b). Sex of specimen unknown . . . . . *9*.

8b. Sex known . . . . *10*.

9a(8a). Tail 30% or less of total length . . . . *Mustela erminea*, p. 133.

9b. Tail 32% or more of total length . . . . *Mustela frenata*, p. 136.

10a(8b). Adult male . . . . *11*.

10b. Adult female . . . . *12*.

11a(10a). Total length less than 340 mm, tail shorter than 106 mm . . . . *Mustela erminea*, p. 133.

11b. Total length more than 360 mm, tail longer than 120 mm . . . *Mustela frenata*, p. 136.

12a(10b). Total length less than 295 mm, tail shorter than 85 mm . . . . *Mustela erminea*, p. 133.

12b. Total length more than 295 mm, tail longer than 95 mm . . . *Mustela frenata*, p. 136.

13a(6b). Ear brown, a narrow light buff border on the edge . . . . *14*.

13b. Ear uniform brown without a buff border . . . . *15*.

14a(13a). Dorsal fur uniform dark to light brown; ears light brown, belly pale brown, a light buff or orange patch or spots on throat and breast; total length less than 675 mm; tail shorter than 225 mm . . . . *Martes americana*, p. 129.

14b. Dorsal fur most often dark brown, sometimes lighter; head and shoulders grizzled; underparts blackish brown; occasionally small white spots on chest and belly; total length more than 675 mm; tail longer than 225 mm. . . . . *Martes pennanti*, p. 131.

15a(13b). Toes webbed; total length over 800 mm; tail longer than 250 mm and strongly tapering from thick base . . . .*Lutra canadensis*, p. 147.

15b. Toes not webbed; total length under 700 mm; tail shorter than 250 mm, not strongly tapering from thick base . . . . *Mustela vison*, p. 137.

16a(5b). Back and sides yellowish gray, feet black, belly whitish, cheeks white with black crescent behind each eye; longest front claw longer than 25 mm . . . . *Taxidea taxus*, p. 141.

16b. Pelage black, generally with white spots or stripes; longest front claw shorter than 25 mm . . . . *17.*

17a(16b). White spot on forehead; four or six broken longitudinal white stripes or rows of spots on sides, turning downward on haunches (pattern sometimes reduced to faint whitish marks on an otherwise black pelt) . . . . *Spilogale putorius*, p. 143.

17b. Median white stripe on forehead and base of nose; top of head white, from which two white stripes diverge over back; white stripes varying in width between specimens but not broken into rows of spots . . . . *Mephitis mephitis*, p. 145.

18a(1b). Claws sharp and retractile; tail less than 200 mm or more than 500 mm long . . . . Felidae, *23.*

18b. Claws nonretractile, usually blunt; tail normally between 200 and 500 mm long . . . . Canidae, *19.*

19a.(18b). Tail more than 33% of total length; hind foot shorter than 185 mm . . . . . *20.*

19b. Tail less than 33% of total length; hind foot longer than 185 mm . . . . *22.*

20a(19a). Tail white-tipped; legs and feet blackish . . . . . . *Vulpes vulpes*, p. 120.

20b. Tail black-tipped; legs and feet rusty or not markedly different from rest of pelage . . . . *21.*

21a(20b). Weight more than 3 kg; legs and feet rusty; back gray, a black middorsal line extending onto tail and expanding to form black tail tip . . . . *Urocyon cinereoargenteus*, p. 122.

21b. Weight less than 3 kg; upper parts and outer surface of legs buffy; belly creamy white. (Not known to occur in Minnesota in 1980. Please see note on rare species in the preface to these keys) . . . . **Vulpes velox*, p. 122.

22a(19b). Total length under 1370 mm; nose pad less than 25 mm wide; hind foot no longer than 220 mm; weight (with stomach empty) no more than 20 kg . .*Canis latrans*, p. 114.

22b. Total length more than 1400 mm; nose pad wider than 30 mm; hind foot at least 225 mm long; weight at least 25 kg . . . . *Canis lupus*, p. 116.

23a(18a). Adults unspotted, upper parts a uniform tan or buff; total length 1.8 m or more; tail more than 30% of total length . . . . **Felis concolor*, p. 149.

23b. Adults more or less spotted or at least grizzled; total length less than 1.25 m; tail less than 20% of total length . . . . *24.*

24a(23b). Tip of tail black above and below, as though dipped in ink; tail less than 14% of total length; hind foot 180 mm or longer; ear tufts longer than 35 mm . . . . *Lynx canadensis*, p. 151.

24b. Tip of tail black only on top; tail typically more than 14% of total length; hind foot usually shorter than 200 mm; ear tufts absent or shorter than 35 mm . . . *Lynx rufus*, p. 153.

VII. Order PERISSODACTYLA . . . . Equidae: horses, asses, and mules, p. 171.

VIII. Order ARTIODACTYLA (wild species only)

1a. Feet with four digits (digits 3 and 4 bearing large hooves, digits 2 and 5 bearing smaller side hooves) . . . . *2.*

1b. Feet with only two digits; back and sides buff; underparts and rump white; white patches on throat; black markings on face and neck . . . . *Antilocapridae, *Antilocapra americana*, p. 167.

2a(1a). A deep cloak or mane of dense woolly hair over head and shoulders, contrasting with short, smooth hair over rest of body; shoulders strongly humped; permanent horns in both sexes . . . . Bovidae, *Bison bison*, p. 168.

2b. No deep woolly cloak or mane over head and shoulders; shoulders not strongly humped; deciduous bony antlers in one or both sexes in season . . . . Cervidae, *3.*

3a(2b). Snout pendulous; dewlap always present; pelage dark brown above and below, lower legs lighter; about the size of a large horse . . . .*Alces alces*, p. 162.

3b. Snout not pendulous; no dewlap; pelage not uniformly dark brown; size distinctly smaller than a horse *or*, if nearly the size of a horse, a large creamy-buff rump patch . . . . . *4*.

4a(3b). Most of pelage dark, hair on neck and throat cream-colored and relatively long; nose haired all the way around nostril; antlers on both sexes in season. (Not known to occur in Minnesota in 1980. Please see note on rare species in the preface to these keys) . . . . . **Rangifer tarandus*, p. 165.

4b. Neck and throat not creamy-white in contrast to back and sides; nose not haired all the way around nostril, nose pad on inside lower edge of nostril bare; antlers present on males in season, rarely occurring on females . . . *5*.

5a(4b). A large buff or creamy rump patch, tail the same color and not over 185 mm long; back and sides pale, head and neck dark; total length of females more than 1.75 m, of males more than 2.2 m . . . . *Cervus elaphus*, p. 156.

5b. Rump patch small or absent, tail either brown above and white below or white with black tip; back and sides not paler than head and neck; total length of females 2.0 m or less, of males 2.3 m or less . . . . *6*.

6a(5b). Tail brown above and white below, with white tip; antlers normally comprising a main beam and vertical, secondary tines . . . .*Odocoileus virginianus*, p. 159.

6b. Tail white and black above, with black tip, light rump patch clearly visible on either side of tail; antlers normally dichotomously branching . . . . *Odocoileus hemionus*, p. 157.

Key to Skulls

An asterisk before a species denotes an unusual find.

Key to the Orders

1a. Thirteen teeth in each side of upper jaw, 12 in each side of lower jaw, tooth formula 5/4, 1/1, 3/3, 4/4 × 2 = 50; angle of lower jaw inflected (fig. 81); jugal bone forming part of mandibular fossa; nasals broadly expanded posteriorly; braincase smaller than rostrum . . . . .POLYPROTODONTA, I.

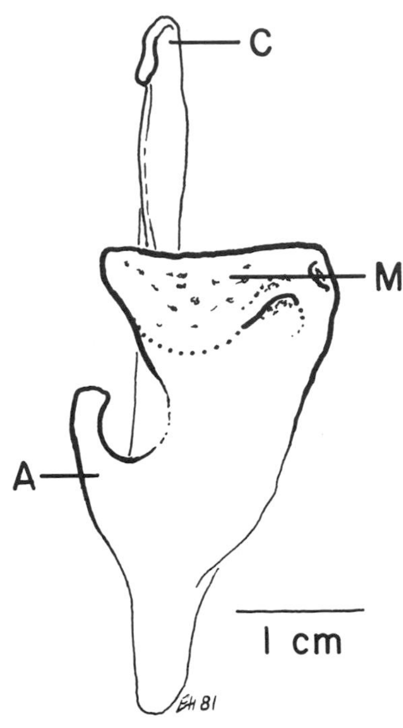

Figure 81. Rear of right mandible of *Didelphis virginiana*. A = angle or angular process. C = coronoid process. M = mandibular condyle.

1b. Never more than 11 teeth in each side of upper and lower jaws, maximum tooth formula 3/3, 1/1, 4/4, 3/3 × 2 = 44; angle of lower jaw rarely inflected; other characters variable . . . . *2*.

2a(1b). A bony ring or bar behind orbit, or orbit completely enclosed by bone to the rear; *or*, if no bony postorbital ring or bar, canines triangular in cross section and pointing outward . . . . *3*.

2b. No complete bony postorbital bar connecting frontal bone above with zygoma below (in some specimens dried cartilage may resemble a bony postorbital bar) . . . . *5*.

3a(2a). Tooth formula 2/2, 1/1, 2/2, 2-3/2-3 ×

2 = 28-32; no large diastema anywhere in normal toothrow; upper and lower toothrows horseshoe-shaped; face nearly vertical; orbits facing forward; foramen magnum facing directly downward; braincase about as large as rest of skull . . . . PRIMATES, *Homo Sapiens*. (Anyone finding human skeletal remains should note the locality carefully and contact a professional anthropologist at a college or university or at a public science or natural history museum.)

3b. Tooth formula not as above; face not vertical (substantial rostrum anterior to orbits); foramen magnum at or near rear of skull; braincase in adults much smaller than rest of skull . . *4*.

4a(3b). Upper incisors present, canines present or not; occlusal surface of cheek teeth a complex of folded loops; nasal bones expanded posteriorly; last three premolars much like first two molars in both upper and lower jaws; dentaries completely fused at anterior suture; alisphenoid canal present (fig. 80) . . . . PERISSODACTYLA, VII.

4b. Upper incisors absent, lower canines incisiform (lower jaw seems to have four incisors); a wide diastema between lower incisors and cheek teeth; occlusal surface of cheek teeth a pattern of crescents; permanent premolars distinctly simpler than molars (but deciduous or milk premolars molariform); nasal bones not expanded posteriorly; mandibular suture not fused; alisphenoid canal absent (ruminant artiodactyls);

OR

Tooth formula 3/3, 1/1, 4/4, 3/3 × 2 = 44; canines triangular in cross section and pointing outward; no postorbital bar; no large diastema in tooth row; occlusal surface of cheek composed of cusps; dentaries fused anteriorly; alisphenoid canal absent (domestic pig) . . . . ARTIODACTYLA, VIII.

5a(2b). Prominent canines extending beyond general surface of tooth row (upper canine is *fourth* tooth unless there is a marked diastema at front of upper jaw between right and left incisors, in which case upper canine is second or third tooth) . . . . *6*.

5b. Canines absent and a large diastema between incisors and cheek teeth; *or* canines not extending beyond general surface of tooth row (third incisor is caniniform in *Condylura*) . . . . *7*.

6a(5a). Premaxillary bones separated in front by a large gap, thus creating a diastema between small upper right and upper left incisors; one or two upper incisors on each side; canines with sharp posterior edges, not round in cross section; skull shorter than 25 mm . . . . CHIROPTERA, III.

6b. Premaxillary bones joined in front, no diastema between left and right upper incisors; incisors always 3/3; canines always somewhat rounded in cross section, no sharp posterior edge; skull longer than 25 mm . . . . CARNIVORA, VI.

7a(5b). At least nine teeth in each upper jaw (some may be visible only with magnification); six or more teeth in each lower jaw; no upper or lower diastema of over 2 mm between incisors and cheek teeth; cheek teeth sharply cuspidate; zygomata delicate or absent . . . . INSECTIVORA, II.

7b. Eight or fewer teeth in each upper jaw, six or fewer in each ramus; upper and lower diastemata of 4 mm or more between incisors and cheek teeth; canines absent; one or two incisors in each upper jaw (if two, the second a small peg behind the first); prominent incisors curved and with chisellike ends; cheek teeth with grinding surfaces or low rounded cusps; zygomata present and sturdy . . . . . . *8*.

8a(7b). Only one incisor in each upper jaw; upper cheek five or fewer, lower cheek teeth four or fewer; sides of rostrum not fenestrated . . . . RODENTIA, V.

8b. Two incisors in each upper jaw, the second a small peg behind the first; upper cheek teeth six, lower cheek teeth five; sides of rostrum fenestrated (fig. 15) . . LAGOMORPHA, IV.

Key to Genera and Species by Orders

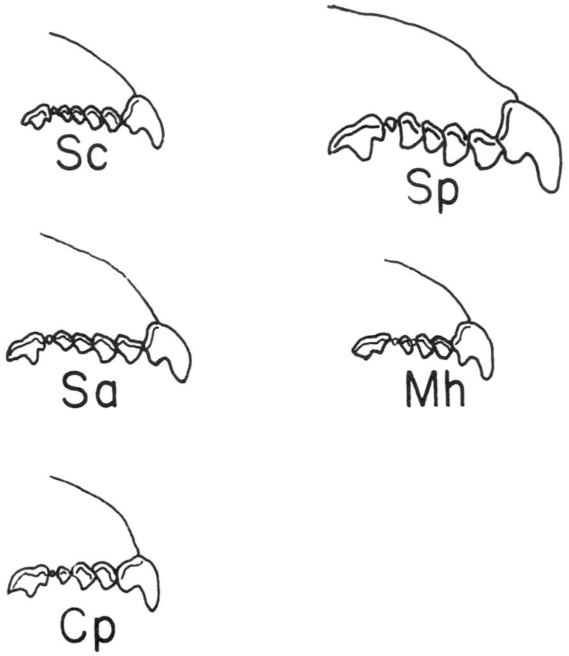

I. Order POLYPROTODONTA . . . . .Didelphidae, *Didelphis virginiana*, p. 17.

II. Order INSECTIVORA

1a. First upper incisor large, protruding, and hooked, with a second cusp behind the larger first cusp; some teeth with brown-tipped cusps; no zygomata; skull shorter than 30 mm . . . . Soricidae, *3*.

1b. First upper incisor without a second cusp and *either* larger than other incisors but neither protruding nor hooked, *or* protruding but neither larger than most other teeth nor hooked; teeth white, not brown-tipped; slender zygomata present; skull longer than 30 mm . . . . .Talpidae, *2*.

2a(1b). Forty-four teeth, 11 in each side of each jaw (some minute unicuspids may be lost in cleaning); anterior end of premaxillary, bearing first incisors, clearly visible from above; angular process of mandible a slender, horizontal rod; skull narrower than 15 mm . . . . *Condylura cristata*, p. 30.

2b. Thirty-six teeth, 10 in each upper jaw and eight in each lower jaw; anterior end of premaxillary and first incisors not visible from above, hidden by nasals; angular process sturdy, not a slender, horizontal rod; skull wider than 15 mm . . . . *Scalopus aquaticus*, p. 28.

3a(1a). Thirty teeth, nine in each upper jaw and six in each lower jaw; four unicuspids (fig. 82), the fourth a tiny peg between the third unicuspid and the first cheek tooth; skull about 16.5 mm long . . . **Cryptotis parva*, p. 27.

3b. Thirty-two teeth (some of the five upper unicuspids minute), 10 in each upper jaw and six in each lower jaw . . . . *4*.

4a(3b). Width of skull 11 mm or more, skull sturdy and angular, noticeably ridged; lower jaw longer than 13.7 mm . . . . *Blarina brevicauda*, p. 25.

4b. Skull less than 11 mm wide; skull delicate and rounded, not prominently ridged; lower jaw shorter than 13.7 mm . . . . *5*.

5a(4b). Skull 15 to 16 mm long . . . . . . *6*.

5b. Skull longer than 17 mm . . . . *7*.

6a(5a). Only three of the five upper unicuspids clearly visible from the side, the third and fifth present but reduced to tiny pegs (fig. 82) . . . . *.Microsorex hoyi*, p. 24.

6b. Four of the five upper unicuspids clearly visible from the side, the first and second larger than the third and fourth, but only the fifth reduced to a tiny peg . . . . *Sorex cinereus*, p. 20.

7a(5b). Skull longer than 21 mm; four upper unicuspids clearly visible from the side, the third smaller than the fourth (fig. 82); mandible longer than 13.3 mm . . . . *Sorex palustris*, p. 22.

7b. Skull shorter than 21 mm; four upper uni-

Figure 82. Diagrammatic outlines of the front upper dentitions of five shrews. Sc = *Sorex cinereus*; Sp = *S. palustris*; Sa = *S. arcticus*; Mh = *Microsorex hoyi*; Cp = *Cryptotis parva*.

cuspids clearly visible from the side, the third and fourth of about equal size (fig. 82); mandible shorter than 13.3 mm . . . . *Sorex arcticus*, p. 23.

III. Order CHIROPTERA

1a. Incisors 1/3; seven teeth in each upper jaw, nine in each lower jaw; first upper premolar minute . . . . *Lasiurus*, *2*.

1b. Incisors 2/3 . . . . *3*.

2a(1a). Skull shorter than 15.5 mm . . . . *Lasiurus borealis*, p. 39.

2b. Skull longer than 15.5 mm . . . . *Lasiurus cinereus*, p. 40.

3a(1b). Thirty-six teeth, eight in each upper jaw and ten in each lower jaw; cheek teeth 5/6 . . . . *Lasionycteris noctivagans*, p. 36.

3b. Total number of teeth 32, 34, or 38; cheek teeth 4/5, 5/5, or 6/6 . . . . *4*.

4a(3b). Thirty-eight teeth; cheek teeth 6/6 . . . . *5*.

4b. Thirty-two or 34 teeth; cheek teeth 4/5, or 5/5 . . . . *6*.

5a(4a). Least width of interorbital constriction at least 4 mm . . . . *Myotis lucifugus*, p. 33.

5b. Least width of interorbital constriction under 4 mm . . . . *Myotis keenii*, p. 35.

6a(4b). Thirty-two teeth; cheek teeth 4/5; skull about 19 mm long . . . . *Eptesicus fuscus*, p. 38.

6b. Thirty-four teeth; cheek teeth 5/5; skull about 12.5 mm long . . . . *Pipistrellus subflavus*, p. 37.

IV. Order LAGOMORPHA, Family Leporidae. (In this group, it is particularly important that the orbits be free of dried connective tissue which can be mistaken for bone.)

1a. Interparietal distinct in adult, usually surrounded by a clear suture line, rarely by a raised bony ridge; postorbital process either fused to braincase or running parallel to it, not diverging from it; zygoma not substantially deeper at center than at either end, either with parallel edges or deep front end . . . . . *2*.

1b. Interparietal not distinct in adult; postorbital process diverging from cranium; zygoma more than 15% deeper at center than at anterior constriction . . . . *Lepus*, *3*.

2a(1a). Interparietal distinctly triangular and marked by a distinct suture, not by a bony ridge; postorbital process fused to, or closely approaching, cranium; zygoma about the same depth for most of its length; skull less than 85 mm long . . . *Sylvilagus floridanus*, p. 44.

2b. Interparietal approximating a transverse ellipse, sometimes ringed by a bony ridge; zygoma 50% or more deeper near anterior end than at narrowest point (at center or farther back); adult skull at least 90 mm long, except in dwarf breeds; (NB: this is the domestic rabbit, and most skulls encountered may be of eight-week old fryers, 70–75 mm long) . . . . *Oryctolagus cuniculus*, p. 170.

3a(1b). Supraorbital processes broad and triangular, anterior ends more than 2 mm higher than adjacent region of frontal bones; skull longer than 90 mm . . . . *Lepus townsendii*, p. 48.

3b. Supraorbital processes narrow, not appreciably higher than adjacent region of frontal bones; skull shorter than 90 mm . . . . *Lepus americanus*, p. 46.

V. Order RODENTIA

1a. Infraorbital foramen smaller than foramen magnum; lower jaw V-shaped in ventral view, the angular process originating medially to a line drawn along outside border of root of lower incisor (fig. 83); skull length variable . . . . *2*.

1b. Infraorbital foramen larger than foramen magnum; lower jaw Y-shaped, the angular process originating laterally to a line drawn along outside border of root of lower incisor (fig. 83); adult skull length at least 80 mm . . . .
Erethizontidae, *Erethizon dorsatum*, p. 109.

2a(1a). Infraorbital foramen on side of rostrum, anterior to zygomatic plate, with opening visible when skull is viewed from the side

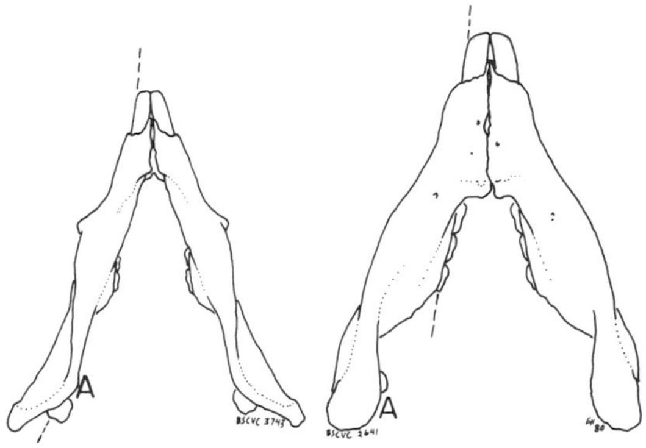

Figure 83. Ventral views of mandibles of *Marmota monax* (left) and *Erethizon dorsatum*. A = angular process.

(fig. 31); cheek teeth 4/4 . . . . *3*.

2b. Infraorbital foramen variable but, if on the side of rostrum anterior to the zygomatic plate, the opening not visible in side view; cheek teeth variable . . . . *5*.

3a(2a). Skull mouselike, zygomata delicate; skull shorter than 24 mm; auditory bullae inflated . . . . *Heteromyidae, *Perognathus flavescens*, p. 78.

3b. Skull massive, zygomata sturdy; skull longer than 30 mm; auditory bullae not inflated . . . . Geomyidae, *4*.

4a(3b). Anterior surface of upper incisor with one light, medial groove . . . . *Thomomys talpoides*, p. 75.

4b. Anterior surface of upper incisor with two deep grooves . . . *Geomys bursarius*, p. 76.

5a(2b). Postorbital processes prominent . . . . Sciuridae, *6*.

5b. Postorbital processes absent *or*, if present, rudimentary, and skull longer than 75 mm . . . . *16*.

6a(5a). Anterior surface of incisors white; skull longer than 70 mm, concave between postorbital processes; posterior edges of postorbital process perpendicular to midline of skull . . . . *Marmota monax*, p. 56.

6b. Anterior surface of incisors orange or yellow; skull shorter than 70 mm, usually moderately convex between postorbital processes; postorbital processes point posterolaterally . . *7*.

7a(6b). Infraorbital foramen a simple opening in zygomatic plate, facing obliquely forward and downward (fig. 17) . . . . *8*.

7b. Infraorbital foramen the anterior opening of an infraorbital canal, on lower part of rostrum below and in front of zygomatic plate, the opening itself clearly visible only from the front . . . . *9*.

8a(7a). Cheek teeth 4/4; skull longer than 39 but shorter than 44 mm . . . . *Tamias striatus*, p. 52.

8b. Cheek teeth 5/4, P3 a small peg less than one-third the transverse diameter of P4; skull shorter than 39 mm . . . .*Eutamias minimus*, p. 54.

9a(7b). Transverse diameter of P3 less than 45% that of P4, or P3 absent or a tiny rudiment covered by crown of P4; if P3 present, adult skull longer than 60 mm or shorter than 40 mm . . . . *12*.

9b. P3 present and its transverse diameter more than 50% that of P4; adult skull longer than 40 mm and shorter than 60 mm . . . . *Spermophilus*, *10*.

10a(9b). Greatest width of skull (across zygomata) more than 60% of skull length; supraorbital ridges higher than roof of skull between them; P3 robust, its transverse diameter at least 70% that of P4 . . . . *Spermophilus richardsonii*, p. 58.

10b. Greatest width of skull less than 60% of skull length; supraorbital ridges lower than roof of skull between them; P3 less than 60% of transverse diameter of P4 . . . . . . *11*.

11a(10b). Skull shorter than 45 mm; least interorbital diameter under 9 mm . . . . *Spermophilus tridecemlineatus*, p. 60.

11b. Skull longer than 50 mm; least interorbital diameter over 10 mm . . . . *Spermophilus franklinii*, p. 62.

12a(9a). Skull longer than 43 mm . . . . . *13*.

12b. Skull shorter than 40 mm . . . . . . *15*.

13a(12a). Skull shorter than 50 mm; as seen from above, zygomata parallel for most of their length, then zygomatic plate curving abruptly inward to short rostrum; nasals less than 30% of skull length; cheek teeth normally 4/4, length of cheek tooth row no greater than 8 mm; P3, if present, tiny and covered by crown of P4 . . . *Tamiasciurus hudsonicus*, p. 69.

13b. Skull longer than 55 mm; as seen from above, zygomata converging slightly anteriorly, and zygomatic plate curving gradually into rostrum; nasals more than 30% of skull length; length of cheek tooth row 11 mm or more; P3 present or absent . . *Sciurus*, *14*.

14a(13b). Cheek teeth usually 5/4, P3 a simple unicuspid less than 40% of transverse diameter of P4 (P3 absent in occasional specimens); infraorbital foramen a vertical slit 3 mm high or less; skull usually shorter than 68 mm, greatest width across zygomata less than 55% of skull length; width of nasals at anterior end of nasal-premaxillary suture 9.5 mm or less; antero-posterior diameter of I1 less than 3.7 mm, of i1 less than 3.8 mm; length of lower cheek tooth row less than 11.5 mm; cooked bones white. . .*Sciurus carolinensis*, p. 63.

14b. Cheek teeth 4/4; infraorbital foramen a vertical slit 3.5 mm high or more; skull usually longer than 60 mm; greatest width across zygomata more than 55% of skull length; rostrum relatively stout, width of nasals at anterior end of nasal-premaxillary suture 10.4 mm or more; antero-posterior diameter of I1 more than 3.7 mm, of i1 more than 3.8 mm; length of lower cheek tooth row more than 11.5; bones turn pink when cooked . . . . *Sciurus niger*, p. 67.

15a(12b). Skull usually shorter than 36 mm, narrower than 22 mm; upper cheek tooth row shorter than 7 mm; free points of postorbital processes shorter than 2.5 mm . . . .*Glaucomys volans*, p. 71.

15b. Skull usually longer than 36 mm, wider than 21.6 mm; upper cheek tooth row longer than 7 mm; free points of postorbital processes longer than 2.5 mm . . . .*Glaucomys sabrinus*, p. 73.

16a(5b). Skull longer than 110 mm . . . . Castoridae, *Castor canadensis*, p. 79.

16b. Skull shorter than 110 mm . . . . . . *17*.

17a(16b). Zygomatic plate horizontal, narrower than infraorbital foramen, and beneath it (fig. 47); front of upper incisors grooved . . . . Zapodidae, *18*.

17b. Zygomatic plate not horizontal, and broader than infraorbital foramen; upper incisors variable . . . . *19*.

18a(17a). Tooth formula 1/1, 0/0, 1/0, 3/3 × 2 = 18 (our only species with 18 teeth) . . . . *Zapus hudsonius*, p. 106.

18b. No upper premolar, 16 teeth . . . . *Napaeozapus insignis*, p. 107.

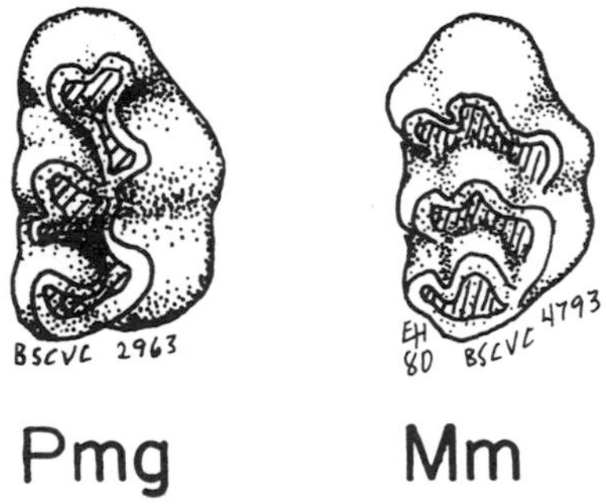

Figure 84. Pmg = left M1 of *Peromyscus maniculatus gracilis* (Cricetidae, Cricetinae); Mm = left M1 of *Mus musculus* (Muridae).

19a(17b). Upper molars with three rows of cusps—one lingual, one medial, and one labial (fig. 84) . . . . Muridae, *20*.

19b. Upper molars with two rows of cusps (fig. 84), or prismatic . . . . . . Cricetidae, *21*.

20a(19a). Skull longer than 25 mm . . . . *Rattus norvegicus*, p. 102.

20b. Skull shorter than 25 mm . . . . *Mus musculus*, p. 104.

21a(19b). Molars prismatic, folded loops of enamel enclosing "lakes" of dentine and forming a flat grinding surface; upper molar row longer than incisive (anterior palatine) foramen . . . . subfamily Microtinae, *26*.

21b. Molars cuspidate, two rows of rounded cusps present except in very old individuals; upper molar row as long as or shorter than incisive foramen . subfamily Cricetinae, *22*.

22a(21b). Deep longitudinal groove down middle of front of each upper incisor . . . . **Reithrodontomys megalotis*, p. 83.

22b. Upper incisors not grooved . . . . . . *23*.

23a(22b). Adult skull longer than 26.5 mm; zygomatic breadth more than 14.5 mm; upper and lower molar rows each longer than 4mm; M3 and m3 much reduced, occlusal surface less than one-third of area of M2 and m2 respectively; coronoid process of mandible projecting more than 2 mm back from front end of notch beneath it . . . . **Onychomys leucogaster*, p. 88.

23b. Skull shorter than 26.5 mm; zygomatic breadth less than 14 mm; upper and lower molar row each shorter than 4 mm; M3 and m3 not greatly reduced, their occlusal surfaces at least one-half of area of M2 and m2 respectively; coronoid process a mere point no more than 1 mm high . . . . *Peromyscus*, *24*.

NB: The three forms of *Peromyscus* in the Upper Midwest are sometimes difficult to distinguish, and authorities differ on the distinguishing characters themselves (compare Burt 1957:211–212 with Banfield 1974:165 and 172 concerning the appearance of the infraorbital foramen). The following two couplets should separate most adults.

24a(23b). Inner and outer edges of incisive foramina roughly parallel for most of length, outer edges gently sloping from palate into foramina except at anterior and posterior ends, lateral maxillary prominences weak (fig. 85) . . . . *25*.

24b. Outer edges of incisive foramina converging inward anteriorly, each outer edge forming an obtuse angle near the premaxillary/maxillary suture, each outer edge sharply turning into the foramen for most of its length, lateral maxillary prominences pronounced (fig. 85) . . . . *Peromyscus leucopus*, p. 86.

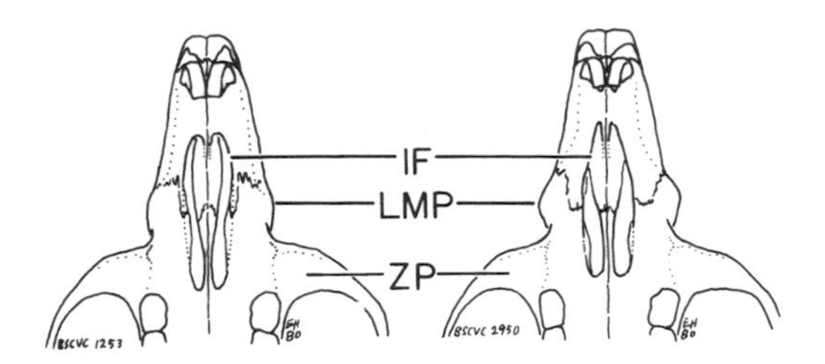

Figure 85. Front of palate of *Peromyscus maniculatus gracilis* (BSCVC 1253), left, and of *P. leucopus* (BSCVC 2950), right. IF = incisive foramen; LMP = lateral maxillary prominence; ZP = zygomatic plate.

25a(24a). Skull shorter than 25.5 mm; zygomatic breadth less than 12.8 mm . . . . *Peromyscus maniculatus bairdii*, p. 84.

25b. Skull longer than 25.5 mm; zygomatic breadth more than 12.2 mm . . . . *Peromyscus maniculatus gracilis*, p. 85.

26a(21a). Anterior faces of upper incisors grooved; roots of lower incisors originating directly beneath lower molar row . . . . tribe Lemmini, lemmings, *27*.

26b. Upper incisors not grooved; roots of lower incisors originating posteriorly and laterally to molar row in base of articular process . . . . tribe Microtini, voles, *28*.

27a.(26a). Outer (labial) edges of lower molars with distinct angles (fig. 86 *Sco*) . . . . *Synaptomys cooperi*, p. 99.

27b. Outer edge of lower molars without distinct angles (fig. 86 *Sbo*) . . . . **Synaptomys borealis*, p. 101.

28a(26b). Skull longer than 50 mm; lower molar row longer than 14 mm . . . . *Ondatra zibethicus*, p. 97.

28b. Skull shorter than 50 mm; lower molar row shorter than 14 mm . . . . *29*.

29a(28b). Posterior border of palate a simple shelf, lacking a median projection, and connecting with palatines only at sides of narial cavity (M1–M3 illustrated in fig. 86 *Cga*) . . . . *Clethrionomys gapperi*, p. 90.

29b. Posterior border of palate not a simple shelf but with a median projection, and connecting with palatines at both center and sides of narial

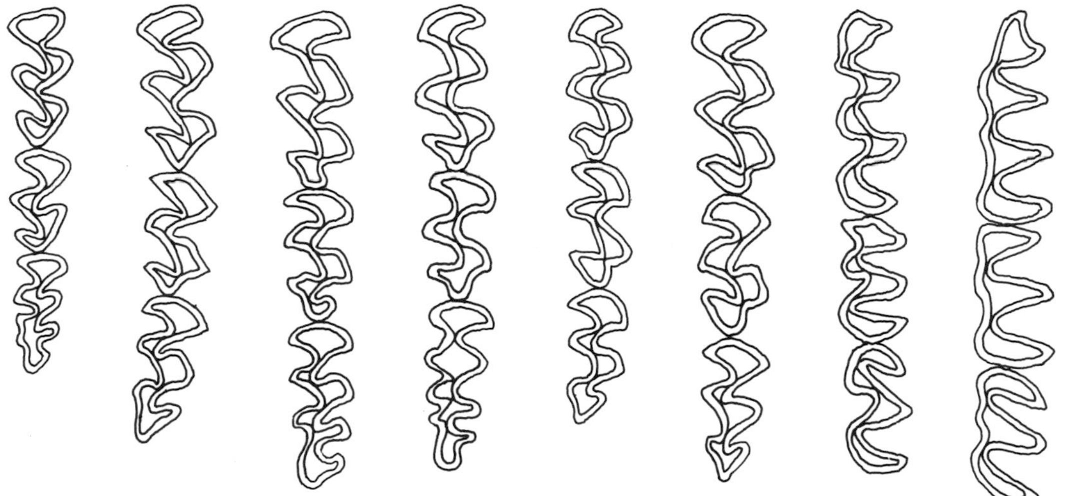

Cga Pin Mpe Mch Moc Mpi Sco Sbo

Figure 86. Diagrams of selected cheek teeth of microtines, excluding *Ondatra*. Cga to Mpi = *upper* molars (M1–M3) of *Clethrionomys gapperi, Phenacomys intermedius, Microtus pennsylvanicus, M. chrotorrhinus, M. ochrogaster,* and *M. pinetorum*, respectively. Sco and Sbo = *lower* molars (m1–m3) of *Synaptomys cooperi* and *S. borealis*.

cavity . . . . *30.*

30a(29b). Skull more than 58.5% as wide as it is long; least interorbital width usually more than 4 mm; M3 with two closed triangles (fig. 86 *Mpi*) . . . **Microtus pinetorum*, p. 96.

30b. Skull less than 58.5% as wide as it is long; least interorbital width usually less than 4 mm; M3 not as above *31*

31a(30b). Three or four reentrant angles on labial surface of M3 (fig. 86 *Mpe or Mch*); height of large, inflated auditory bulla 6 mm or more . . . . *32.*

31b. Two reentrant angles on labial surface of M3; bulla less than 6 mm high, somewhat flattened, only slightly inflated (fig. 86 *Pin* or *Moc*) . . . . *33.*

32a(31a). A fifth loop at posterior end of M2, three closed triangles on M3 (fig. 86 *Mpe*); incisive foramen longer than 5 mm . . . . *Microtus pennsylvanicus*, p. 92.

32b. No fifth loop on M2, five closed triangles on M3; incisive foramen shorter than 5 mm (fig. 86 *Mch*) . . . . **Microtus chrotorrhinus*, p. 94.

33a(31b). Upper molar row shorter than 6 mm . . . . **Microtus ochrogaster, p. 95.*

33b. Upper molar row longer than 6 mm . . . **Phenacomys intermedius*, p. 91.

VI. Order CARNIVORA

1a. Tooth formula 3/3, 1/1, 2-3/2, 1/1, × 2 = 28-30; M1 (last upper cheek tooth) reduced, no larger than I3; rostrum short and rounded . . . . Felidae, *2.*

1b. Thirty-four or more teeth, at least eight in each upper jaw and at least nine in each lower jaw; last upper molar variable, but over twice the size of I3 . . . . *5.*

2a(1a). Four cheek teeth behind canines in each upper jaw, the first (P2) much reduced; adult skull longer than 160 mm *or* shorter than 110 mm . . . . *Felis, 3.*

2b. P2 absent; adult skull between 115 and 155 mm long . . . . *Lynx, 4.*

3a(2a). Adult skull longer than 160 mm . . . . **Felis concolor*, p. 149.

3b. Adult skull shorter than 110 mm . . . . *Felis catus*, p. 171.

4a(2b). Least interorbital breadth 28 mm or more; greatest width of presphenoid 6 mm or more; anterior condyloid and posterior lacer-

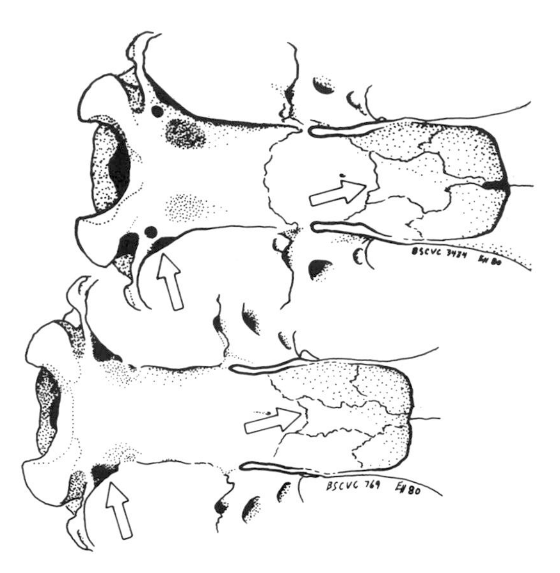

Figure 87. Underside of skulls of *Lynx canadensis*, above, and *L. rufus*, below. Note broad presphenoid and separate openings of anterior condyloid and posterior lacerate foramina in *L. canadensis*, and narrow presphenoid and confluent opening of the two foramina in *L. rufus*. Most *L. rufus* presphenoids are narrower than the one illustrated. Other differences in the figures relate to age of the animals and slight differences in angle of view.

Figure 88. Oblique ventral view of *Canis lupus* skull. The alisphenoid canal is the passage indicated by the hatched arrow below A.

ate foramina opening separately in basioccipital region (fig. 87); upper carnassial (P4) longer than 16.6 mm; lower carnassial (m1) longer than 13.5 mm . . . . *Lynx canadensis*, p. 151.

4b. Least interorbital breadth under 28 mm; presphenoid not widened posteriorly, narrower than 6 mm; anterior condyloid and posterior lacerate foramina opening into a common depression (fig. 87); P4 shorter than 16.6 mm; m1 shorter than 13.5 mm . . . . *Lynx rufus*, p. 153.

5a(1b). Larger, posterior cheek teeth with low, flattened or rounded cusps; skull longer than 200 mm . . . . Ursidae, *6*.

5b. Larger cheek teeth with relatively distinct cutting or shearing edges, *or*, if with flattened or rounded cusps, then skull shorter than 140 mm . . . . *7*.

6a(5a). Length of m1 more than 20.4 mm, width more than 10.5 mm; (will be found in Minnesota, if at all, only as old bones) . . . . **Ursus arctos*, p. 126.

6b. Length of m1 less than 20.4 mm, width less than 10.5 mm . . *Ursus americanus*, p. 124.

7a(5b). Alisphenoid canal present (fig. 88); 10 teeth in each upper jaw, tooth formula 3/3, 1/1, 4/4, 2/3 × 2 = 42; carnassials well developed . . . . Canidae, *8*.

7b. Alisphenoid canal absent; nine or fewer teeth in each upper jaw, *or*, if ten, carnassials not well developed, last two teeth in upper and lower jaws with only moderately high cusps . . . . *13*.

8a(7a). A distinct notch in posterior lower border of dentary (fig. 57); parietal ridges well separated and somewhat parallel for most of their length, forming a U-shaped pattern as seen from above (fig. 57); upper incisors not lobed . . *Urocyon cinereoargenteus*, p. 122.

8b. No distinct notch in lower border of dentary; parietal ridges, if present, joining in a V-shaped configuration and forming a single, median, sagittal crest; upper incisors lobed . . . . *9*.

9a(8b). Skull shorter than 160 mm; postorbital processes not inflated, slightly concave above; anteroposterior length of p1-m3 less than 70 mm . . . . *Vulpes*, *10*.

9b. Skull usually longer than 160 mm; postorbital processes inflated, slightly convex above; anteroposterior length of p1-m3 usually more than 70 mm . . . . *Canis, 11.*

10a(9a). Adult skull longer than 120 mm . . . . *Vulpes vulpes*, p. 120.

10b. Adult skull shorter than 120 mm; (a possible but unlikely find in southwestern Minnesota) . . . . **Vulpes velox*, p. 122.

11a(9b). Skull length variable (may be under 160 mm in small breeds); width from labial surface of left M1 to labial surface of right M1 over 31% of skull length and minimum width of rostrum over 18% of skull length; palatal width 37 to 91% as long as upper cheek tooth row (fig. 89) (except in some small-

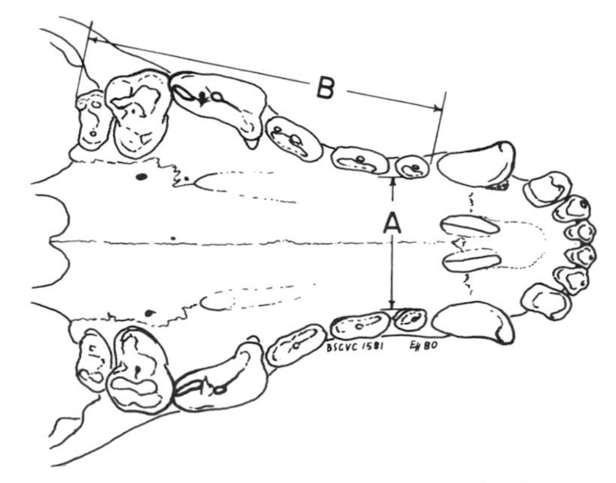

Figure 89. Palate of *Canis lupus*. A = palatal width, the distance between the *alveoli* of the left and right P1s. B = length of the upper cheek tooth row. Adapted from Howard (1949).

skulled breeds); rostrum usually somewhat concave and interorbital sinuses highly inflated, producing a pronounced "brow" (except in collies and similar breeds) . . . . *Canis familiaris*, p. 171.

11b. Width between labial surfaces of right and left M1 under 31% of skull length and minimum width of rostrum under 18% of skull length; palatal width less than 37% as long as upper cheek tooth row (fig. 89); little development of a "brow" . . . . *12.*

12a(11b). Skull shorter than 225 mm . . . . *Canis latrans*, p. 114.

12b. Skull longer than 230 mm . . . . *Canis lupus*, p. 116.

13a(7b). Forty teeth, 10 in each side of each jaw, tooth formula 3/3, 1/1, 4/4, 2/2 × 2 = 40; cheek teeth with low to medium cusps, carnassials weakly developed; auditory bullae rounded, not flattened . . . . Procyonidae, *Procyon lotor*, p. 127.

13b. Thirty-eight or fewer teeth, nine or fewer in each upper jaw, 10 or fewer in each lower jaw; carnassials variable, but always well developed in species with 10 teeth in lower jaw . . . . .Mustelidae, *14.*

14a(13b). Bony palate terminating near back of last cheek tooth, rarely as much as 2 mm farther back; last cheek tooth blocky, noticeably larger than preceding tooth; bulla extremely flattened, a small portion projecting no more than 3 mm below surrounding base of skull . . . . subfamily Mephitinae, skunks, *15.*

14b. Palate terminating well back of last cheek tooth, more or less opposite the midpoints of zygomata; last cheek tooth variable, as large as or smaller than tooth preceding it; bulla variable . . . . *16.*

15a(14a). Skull shorter than 60 mm; almost flat between orbits . . . . *Spilogale putorius*, p. 143.

15b. Skull longer than 60 mm, highly convex between orbits . . . . *Mephitis mephitis*, p. 145.

16a(14b). Five cheek teeth in each upper jaw . . . . *17.*

16b. Four cheek teeth in each upper jaw . . *20.*

17a(16a). Skull longer than 130 mm; tooth formula 3/3, 1/1, 4/4, 1/2 × 2 = 38 . . . . **Gulo gulo*, p. 139.

17b. Skull shorter than 130 mm; tooth formula as above *or* 3/3, 1/1, 4/3, 1/2, × 2 = 36 . . . . *18.*

18a(17b). Infraorbital foramen under 6 mm in any diameter; rostrum longer than broad; bulla elongate but rounded; 38 teeth . . . . *.Martes, 19.*

18b. Infraorbital foramen wider than 6 mm; rostrum broader than long; bulla flattened; 36 teeth . . . .*Lutra canadensis*, p. 147.

19a(18a). Skull shorter than 95 mm . . . . *Martes americana*, p. 129.

19b. Skull longer than 95 mm . . . . *Martes pennanti*, p. 131.

20a(16b). Mastoidal breadth about as great as width across zygomata; occlusal surface of last upper cheek tooth (M1) roughly triangular; skull longer than 80 mm . . . . *Taxidea taxus*, p. 141.

20b. Mastoidal breadth less than 95% of zygomatic width; M1 occlusal surface more or less dumbbell-shaped . . . . *Mustela*, *21*.

21a(20b). Skull longer than 55 mm . . . . *22*.

21b. Skull shorter than 55 mm . . . . . . *23*.

22a(21a). Rostrum short (broader than long), sides converging as seen from above; zygomatic width more than 56% of skull length; mastoidal width more than 51.5% of skull length; lingual end of dumbbell-shaped M1 only moderately larger than labial end (rare species, occasionally escaping from captivity) . . . . .*Mustela putorius*, p. 171.

22b. Rostrum average for a mustelid, about as broad as long, sides parallel behind canines; zygomatic width less than 56% of skull length; mastoidal width less than 51.5% of skull length; lingual end of dumbbell-shaped M1 about three times the area of labial end . . . . *Mustela vison*, p. 137.

23a(21b). Skull shorter than 33 mm . . . . *Mustela nivalis*, p. 134.

23b. Skull longer than 33 mm . . . . *24*.

24a(23b). Zygomatic width at least 23.5 mm; postorbital processes sharply pointed . . . . male *Mustela frenata*, p. 136.

24b. Zygomatic width under 23.5 mm . . . *25*.

25a(24b). Zygomatic width more than 19 mm . . . . *26*.

25b. Zygomatic width less than 19 mm; postorbital processes bluntly pointed . . . . female *Mustela erminea*, p. 133.

26a(25a). Postorbital processes sharply pointed . . .female *Mustela frenata*, p. 136.

26b. Postorbital processes bluntly pointed . . . . male *Mustela erminea*, p. 133.

VII. Order PERISSODACTYLA . . . .
Equidae: horses, asses, and mules, p. 171.

VIII. Order ARTIODACTYLA

1a. Upper incisors present, tooth formula 3/3, 1/1, 4/4, 3/3 × 2 = 44; canines triangular and pointing outward; cheek teeth bunodont; no complete postorbital bar; no horns or antlers . . . . . Suidae, *Sus scrofa*, p. 171.

1b. No upper incisors; tooth formula 0/3, 0-1/1, 3/3, 3/3 × 2 = 32-34 (lower canine looks like a fourth incisor); cheek teeth selenodont; postorbital bar complete; horns or antlers may be present . . . . *2*.

2a(1b). Gap between lacrimal and nasal bones, usually conspicuous but reduced to a 6–8 mm wide oblong, best seen from above, in *Rangifer* . . . . *3*.

2b. No gap separating the lacrimal from the nasal . . . . Bovidae, *8*.

3a(2a). Deciduous, seasonal, black horny sheaths on permanent bony horn cores which arise directly above orbits in both sexes; no conspicuous depression in lacrimal bone just anterior to orbit . . . . *Antilocapridae, *Antilocapra americana*, p. 167.

3b. Bony antler bases present or absent, if present arising on cranium well behind orbits; conspicuous depression (lacrimal pit) in lacrimal bone just anterior to orbit . . . Cervidae, *4*.

4a(3b). Nasal bones short, less than half as long as space between their anterior tips and anterior tips of premaxillae; skull longer than 530 mm (skull without premaxillae longer than 435 mm) . . . .*Alces alces*, p. 162.

4b. Nasal bones more than half as long as space between their anterior tips and anterior tips of premaxillae; skull shorter than 530 mm (skull without premaxillae shorter than 435 mm) . . . . *5*.

5a(4b). As seen from below, internal nares not separated by median partition (vomer) . . . . *Cervus elaphus*, p. 156.

5b. Right and left internal nares separated by vomer . . . . *6*.

6a(5b). Skull longer than 350 mm (305 mm without premaxillae); auditory bulla about half as wide as length of bony tube leading to meatus; seasonal, slightly palmate antlers present in both sexes; gap between lacrimal and nasal bones reduced to a slit narrower than 8 mm and best seen from above (likely to be found in Minnesota only as old bones) . . . . *Rangifer tarandus*, p. 165.

6b. Skull shorter than 355 mm (310 mm without premaxillae); bulla about as wide as length of bony tube leading to meatus; females normally without antlers; lacrimal gap wider (higher) than 10 mm . . . . *7*.

7a(6b). Lacrimal pit deep, will normally conceal at least 5 mm of a probe, as viewed from directly above; seasonal antlers in males usually branching dichotomously . . . . *Odocoileus hemionus*, p. 157.

7b. Lacrimal pit shallow, will normally conceal only 0-4 mm of a probe, as viewed from directly above; seasonal antlers in males normally with one main beam and unbranched tines (tines may branch in old males; see fig. 76) . . . . .*Odocoileus virginianus*, p. 159.

8a(2b). Upper cheek tooth row (P2-M3) shorter than 110 mm . . . . *9*.

8b. P2-M3 longer than 110 mm . . . . . . *10*.

9a(8a). Mastoid processes long, extending more than 8 mm below lowest point of occipital condyles; maxillary gum line only slightly convex, its midpoint less than 3 mm below a line drawn from anterior base of P2 to posterior base of M3; sides of basioccipital divergent, posterior end less than 80% as wide as greatest diameter; horns, if present, tight spirals close to skull . . . . *Ovis aries*, p. 171.

9b. Mastoid processes short, extending less than 4 mm below lowest point of occipital condyles; maxillary gum line strongly convex, its midpoint more than 6 mm below a line from P2 to M3; sides of basioccipital nearly parallel, four protuberances anterior to occipital condyles roughly forming a square; horns, if present, usually backward-curving spikes . . . .*Capra hircus*, p. 171.

10a(8b). Upper posterior process of premaxillary not extending to nasal; zygomata not visible from directly above, hidden by frontals; permanent horns consisting of a bony core covered by keratinaceous horn in both sexes; back of skull rounded, as seen from above, extending at least 30 mm beyond horn cores . . . . *Bison bison*, p. 168.

10b. Upper posterior process of premaxillary in contact with nasal; zygomata visible from directly above; horns present or absent; back of skull square as seen from above and from the side, horn cores, if present, at back corners of skull . . . . *Bos taurus*, p. 171.

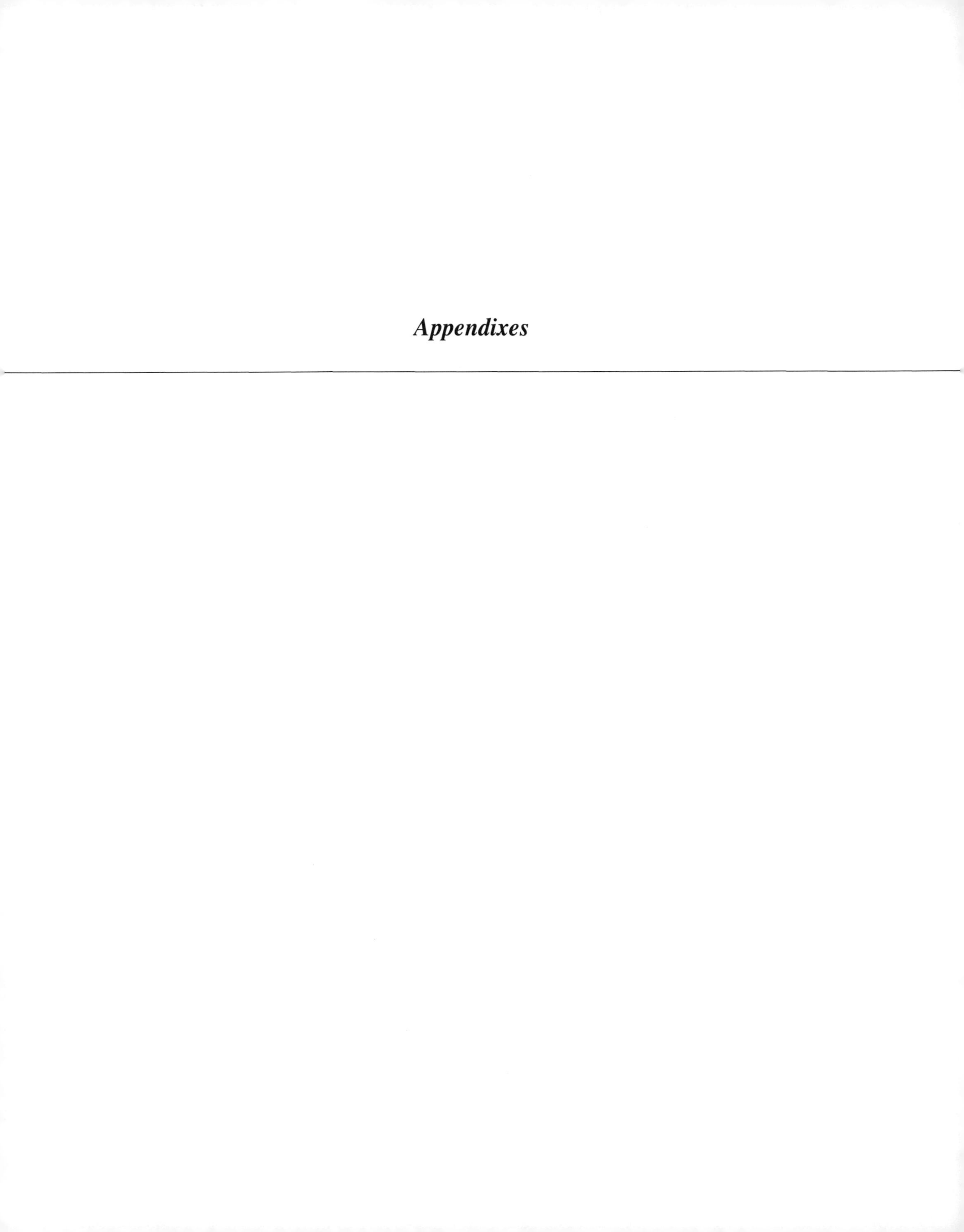

Appendixes

Subspecies List

I have not considered subspecies except in the accounts on *Peromyscus maniculatus*. The following list of subspecies is based on Hall and Kelson (1959). If only one subspecies is listed, that is the only one recognized in Minnesota. When more than one subspecies has been recognized in the state, I give a general indication of the range of each.

Order Polyprotodonta
- Family Didelphidae
  - *Didelphis virginiana virginiana*

Order Insectivora
- Family Soricidae
  - *Sorex cinereus haydeni*—western border
  - *S. c. leseurii*—extreme southeast
  - *S. c. cinereus*—remainder
  - *S. palustris hydrobadistes*—central
  - *S. p. albibarbis*—north
  - *S. arcticus laricorum*
  - *Microsorex hoyi intervectus*—northeast
  - *M. h. hoyi*—remainder
  - *Blarina brevicauda brevicauda*
  - *Cryptotis parva parva*
- Family Talpidae
  - *Scalopus aquaticus machrinus*—southeast
  - *S. a. machrinoides*—central, southwest
  - *Condylura cristata cristata*

Order Chiroptera
- Family Vespertilionidae
  - *Myotis lucifugus lucifugus*
  - *M. keenii septentrionalis*
  - *Lasionycteris noctivagans*
  - *Pipistrellus subflavus obscurus*
  - *Eptesicus fuscus fuscus*
  - *Lasiurus borealis borealis*
  - *L. cinereus cinereus*

Order Lagomorpha
- Family Leporidae
  - *Sylvilagus floridanus similus*—western border
  - *S. f. mearnsii*—remainder
  - *Lepus americanus phaeonotus*
  - *L. townsendii campanius*

Order Rodentia
- Family Sciuridae
  - *Tamias striatus griseus*
  - *Eutamias minimus neglectus*
  - *Marmota monax canadensis*—northcentral, northeast
  - *M. m. monax*—southern tier
  - *M. m. rufescens*—remainder
  - *Spermophilus richardsonii richardsonii*
  - *S. tridecemlineatus tridecemlineatus*
  - *S. franklinii*
  - *Sciurus carolinensis pennsylvanicus*—southcentral, southeast
  - *S. c. hypophaeus*—remainder
  - *S. niger rufiventer*
  - *Tamiasciurus hudsonicus loquax*—extreme southeast
  - *T. h. hudsonicus*—northcentral, northeast
  - *T. h. minnesota*—remainder

Glaucomys volans volans
G. s. sabrinus
Family Geomyidae
Thomomys talpoides rufescens
Geomys bursarius bursarius
Family Heteromyidae
Perognathus flavescens perniger
Family Castoridae
Castor canadensis missouriensis—southwest, or southern tier (Bowles 1975)
C. c. canadensis—remainder
Family Cricetidae
Reithrodontomys megalotis dychei
Peromyscus maniculatus bairdii—see page 84
P. m. gracilis—see page 85
P. leucopus aridulus—Browns Valley
P. l. noveboracensis—remainder
Onychomys leucogaster leucogaster
Clethrionomys gapperi gapperi—northcentral, northeast
C. g. loringi—remainder
Phenacomys intermedius celatus
Microtus pennsylvanicus drummondi—extreme northwest
M. p. pennsylvanicus—remainder
M. chrotorrhinus chrotorrhinus
M. ochrogaster ochrogaster—southern tier, southeast
M. o. minor—remainder
M. pinetorum nemoralis
Ondatra zibethicus cinnamominus—western border, southwest
O. z. zibethicus—remainder
Synaptomys cooperi cooperi—northern two-thirds
S. c. gossii—southern one-third
S. borealis smithi
Family Muridae
Rattus norvegicus norvegicus
Mus musculus domesticus
Family Zapodidae
Zapus hudsonius hudsonius—northcentral, northeast
Z. h. intermedius—remainder
Napaeozapus insignis frutectans

Family Erethizontidae
Erethizon dorsatum dorsatum

Order Carnivora
Family Canidae
Canis latrans thamnos
Canis lupus lycaon
Vulpes velox hebes
Vulpes vulpes regalis
Urocyon cinereoargenteus ocythous
Family Ursidae
Ursus americanus americanus
Family Procyonidae
Procyon lotor hirtus
Family Mustelidae
Martes americana americana
M. pennanti pennanti
Mustela erminea bangsi
M. nivalis campestris—extreme southwest
M. n. rixosa—remainder
M. frenata spadix
M. vison lacustris—northern border (but see Timm 1975)
M. v. letifera—remainder
Gulo gulo luscus
Taxidea taxus taxus
Spilogale putorius interrupta
Mephitis mephitis hudsonica
Lutra canadensis canadensis
Family Felidae
Felis concolor schorgeri
Lynx canadensis canadensis
L. rufus superiorensis

Order Artiodactyla
Family Cervidae
Cervus elaphus canadensis
Odocoileus hemionus hemionus
O. virginianus macroura—southeast
O. v. borealis—east-central, northeast
O. v. dacotensis—west
Alces alces andersoni
Rangifer tarandus sylvestris
Family Antilocapridae
Antilocapra americana americana
Family Bovidae
Bison bison athabascae

Glossary

acrid. Sharply pungent, irritating.

adaptive radiation. The evolution, from a single ancestral species (or higher taxon), of many descendent species (or higher taxa) that are adapted to diverse ways of life.

alisphenoid canal. A tunnel in the base of the alisphenoid bone of some mammals; a blood vessel passes through it. See fig. 88.

alisphenoids. Paired bones, anterior to the squamosals, contributing to the lateral walls of the braincase.

allopatric. Occupying different geographic areas. Cf. **sympatric.**

altricial. Helpless at birth, relatively hairless, with closed eyelids, requiring intensive parental care. Cf. **precocial.**

angular process. Projection at posteroventral angle of the lower jaw, below the mandibular condyle.

anterior condyloid foramen. One of a pair of foramina located on each side of the underside of some skulls posteromedially to the bulla. See **posterior lacerate foramen** and fig. 87.

anterior palatine foramen. See **incisive foramina.**

anticoagulant. A substance that inhibits the clotting reactions of blood.

antlers. Paired outgrowths of the frontal bones of deer; normally grown and shed annually; variously branched; typically characteristic of males. Cf. **horns.**

articulare. Bone at the back of the lower jaw of reptiles and birds that articulates with the quadrate of the skull; becomes the malleus in mammals.

astragulus. One of the tarsal bones, articulating with the base of the tibia and with the calcaneum.

atlas. First cervical vertebra.

auditory bulla (pl. bullae). A bony chamber, usually more or less rounded, which encloses the middle ear.

axis. Second cervical vertebra.

baculum (pl. bacula, syn. os penis). Bone in the penis of some male mammals; distinctive morphology often useful in classification. Cf. **os clitoridis.**

basal metabolic rate. Rate of energy expenditure in a resting mammal that is not subjected to thermal or other stress.

basioccipital. Medial bone at the rear of the skull, forming the lower border of the foramen magnum. See **occipital.**

basisphenoid. A median bone on the underside of the skull, in front of the basioccipital and behind the presphenoid.

beam. Basal or main segment of an antler.

Beringia. Region of northwestern North America and northeastern Asia, intermittently a continuous land bridge up to 2,000 km wide during the Cenozoic.

binomial. A two-word name; in biology a genus (a noun) followed by a specific modifier (an adjective or a noun in adjectival form), e.g., *Sciurus carolinensis* (squirrel, of Carolina).

biome. One of several major ecologic communities of the world, typically characterized by their dominant vegetation (e.g., taiga, tropical rainforest).

blastocyst. An early stage in the embryology of placental mammals comprising a hollow sphere of cells (the trophoblast) and an internal cell mass that will become the definitive embryo.

boar. Adult male pig or bear.

boreal. Pertaining to the north.

boreal forest. See **taiga**.

browse. v. To feed on leaves and twigs of woody plants; n. (collective) such food plants.

buck. Adult male of medium-sized deer, of pronghorns, and of rabbits (but not of wapiti or moose). Cf. **bull.**

bull. Adult male of cattle, bison, wapiti, moose, Old World buffalo, elephants, seals, and others. Cf. **buck.**

bulla. See **auditory bulla.**

bunodont. Having rounded tubercles or cusps on the cheek teeth.

buteo (or buteo hawk). Any broad-winged soaring hawk of the family Accipitridae, subfamily Buteoninae.

cache. n. A large store of food in a single place; v. to store food in a cache. Cf. **scatter-hoard.**

caecotrophy. See **reingestion.**

caecum. A blind pouch at the beginning of the large intestine.

calcaneum. The hindmost or uppermost of the tarsal bones, articulating with the astragulus. This is the heel bone, the point of attachment for the Achilles tendon.

canine. a. Pertaining to dogs; n. a unicuspid tooth immediately behind the incisors, often with a prominent conical crown; the upper canine is rooted in the maxilla but typically emerges at the maxillary-premaxillary suture.

carnassials. A pair of sectorial cheek teeth (P4 and m1 in adult carnivore dentition) which serve to shear meat and crack bones.

carnivore. 1. Any meat-eating animal, particularly one that feeds on vertebrates or large invertebrates (Cf. **insectivore**). 2. Any member of the order Carnivora, regardless of its diet.

carrying capacity. The population density of a given species which a habitat can support without deterioration of the habitat.

Cenozoic. The current geologic era. See table 1, p. 5.

cheek teeth. Teeth behind the canines, commonly multicusped or otherwise modified for grinding, crushing, and shearing food; premolars and molars.

chemosterilant. A chemical used to sterilize animals, e.g., as a pest-control measure.

chew the cud. See **rumination.**

class. An obligatory Linnaean category, within a phylum and comprising one or more orders.

climax. The last biotic community in an ecologic succession, one that is commonly able to perpetuate itself over a long time in a given climate.

cochlea. A tapering tube (also coiled in mammals) in the inner ear, in which vibrations transmitted to the cochlear fluid by the middle ear ossicles stimulate the transmission of auditory nerve impulses.

colony. Usually a social group that is somewhat sedentary, not roaming, because of attachment to and dependence upon a discrete habitat or a burrow system.

competitive exclusion. The exclusion from a given habitat of one species by a closely competing species (one with a similar ecologic niche).

condyloid process. See **mandibular condyle.**

conspecific. Belonging to the same species.

continental drift. The horizontal movement of continental masses (the emergent continents plus submerged continental shelves) with respect to one another. See **plate tectonics.**

convergence, convergent evolution. The evolution of similar adaptations to similar ecologic conditions in distantly related forms (e.g., prehensile tails in opossums and kinkajous). Cf. **parallelism.**

coprophagy. See **reingestion.**

coronoid process. A posterior projection of the dentary, above the mandibular condyle.

cosmopolitan. Occurring in all major land areas; worldwide in distribution or nearly so.

cranium. The portion of the skull housing the brain.

crepuscular. Active at morning and/or evening twilight. Cf. **diurnal, nocturnal.**

Cretaceous. The last period of the Mesozoic. See table 1, p. 5.

critical period. See **imprinting.**

cuspidate. Referring to teeth that have more or less pointed projections called cusps.

deciduous. Pertaining to the milk or baby teeth of most mammals, later replaced by adult or permanent dentition; also to trees of which the leaves are all shed annually, usually in the fall.

delayed fertilization. Reproductive pattern in which sperm are stored in the female's reproductive tract after copulation, and fertilize the egg weeks or months later. Cf. **delayed implantation.**

delayed implantation. Reproductive pattern in which the sperm fertilizes the egg shortly after copulation, the zygote develops to the blastocyst stage, but the blastocyst then ceases development for several weeks or months before implanting in the uterine wall and resuming development. Cf. **delayed fertilization.**

den up. To retire into a den for a long period of winter sleep, but without the large drop in internal temperature typical of "true" hibernation.

dentary. Bone forming the entire left or right ramus of the mammalian mandible, or lower jaw; in most mammals, a distinct medial suture separates the two dentaries, but in some (e.g., pigs, horses, and people) the two fuse to form a single rigid bone.

development. 1. The bringing of something into being or into full realization. 2. A euphemism for the modification of relatively natural, diverse ecosystems into drastically simplified, artificial ecosystems for (often short-term) monetary gain.

diastema. A space between consecutive teeth, generally a large space in relation to tooth diameter.

dichotomous. Branching successively into two-part forks, as in a plant stem, an antler, or a taxonomic key.

dicot. A member of the Dicotyledones, those flowering plants typically having net-veined leaves, flower parts in multiples of four or five, vascular tissue in a ring of bundles or in a continuous circle, and two cotyledons in the seed. Cf. **monocot.**

digitigrade. Walking with only the toes touching the ground. Cf. **plantigrade, unguligrade.**

digits. Fingers or toes.

dimorphic. Having two distinct forms or appearances: 1. sexual—e.g., antlers in male but not in female mule deer; 2. seasonal—e.g., white winter and brown summer coats of ermine.

diphyodont. Having two successive sets of teeth, a milk or deciduous dentition, and a permanent dentition. A distinctive and primitive condition of the class Mammalia which has been lost or modified in many groups.

disjunct distribution. Occurrence of a species or other taxon in two noncontiguous geographic areas.

diurnal. Active during the daylight hours. Cf. **crepuscular, nocturnal.**

domestic, domesticated. Of animals raised and maintained by people for use as a source of food, work, raw materials, or enjoyment. Cf. **exotic, feral, native, wild.**

duff. Partially decayed organic litter on the ground; humus.

ear ossicles. Three small middle ear bones, the malleus, incus, and stapes, that transduce sound vibrations from the eardrum to the cochlea. See **articulare** and **quadrate.**

echolocation. The detection of the position and characteristics of an object by emitting high-frequency sounds and receiving and interpreting the echoes; highly developed in most bats and whales.

ecologic niche. The sum total of the environmental factors that affect a species, especially those that are necessary for its survival; may be defined narrowly to include the resources the species requires, or broadly to include its entire role in the economy of the biotic community.

ecotone. A transition zone between two recognized biotic communities.

ectotherm. An animal whose internal temperature, whether it is stable or fluctuating, is determined by the external temperature, not by the regulation of an internal heat source. Cf. **endotherm.**

embryonic reabsorption. Cessation of development, breakdown, and absorption of the remains of an embryo by the prospective mother, most commonly under conditions of stress.

endotherm. An animal that maintains a relatively constant body temperature by regulating its rate of internal heat production and its rate of heat loss. Cf. **ectotherm.**

Eocene. The second epoch in the Cenozoic. See table 1, p. 5.

ericaceous. Belonging to or pertaining to shrubs of the heath family, Ericaceae.

evergrowing (rootless). Pertaining to teeth with widely open roots that continue to grow after eruption for all or a large part of the life of a mammal.

exoskeleton. A skeleton on the outside of an animal, as found in insects, mollusks, and other invertebrates, in contrast to the endoskeleton typical of vertebrates.

exotic. Foreign, not native; introduced from elsewhere. Cf. **domestic, feral, native, wild.**

extant. Of a species or other taxon: having living members at present. Cf. **extinct.**

extinct. Of an entire species or other taxon: no longer having living members; the verb form "to go extinct" is now widely used in the literature. Cf. **extant.**

family. 1. An obligatory Linnaean category, within an order and comprising one or more genera. Families of animals all end in "-idae." 2. A social unit of parents, offspring, etc.

fecundity. Capacity for producing young.

fenestration. A large, often irregular opening or openwork area in a skull, exposing other bony tissue beneath, rather than a relatively circular hole leading into a cavity. Cf. **foramen.**

feral. Having reverted to a wild state from one of domestication. Cf. **domestic, exotic, native, wild.**

fetid. Having a strong, foul odor.

fibrinogen. A blood protein involved in the formation of clots.

flatworm. Any member of the invertebrate phylum Platyhelminthes. With reference to mammals, usually a parasitic fluke (Class Trematoda), such as *Parelaphostrongylus tenuis* in deer, or a parasitic tapeworm (Class Cestoidea).

food chain. A series of kinds of organisms, usually starting with green plants, each of which serves as food for the next. A useful concept, but food interactions are generally more complex than the term suggests.

foramen (pl. foramina). A discrete, roughly circular or elliptical opening in a bone. Cf. **fenestration.**

foramen magnum. Large opening at the rear or base of the cranium through which the spinal cord connects to the brain.

forb. A broad-leaved herb (e.g., clover), in contrast to a **graminoid.**

form. A little-modified depression in the ground or ground vegetation, used as a retreat or shelter (usually by a rabbit or hare.)

fossorial. Subterranean; generally with reference to adaptations for digging or other aspects of life underground.

genus (pl. genera). An obligatory Linnaean category, within a family and comprising one or more species; the first part of a binomial.

gestation. Broadly, the period from copulation *or* from union of sperm and egg (fertilization) until birth (parturition). Narrowly, the period of active development after implantation until birth. See **delayed fertilization, delayed implantation.**

glacial drift. Material (rocks, gravel, sand, etc.) deposited by a glacier. Most Minnesota bedrock is covered with this.

glottis. The opening from the pharynx into the trachea.

graminoid. A narrow-leaved, grassy herb (e.g., a grass, rush, or sedge), in contrast to a **forb.**

gregarious. Tending to socialize, to form social groups; opposite of solitary.

guard hairs. Long, relatively coarse hairs that project beyond and therefore lie over the underfur.

habitat. Kind of ecologic community where an individual or a population or a species resides, within the geographic range of that species.

hallux. The big toe; innermost digit of the primitive pentadactyl (five-toed) hind foot. Cf. **pollex.**

hectare. (abbrev. ha). A square 100m on a side, or any area of 10,000m^2.

herb. Any soft-bodied (nonwoody) vascular plant. See **forb, graminoid.**

herbaceous. Like an herb; nonwoody.

herbivore. An animal that feeds on plants.

heterodont. Having teeth that are of a variety of distinct shapes (as most mammals do). Cf. **homodont.**

heterotherm. An animal that is endothermic during the active months (e.g., woodchucks), or hours (e.g., bats), but more or less ectothermic during inactive times, which are generally called hibernation.

hibernaculum (pl. hibernacula). The place where a "true" hibernator hibernates, or a "winter sleeper" dens up.

hibernation. See **heterotherm.**

hierarchy, taxonomic. A group of ranked categories used in biological taxonomy.

Holarctic. The Nearctic and Palearctic biogeographic regions combined.

Holocene. See **Recent.**

home range. The area that an *individual* animal (or a closely knit social group of animals) customarily uses and is familiar with. Not synonymous with range or territory (though in some species it may coincide with the latter).

homodont. Having teeth that are essentially all of the same shape, an uncommon condition in mammals. Cf. **heterodont.**

horns. In mammals, typically paired, unbranched, bony projections from the skull, covered by (usually) permanent keratin sheaths. **Antlers** are different. Pronghorns are a special case (see p. 167).

hydrophobia. See **rabies.**

hypogeous fungi. Fungi that are entirely subterranean, including their reproductive structures.

hypsithermal interval. A period, about 4,000 to 6,000 years ago, when North American climates were warmer and drier than at present.

imprinting. A behavior pattern in which an animal becomes able to recognize another individual, or to perform a particular act, by being appropriately stimulated during a restricted time, called the critical period.

incisiform. Shaped like an incisor (usually of a tooth that is not an incisor—e.g., a canine—or of a tooth of uncertain homology).

incisive foramina (anterior palatine foramina). Paired openings in the front of the palate at the junction of the premaxillae and maxillae.

incisors. Simple, often chisellike or adzlike teeth in the front of the tooth rows of most mammals. The upper incisors are borne on the premaxillary.

incus. The middle of the three mammalian ear ossicles. See **quadrates.**

infraclass. An optional Linnaean category, a division of a subclass.

infraorbital canal. A passage from the anterior wall of the orbit to the side of the rostrum, emerging as the **infraorbital foramen** on or in front of the zygomatic process of the maxilla.

insectivore. 1. Any animal that eats insects or, more generally, small terrestrial or aquatic arthropods or other invertebrates. 2. A member of the mammalian order Insectivora, regardless of its diet. Cf. **carnivore.**

interfemoral membrane (uropatagium). The part of a bat's flight membrane that connects the hind legs and which usually encompasses at least part of the tail.

internal nares (sing. naris). The openings of the nasal passages at the back of the secondary palate (roof of the mouth).

interorbital breadth, interorbital width. The minimum distance between the orbits, measured across the roof of the skull.

interparietal. A median, often triangular bone, at the back of the roof of the skull, between the occipital and the posterior borders of the parietals.

jugal (syn. malar). A bone that connects the squamosal with the maxilla and which usually lies wholly within the zygoma; absent in the zygomata of the extinct Multituberculata.

Jurassic. The middle of the three periods of the Mesozoic. See table 1, p. 5.

keratinaceous. Composed of the protein keratin, the major constituent of nails, claws, callouses, and horns (but not antlers). See **horn.**

labial. Pertaining to the lips; of structures in the mouth: outward, toward the lips. Cf. **lingual.**

lability, thermal. See **thermal lability.**

lacrimal. A bone, usually small, in the anterior wall of the orbit, emerging slightly or extensively onto the rostrum, usually pierced by the lacrimal duct, which drains tears from the lacrimal gland to the nasal cavity.

lactation. The production of milk by the mammae (sing. mamma), or mammary glands.

lingual. Pertaining to the tongue; of structures in the mouth: inward, toward the tongue. Cf. **labial.**

Linnaean hierarchy. See **hierarchy, taxonomic.**

lipids. Fats and related organic chemicals. Includes oils in this family of biochemicals, but not hydrocarbon oils derived from petroleum.

lophodont. Of teeth with occlusal surfaces consisting of several transverse ridges.

malar. See **jugal.**

malleus. The outermost of the three mammalian middle ear ossicles. See **articulare.**

mammary glands. Paired ventral skin glands of female mammals which, during early postnatal development, produce milk. Present but vestigial in males.

mandible. The entire lower jaw (both sides) in any vertebrate, regardless of how many bones it comprises. Each side is a **ramus.** See **dentary.**

mandibular condyle. The convex articulating surface at the rear of each ramus.

mandibular fossae. The (usually) convex surfaces on the sides of the skull against which the mandibular condyles articulate; in mammals, always formed at least in part from the squamosal bone.

marsupial. 1. Having a pouch. 2. Referring to any member of the order or group of orders called the Marsupialia (includes the Polyprotodonta as used here). See **marsupium.**

marsupium. A pouch; specifically, the pouch present in most female marsupials, into which the teats open and in which the young are housed and fed for a considerable time after birth.

mast. Nuts, collectively; acorns, hickory nuts, beechnuts, etc.

mastoid process. A ventral projection of the petromastoid bone, just behind the auditory bulla. Absent in some mammals.

mastoidal breadth. The distance between the outer surfaces of the right and left mastoidal processes.

maxilla(e), maxillary(ies). Paired bones forming (usually) most of the sides of the rostrum and of the roof of the mouth (and therefore of the floor of the nasal cavities) and bearing the canines and cheek teeth.

meatus. An opening, in particular the outer ear opening (external auditory meatus).

mesaxonic. Having the axis or plane of symmetry of the feet pass through the middle (third) digit, as in horses. Cf. **paraxonic.**

mesic. Of a habitat that is moderately moist or humid, neither arid nor wet. Cf. **xeric.**

Mesozoic. The geologic era before the present one (the Cenozoic). See table 1, p. 5.

mima mounds. Repeated mounds that occur in some prairies, about a meter high and many meters across, usually associated with, and probably in part created by, pocket gophers or other fossorial mammals.

Miocene. The fourth epoch in the Cenozoic. See table 1, p. 5.

molar. The most posterior of the kinds of teeth of most mammals, usually with multiple roots and often with complex occlusal surfaces. Molars are permanent teeth with no milk predecessors.

molariform. Like a molar; most often, of premolars that resemble the molars of the same species.

monocot. A member of the Monocotyledones, those flowering plants typically having parallel-veined leaves, flower parts in multiples of three, vascular bundles scattered through the stem, and one cotyledon in the seed. Cf. **dicot.**

monogamous. Of reproductive behavior in which one male consorts with one female for one or more breeding seasons. Cf. **polygamous, promiscuous.**

monotypic. Of a taxonomic group that contains only one taxon of the next lower rank; *Condylura* is a monotypic genus containing only one species, *C. cristata*.

multiparous. Of a female that has had more than one litter (or more than one offspring, in species that typically bear young one at a time).

mutualism. A symbiotic relationship between individuals of two species that benefits each of them. Cf. **symbiosis.**

nares. Openings into the nasal (or narial) canals; the openings to the outside are external nares, or nostrils; those into the back of the mouth are internal nares, or choanae.

nasals. A pair of bones on the upper surface of the rostrum, between the maxillae, forming a roof over much of the nasal cavity.

native. Literally, born in the area in which one resides. Of species of mammals, native generally means occurring in the area in question without having been brought there by people, intentionally or not. Cf. **domestic, exotic, feral, wild.** How long it will be useful to maintain some of these labels remains to be seen.

Nearctic. A biogeographic region comprising most of North America, south to the Isthmus of Tehuantepec. See **Holarctic, Neotropical.**

Neotropical. A biogeographic region comprising South America, Central America, and southeast Mexico to the Isthmus of Tehuantepec and a bit farther along both coasts of Mexico. See **Nearctic.**

niche. See ecologic niche.

nocturnal. Active at night. Cf. **crepuscular, diurnal.**

nuchal crest. A transverse ridge along the posterodorsal border of the skull.

occipital. Composite bone surrounding the foramen magnum at the back or base of the skull. Its lower component, the basioccipital, extends forward in the floor of the skull and is usually separately labeled.

occipital condyles. A pair of projections, ventrolateral to the foramen magnum, which articulate with the atlas.

occlusal surface. The working (cutting, shearing, grinding, etc.) surface of the teeth.

old field succession vegetation. Vegetation in an abandoned field, which naturally replaces the vegetation maintained there previously. This varies with locality but commonly includes various perennial herbs and sun-loving shrubs and trees, and, eventually, shade-tolerant trees.

Oligocene. The third eopch in the Cenozoic. See table 1, p. 5.
omnivore. An animal that eats both plant and animal matter.
oogenesis. The development and maturation of eggs (ova, sing. ovum) in the ovaries.
orbit. Region of the skull that houses the eye, sometimes forming a well-defined socket.
orbitosphenoid. See **presphenoid.**
order. An obligatory Linnaean category, within a class and comprising one or more families.
organochlorine compounds. Organic compounds containing chlorine, of which some are toxic. See **PCB's.**
os clitoridis. A bone in the clitoris of some female mammals, homologous to the **baculum** of males.
overstory. In a forest, the upper levels of foliage, the tops of mature trees; the canopy. Cf. **understory.**
ovulation. The release of mature eggs from the ovary, normally into the oviducts.

palate. See secondary palate.
palatines. Paired bones on either side of the midline of the secondary palate, behind the maxillae and in front of the internal nares.
Palearctic. A biogeographic region comprising Africa north of the Sahara and all of Eurasia except for an area roughly south of the Himalayas and contiguous mountain ranges. Cf. **Holarctic.**
Paleocene. The first epoch in the Cenozoic. See table 1, p. 5.
papilla (pl. papillae). Small, fleshy protuberances or projections, such as taste buds on the tongue.
parallelism, parallel evolution. The evolution of similar adaptations to similar ecologic conditions in more or less closely related forms. Differs from **convergence** in degree.
paraxonic. Having the axis or plane of symmetry of the feet pass between digits three and four, as in deer. Cf. **mesaxonic.**
parietal ridge. Ridge on the dorsolateral surface of the skull, marking the upper edge of the origin of the temporalis muscle. See **sagittal crest.**
parietals. Paired bones meeting in the midline of the skull behind the frontals, generally forming a major part of the roof of the skull.
patagium. Flight membrane.
PCB's (polychlorinated biphenyls). Organochlorine compounds, useful in the plastics industry and in electric power installations, which are toxic to many species, and which degrade to less harmful compounds rather slowly.
permafrost. A layer of permanently frozen soil, reaching the surface in winter but not in summer, characteristic of the tundra.
pesticide. A poison used to kill unwanted organisms but often capable of killing other organisms. Often a euphemism for a substance more accurately termed a biocide.
phalanges (sing. phalanx). Bones in the digits.
pharynx. In mammals, the region behind the fleshy internal nares which is common to both the digestive and respiratory tracts.
phylogeny. 1. The history of a particular taxon (e.g., of the family Equidae or the tribe Microtini). 2. A "family tree" illustrating such probable relationships.
phylum. An obligatory Linnaean category, within a kingdom and comprising one or more classes.
pinna. External ear.
placental. 1. Having a chorioallantoic placenta, a composite organ of maternal and embryonic or fetal circulatory tissues across which materials in solution—nutrients, wastes, respiratory gases—can pass. 2. Any species of mammal, except for one family of marsupials, that possesses such a structure; a member of the infraclass Eutheria.
plantigrade. Walking with the entire sole of the foot on the ground. Cf. **digitigrade, unguligrade.**
plate tectonics. The study of the movements of the geologic plates that make up the earth's crust, and of the causes of these movements.

Continental drift occurs because the continents ride on some of these plates.

Pleistocene. The sixth epoch in the Cenozoic. See table 1, p. 5.

Pliocene. The fifth epoch in the Cenozoic. See table 1, p. 5.

plunge hole. An emergency entrance into an underground retreat.

pollex. The thumb; innermost digit of the primitive pentadactyl (five-toed) hand or forefoot. Cf. **hallux.**

polygamous. Of reproductive behavior in which an individual of one sex consorts with several of the other sex. **Polygyny** (one male and several females) is more common in mammals than the reverse arrangement, **polyandry.** Cf. **monogamous, promiscuous.**

polymorphic. Having several distinct forms.

posterior lacerate foramen. One of a pair of foramina located on each side of the underside of some skulls posteromedially to the bulla. See **anterior condyloid foramen** and fig. 87.

postorbital bar. See **temporal fossa.**

postorbital process. A projection of the frontal bone of some mammals above and behind the orbit.

prairie. Grassland; usually implies naturally occurring grassland.

Prairie Provinces. Alberta, Saskatchewan, and Manitoba.

preadaptation. A characteristic of a species that is useful (adaptive) in its original environment and which later turns out also to be useful, usually in a different way, in a new or changed environment.

Pre-Cambrian. An informal term for all of geologic time from about 4,600 million to about 600–570 million years ago, when the Cambrian began.

precocial. Relatively mature at birth, somewhat furred, with eyes open, capable of some locomotion. Cf. **altricial.**

premaxillae (syn. premaxillaries). Paired bones at the front of the rostrum, forming part of the secondary palate and of the rostrum, and usually bearing upper incisors.

premolars. Cheek teeth behind the canine (if present) and in front of the molars, often with somewhat complex occlusal surfaces. Unlike molars, premolars typically have milk predecessors.

presphenoid. A median bone that appears in the underside of the skull in front of the basisphenoid, and lateral extensions of which emerge in the orbit as the orbitosphenoids, anterior to the alisphenoids.

primary production. *Gross* primary production is the total amount of organic matter (OM) produced by photosynthesis by the plants in a habitat. *Net* primary production is gross production minus the OM that the plants consume in respiration, and thus is what is left for herbivores (some mammals but mostly insects) to feed upon.

primiparous. Of a female that is about to give, or has given, birth for the first time.

prismatic. Of cheek teeth, having an occlusal surface with more or less triangular folds of enamel. See fig. 86.

promiscuous. Of reproductive behavior in which males mate with receptive females as they encounter them, with no subsequent pair bond. Cf. **monogamous, polygamous.**

quadrates. Paired bones flexibly or rigidly articulating with the back of the skull in reptiles and birds, with which the lower jaw articulates via the articulare. The embryonic quadrate becomes the incus in mammals.

Quaternary. Classically, the second period in the Cenozoic, comprising the Pleistocene epoch only. See table 1, p. 5.

rabies. A viral disease of the central nervous system, which infected (rabid) dogs, skunks, foxes, bats, etc., can transmit to one another and to people, most commonly by biting. People can be cured if treated promptly after exposure.

radio-tracking (radiotelemetry). Tracing the location of (and perhaps other information about) an animal by receiving signals from a small transmitter affixed to it.

ramus. The left or right half of the lower jaw, or **mandible.**

range. The entire geographic area within which members of a species or other taxonomic group occupy suitable habitats.

Recent. The present geologic epoch, also called the Holocene. See table 1, p. 5.

refection. See *reingestion*.

reingestion. The consumption of fecal pellets, of a particular kind, by the individual that produces them.

retractile. Of the claws of all cats except the cheetah, able to be pulled back so as not to touch the surface that the paw is contacting.

rhizome. An underground stem, often horizontal, used by certain plants for food storage.

rootless. See **evergrowing.**

rostrum. The part of a skull in front of the orbits.

rumination. Chewing of the cud; in ruminant artiodactyls with composite stomachs, the regurgitation of food from the first stomach compartment to the mouth for further chewing and subsequent return to the stomach.

rut. Of large-horned or antlered artiodactyls (usually), the belligerent and lusty state of males during the mating season.

sagittal crest. A ridge down the midline of the roof of skull, providing additional space for the origin of the temporalis muscle. Essentially a joining of the two parietal ridges.

scat. A discrete piece of feces, especially of a carnivorous mammal. Foxes produce scats; cows do not.

scatter-hoard. To store items individually over a wide area, as a fox squirrel does acorns.

secondary palate. In mammals and in crocodilians (in which it arose independently), a shelf of bone that separates the nasal cavity from the mouth and thus forms the roof of the mouth.

second-growth forest. Forest that has grown up on land previously cut over.

sectorial. Of teeth especially adapted to cutting, as in cats.

selenodont. Of teeth having longitudinally oriented, crescentic cusps, as in deer and cattle.

sinusoid. Shaped like a sine curve; wavy, sinuous.

species (pl. species). The smallest obligatory Linnaean category, within a genus. The *name* of a species is a binomial, not just the specific epithet (e.g., *Geomys bursarius*, or *G. bursarius*, but not simply *bursarius*).

spermatogenesis. The development and maturation of sperm in the testes.

sphagnum. A moss that forms characteristic spongy hummocks in northern bogs.

squamosals. Paired bones comprising much of the lateral walls of the cranium, the posterior portion of the zygoma, and all or part of the mandibular fossa.

stapes. The innermost of the three mammalian ear ossicles and the only ear ossicle of reptiles and birds.

subclass. An optional Linnaean category, a division of a class.

subcutaneous. Immediately beneath the skin. Subcutaneous fat is a common adaptation for food storage, insulation, physical protection, and sexual recognition.

subfamily. An optional Linnaean category, a division of a family.

suborder. An optional Linnaean category, a division of an order.

subphylum. An optional Linnaean category, a division of a phylum.

subspecies (pl. subspecies). An optional Linnaean category, a division of a species. See **trinomial.**

succession, ecologic. A replacement process in which certain biotic communities create conditions that favor their replacement by other biotic communities, e.g., replacement of prairie by shrubland. In some situations, the process ends in the establishment of a self-perpetuating, relatively long-lived **climax** community.

supraorbital process. A projection of the frontal bone above the orbit.

suture. The junction of two bones, or two parts of a bone, forming a relatively immovable joint. With age, suture lines may disappear entirely.

symbiosis. An intimate ecological relationship between individuals of two different species (as used here, includes parasitism and commensalism, as well as mutualism).

sympatric. Of two species or other groups living in the same region. Cf. **allopatric.**

taiga. A biome dominated by needle-leaved evergreen conifers and characterized by cool to cold, often long winters, relatively cool summers, and moderate to high annual precipitation. In northern North America, it occurs generally south of the **permafrost** line.

taxonomy. Classification, particularly of organisms. See **hierarchy, taxonomic.**

teats. Nipples.

temporal fossa. Space behind the orbit and mostly medial and dorsal to the zygoma, occupied by jaw muscles. In some mammals, a postorbital bar separates it from the orbit.

territory. An area, within or coextensive with the home range, that an individual or a social group defends against at least some other individuals or groups of the same species.

Therapsida. The order of reptiles, in the subclass Synapsida, that included the dominant land carnivores of most of the Triassic and which gave rise to the early mammals in the Upper Triassic.

thermal lability. Ability to vary the internal temperature within a range of several degrees without causing physiological harm.

tine. A branch or prong on an antler.

torpid. Dormant, incapable of locomotion and of most sensation, and with a body temperature well below that of an active mammal of the same species.

trachea. Cartilage-strengthened tube leading from the pharynx to the paired bronchi (sing. bronchus), one of which goes to each lung; windpipe.

tragus. Fleshy projection from the base of the pinna in most bats.

tree line. In the northern hemisphere, the northernmost limit of trees. Roughly coincides, at the north edge of the taiga, with the southern limit of permafrost.

Triassic. The first period in the Mesozoic. See table 1, p. 5.

trematode. A parasitic flatworm of the Class Trematoda; a fluke.

tribe. An optional Linnaean category, a division of a subfamily.

tribosphenic. Of tritubercular upper molars with three principal cusps (a trigon); and of corresponding tuberculosectorial lower molars with three cusps (a trigonid) and a relatively low posterior platform (a talonid). Many Mesozoic mammals had tribosphenic molars, and *Didelphis* has them.

trinomial. The name of a subspecies; e.g., *Marmota monax rufescens*.

tritubercular. See **tribosphenic.**

tubercle. A small, rounded or knoblike protuberance or process.

tuberculosectorial. See **tribosphenic.**

tundra. A high latitude (or high altitude) treeless biome, supporting low or dwarf woody vegetation and a variety of herbs. At high latitudes, typically underlain by permafrost.

underfur. A dense, often woolly, insulating coat of fine hair, in many mammals overlain by longer, heavier guard hairs.

understory. In a forest, the lower level of woody vegetation, generally comprising shrubs, small trees, and young members of canopy or overstory species.

ungulate. 1. a. Hoofed. 2. n. A mammal with hooves.

unguligrade. Walking on the toenails (that is, on hooves). Cf. **digitigrade, plantigrade.**

unicuspid. A tooth with a single cusp.
uropatagium. See **interfemoral membrane.**
uterus (pl. uteri). That part of the female reproductive tract within the wall of which embryonic and fetal development occurs.

vector. An organism that transmits a parasite.
velvet. The furred skin that covers antlers during their growth.
vole. A small microtine rodent, especially one in the tribe Microtini. In our area, includes species of *Clethrionomys*, *Phenacomys*, and *Microtus*, but not *Ondatra*.

weathering. The combination of physical, chemical, and biological processes by which rock or other parent material becomes soil.
wild. Not domesticated. As used here, wild includes native species and exotic undomesticated species (e.g., *Mus musculus*). Cf. **domestic, exotic, feral, native.**
Wisconsin (also Wisconsonian). The last of four major glacial advances during the Pleistocene.

xeric. Arid, dry, deficient in moisture. Cf. **mesic.**

zygoma (pl. zygomata, syn. zygomatic arch). A bony arch at the side of the skull, usually comprising a process of the maxilla, the entire jugal bone, and a process of the squamosal, and enclosing the orbit and the temporal fossa.
zygomatic plate. In most rodents, an expanded, flat region in the zygomatic process of the maxilla.

Bibliography

Bibliography

Since a major aim of this book is to introduce the reader to the professional literature, I have cited many items, listed additional references at the end of each account, and listed all the references in the Bibliography.

Original research results, often called the "primary literature," normally appear in professional journals, usually published by professional societies, or in serials published at irregular intervals by museums or other scientific institutions. For example, the American Society of Mammalogists publishes the Journal of Mammalogy. (Anyone with a serious interest in mammalogy should consider joining the society. For information, contact Elmer Birney, Bell Museum, 10 Church Street S.E., Minneapolis, Minn. 55455.)

The "secondary literature," consisting largely of compilations of information from the primary literature, often appears in book form but may also be published in journals or serials. Texts, field guides, laboratory manuals, and state or regional handbooks are typical examples of secondary literature.

Among the general books (cited fully in the Bibliography) of interest to the beginning mammalogist are two recent texts (Gunderson 1976; Vaughan 1978); a classic text (Hamilton 1939); a laboratory manual (DeBlase and Martin 1974); a short guide to specimen preparation (Hall 1962); a field guide (Burt and Grossenheider 1976); and various state and regional handbooks and manuals (Banfield 1974; Bowles 1975; Burt 1957; Godin 1977; Hall and Kelson 1959; Hamilton and Whitaker 1979; Jackson 1961; Peterson 1966). Books as well as articles concerning particular species or groups are listed in the text.

Articles on mammals appear in hundreds of periodicals and other serials, including:

Acta Theriologica
American Journal of Anatomy
American Journal of Physiology
American Midland Naturalist
American Museum of Natural History, Bulletin
American Naturalist
American Scientist
American Zoologist
Anatomical Record
Animal Behaviour
Animal Behaviour Monographs
Behavioral Ecology and Sociobiology
Behaviour
Biological Reviews of the Cambridge Philosophical Society
Canadian Field-Naturalist
Canadian Journal of Zoology
Carnegie Museum, Annals
Carnivore
Comparative Biochemistry and Physiology
Ecological Monographs
Ecology
Evolution
Evolutionary Biology
Evolutionary Theory
Illinois Natural History Survey
International Union for Conservation of Nature and Natural Resources Publications
Iowa Academy of Science, Proceedings
Journal of Comparative Physiology
Journal of Experimental Zoology
Journal of Forestry
Journal of Mammalogy
Journal of Morphology
Journal of Parasitology
Journal of Wildlife Management
Journal of Zoology (London)
Kansas Academy of Sciences, Transactions
Loon (formerly Flicker)
Mammal Review (England)
Mammalia (France)
Mammalian Species
Michigan State Univ., Museum Publications
Minnesota Academy of Science, Journal (formerly Proceedings)
Minnesota Department of Natural Resources, Technical Bulletin
Minnesota Volunteer

Museum of Comparative Zoology (Harvard), Bulletin
Natural History
Naturaliste Canadien
New York Academy of Science, Transactions
New York Fish and Game Journal
North American Fauna
North American Wildlife and Natural Resources Conference, Transactions
Paleobiology
Physiological Zoology
Quarterly Review of Biology
Science
Scientific American
Southwestern Naturalist
Systematic Zoology
Texas Tech Univ., Museum, Occasional Papers and Special Publications
USDA Forest Service Research Papers
U.S. National Museum, Bulletin
Univ. Connecticut Occasional Papers, Biol. Sciences Series
Univ. Kansas Publications, Museum of Natural History
Univ. Michigan Museum of Zoology, Occasional Papers and Miscellaneous Publications
Univ. Minn., Agricultural Experiment Station, Technical Bulletin
Univ. Wisconsin Stevens Point, Museum of Natural History Reports
Wildlife Monographs

How does one find all the articles on a subject? There is no single answer. The number of periodicals and serials, and the total number of articles published per year, have become too large for an individual to search all the annual indexes and to go from the citations of references in one article to other articles until all the articles on a given subject have been found, though this approach helps. *The Readers' Guide to Periodical Literature* will not do because it indexes few scientific journals. *Biological Abstracts* is the most comprehensive source for biological articles in general; the uninitiated student will, however, need some training in its use. *Zoological Record* publishes separate sections on various major animal groups, and professional mammalogists often find its section *Mammalia* more useful than *Biological Abstracts*. A most useful source is the *Recent Literature of Mammalogy* supplement that accompanies each quarterly issue of the *Journal of Mammalogy*. It lists almost every article of interest to any mammalogist. Many of the same articles are indexed in *Wildlife Review*, a regular publication of the U.S. Fish and Wildlife Service. None of these sources is free from mistakes in spelling, wording, dates, and pagination. You should always check the original article, which you will want to read anyway. Even there, errors in the title or author's name appear occasionally.

Ables, E. D. 1969a. Activity of red foxes in southern Wisconsin. J. Wildl. Mgt. 33:145–153.

———. 1969b. Home range studies of red foxes (*Vulpes vulpes*). J. Mamm. 50:108–120.

———. 1975. Ecology of the red fox in North America. Pp. 216–236 *in* M. W. Fox, ed. The wild canids. Van Nostrand Reinhold, New York.

Ackerman, R., and P. D. Weigl. 1970. Dominance relationships of red and grey squirrels. Ecology 51:332–334.

Adamcik, R. S., and L. B. Keith. 1978. Regional movements and mortality of great horned owls in relation to snowshoe hare fluctuations. Can. Field-Nat. 92: 228–234.

Adamcik, R. S., A. W. Todd, and L. B. Keith. 1978. Demographic and dietary responses of great horned owls during a snowshoe hare cycle. Can. Field-Nat. 92: 156–166.

———. 1979. Demographic and dietary responses of red-tailed hawks during a snowshoe hare fluctuation. Can. Field-Nat. 93:16–27.

Adams, D. B. 1979. The cheetah: native American. Science 205:1155–1158.

Agenbroad, L. D. 1973. A Paleo-Indian bison kill in the Panhandle of Nebraska. Trans. Neb. Acad. Sci. 2:57–61.

Aldous, C. M. 1936. Food habits of *Lepus americanus phaenotus*. J. Mamm. 17:175–176.

———. 1937. Notes on the life history of the snowshoe hare. J. Mamm. 18:46–57.

Aldous, S. E. 1938. Beaver food utilization studies. J. Wildl. Mgt. 2:215–222.

———. 1941. Food habits of chipmunks. J. Mamm. 22:18–24.

Aldous, S. E., and J. Manweiler. 1942. The winter food habits of the short-tailed weasel in northern Minnesota. J. Mamm. 23:250–255.

Aleksiuk, M. 1968. Scent-mound communication, territoriality, and population regulation in beaver (*Castor canadensis* Kuhl). J. Mamm. 49:759–762.

Aleksiuk, M., and A. Frohlinger. 1971. Seasonal metabolic organization in the muskrat (*Ondatra zibethica*). I. Changes in growth, thyroid activity, brown adipose tissue, and organ weights in nature. Can. J. Zool. 49:1143–1154.

Allan, P. F. 1936. *Microtus ochrogaster* in Minnesota. J. Mamm. 17:291.

———. 1947. Blue jay attacks red bats. J. Mamm. 28:180.

Allen, D. L. 1939. Winter habits of Michigan skunks. J. Wildl. Mgt. 3:212–228.

———. 1943. Michigan fox squirrel management. Mich. Dept. Cons., Game Div. Publ. 100:1–404.

———. 1974. Of fire, moose, and wolves. Audubon 76(6):38–49.

Allen, D. L., and L. D. Mech. 1963. Wolves versus moose

on Isle Royale. Natl. Geogr. 123:200–219.

Allen, D. L., and W. W. Shapton. 1942. An ecological study of winter dens, with special reference to the eastern skunk. Ecology 23:59–68.

Allen, E. G. 1938. The habits and life history of the eastern chipmunk, *Tamias striatus lysteri*. N.Y. State Mus., Bull. 314:1–122.

Allen, Glover. 1933. The least weasel a circumboreal species. J. Mamm. 14:316–319.

Allen, J. M. 1954. Gray and fox squirrel management in Indiana. 2d ed. Ind. Dept. Cons., P-R Bull. 1:1–112.

Allen, R. E., and D. R. McCullough. 1976. Deer-car accidents in southern Michigan. J. Wildl. Mgt. 40:317–325.

Allin, E. F. 1975. Evolution of the mammalian middle ear. J. Morph. 147:403–438.

Ames, A. E. 1873. Mammalia of Minnesota. Minn. Acad. Sci., Bull. 1:68–71.

Amstrup, S. C., and John Beecham. 1976. Activity patterns of radio-collared black bears in Idaho. J. Wildl. Mgt. 40:340–348.

Andelt, W. F., D. P. Althoff, and P. S. Gipson. 1979. Movements of breeding coyotes with emphasis on den site relationships. J. Mamm. 60:568–575.

Andelt, W. F., and P. S. Gipson. 1979a. Domestic turkey losses to radio-tagged coyotes. J. Wildl. Mgt. 43:673–679.

———. 1979b. Home range, activity, and daily movements of coyotes. J. Wildl. Mgt. 43:944–951.

Andelt, W. F., and B. R. Mahan. 1977. Ecology of an urban coyote. Proc. Neb. Acad. Sci. 80:5.

Andersen, D. C. 1978. Observations on reproduction, growth, and behavior of the northern pocket gopher (*Thomomys talpoides*). J. Mamm. 59:418–422.

Andersen, K. W., and J. K. Jones, Jr. 1971. Mammals of northwestern South Dakota. Univ. Kan. Publ., Mus. Nat. Hist. 19:361–393.

Anderson, A. E., D. E. Medin, and D. C. Bowden. 1974. Growth and morphometry of the carcass, selected bones, organs, and glands of mule deer. Wildl. Monogr. 39:6–115.

Anderson, P. K. 1978. The serendipitous mouse. Nat. Hist. 87(4):38–43.

Anderson, R. K., and M. Davis. 1978. Review status of eastern pine marten (Wisconsin). Pp. 100–101 *in* M. K. Brown, ed. The pine marten. N.Y. State Dept. Env. Cons., unnumb. rep.

Anderson, R. M. 1965. Methods of collecting and preserving vertebrate animals. Nat. Mus. Can., Bull. 69, Biol. Ser. 18.

Anderson, S. 1959. The baculum of microtine rodents. Univ. Kan. Publ., Mus. Nat. Hist. 12:181–216.

———. 1967. Introduction to the rodents. Pp. 206–209 *in* S. Anderson and J. K. Jones, Jr., eds. Recent mammals of the world. Ronald Press, New York.

Andrews, R. D. 1974. An irruption of least shrews, *Cryptotis parva*, in Illinois. Trans. Ill. State Acad. Sci. 67:5–7.

Andrews, R. V., et al. 1971. Capture and handling of wild Norway rats used for endocrine studies. J. Mamm. 52:820–824.

Anisko, J. J. 1976. Communication by chemical signals in Canidae. Pp. 283–293 *in* R. L. Doty, ed. Mammalian olfaction, reproductive processes, and behavior. Academic Press, New York.

Anonymous. 1965. Marten (*Martes americana*). Sask. Dept. Nat. Resour., Wildl. Resour. Reader 139:1–2.

Anthony, E. L. P., and T. H. Kunz. 1977. Feeding strategies of the little brown bat, *Myotis lucifugus*, in southern New Hampshire. Ecology 58:775–786.

Arata, A. A. 1967. Muroid, gliroid, and dipodoid rodents. Pp. 226–253 *in* S. Anderson and J. K. Jones, Jr., eds. Recent mammals of the world. Ronald Press, New York.

Arimond, Sheila. 1980. Black bears of northeast Minnesota: ecology and management. Proc. Symp. Mammal. Ecol. Hab. Mgt. Minn., Bemidji State Univ. 67–69.

Arlton, A. V. 1936. An ecological study of the mole. J. Mamm. 17:349–371.

Armstrong, D. M. 1972. Distribution of mammals in Colorado. Univ. Kan. Mus. Nat. Hist., Monogr. 3:x + 415.

Asdell, S. A. 1964. Patterns of mammalian reproduction. 2d. ed. Comstock, Ithaca, N.Y.

———. 1966. Evolutionary trends in physiology of reproduction. Symp. Zool. Soc. London 15:1–13.

Auerlich, R. J., R. K. Ringer, and D. Polin. 1972. Rate of accumulation of chlorinated hydrocarbon pesticide residues in adipose tissue of mink. Can. J. Zool. 50:1167–1173.

Austin, C. R., and R. V. Short, eds. 1976. Reproduction in mammals. Vol. 6. The evolution of reproduction. Cambridge Univ. Press, New York.

Autenrieth, R. E., and E. Fichter. 1975. On the behavior and socialization of pronghorn fawns. Wildl. Monogr. 42:1–111.

Bahnak, B. R., et al. 1979. Seasonal and nutritional effects on serum nitrogen constituents in white-tailed deer. J. Wildl. Mgt. 43:454–460.

Bailey, B. 1923. Meat-eating propensities of some rodents of Minnesota. J. Mamm. 4:129.

———. 1929. Mammals of Sherburne County, Minnesota. J. Mamm. 10:153–164.

Bailey, E. D. 1965a. The influence of social interaction and season on weight change in woodchucks. J. Mamm. 46:438–445.

———. 1965b. Seasonal changes in metabolic activity of

non-hibernating woodchucks. Can. J. Zool. 43:905–909.

Bailey, J. A. 1969. Exploratory study of nutrition of young cottontails. J. Wildl. Mgt. 33:346–353.

Bailey, R. G. 1976. Ecoregions of the United States. USDA, U.S. For. Serv., Ogden, Utah.

Bailey, T. N. 1971. Biology of striped skunks on a southwestern Lake Erie marsh. Am. Midl. Nat. 85:196–207.

———. 1974. Social organization in a bobcat population. J. Wildl. Mgt. 38:435–446.

Bailey, Vernon. 1893. The prairie ground squirrels or spermophiles of the Mississippi Valley. USDA Div. Ornith. Mamm., Bull. 4:1–69.

———. 1926. A biological survey of North Dakota. N. Am. Fauna 49.

Baker, Ann E. M. 1974. Interspecific aggressive behavior of pocket gophers *Thomomys bottae* and *T. talpoides* (Geomyidae: Rodentia). Ecology 55:671–673.

Baker, D. L. et al. 1979. Energy requirements of mule deer fawns in winter. J. Wildl. Mgt. 43:162–169.

Baker, R. H. 1944. An ecological study of tree squirrels in eastern Texas. J. Mamm. 25:8–24.

———. 1971. Nutritional strategies of myomorph rodents in North American grasslands. J. Mamm. 52:800–805.

Baker, R. J., R. K. Barnett, and I. F. Greenbaum. 1979. Chromosomal evolution in grasshopper mice (*Onychomys*: Cricetidae). J. Mamm. 60:297–306.

Baker, R. J., and J. L. Patton. 1967. Karyotypes and karyotypic variation of North American vespertilionid bats. J. Mamm. 48:270–286.

Baker, R. J., S. L. Williams, and J. C. Patton. 1973. Chromosomal variation in the plains pocket gopher, *Geomys bursarius major*. J. Mamm. 54:765–769.

Bakker, R. T. 1971. Dinosaur physiology and the origin of mammals. Evolution 25:636–658.

———. 1975a. Dinosaur renaissance. Sci. Am. 232(4):58ff.

———. 1975b. Experimental and fossil evidence for the evolution of tetrapod bioenergetics. Pp. 365–399 *in* D. Gates and R. Schmerl, eds. Perspectives in biophysical ecology. Springer Verlag, New York.

Bakko, E. B. 1975. A field water balance study of gray squirrels (*Sciurus carolinensis*) and red squirrels (*Tamiasciurus hudsonicus*). Comp. Biochem. Physiol. 51A:759–768.

———. 1977. Influence of collecting techniques on estimate of natural renal function in red squirrels. Am. Midl. Nat. 97:502–504.

Balph. D. F. 1961. Underground concealment as a method of predation. J. Mamm. 42:423–424.

Balser, D. S. 1964. Management of predator populations with antifertility agents. J. Wildl. Mgt. 28:352–358.

Balser, D. S. and W. H. Longley. 1966. Increase of the fisher in Minnesota. J. Mamm. 47:547–550.

Banfield, A. W. F. 1961. A revision of the reindeer and caribou, genus *Rangifer*. Nat. Mus. Can., Bull. 177.

———. 1974. The mammals of Canada. Univ. Toronto Press, Toronto.

Banfield, A. W. F., and J. S. Tener. 1958. A preliminary study of the Ungava caribou. J. Mamm. 39:560–573.

Barash, D. P. 1974a. Mother-infant relations in captive woodchucks (*Marmota monax*). Anim. Behav. 22:446–448.

———. 1974b. The evolution of marmot societies: a general theory. Science 185:415–420.

———. 1974c. Neighbor recognition in two "solitary" carnivores: the raccoon (*Procyon lotor*) and the red fox (*Vulpes fulva*). Science 185:794–796.

———. 1975. Marmot alarm calling and the question of altruistic behavior. Am. Midl. Nat. 94:468–470.

Barbehenn, K. R. 1958. Spatial and population relationships between *Microtus* and *Blarina*. Ecology 39:293–304.

———. 1973. A comment on predator-prey species proportions. Am. Nat. 107:152–154.

Barbour, R. W., and W. H. Davis. 1969. Bats of America. Univ. Press Ky., Lexington.

Barbour, R. W., and W. H. Davis, and M. D. Hassell. 1966. The need of vision in homing *Myotis sodalis*. J. Mamm. 47:356–357.

Barfield, R. J., and L. A. Geyer. 1972. Sexual behavior: ultrasonic postejaculatory song of the male rat. Science 176:1349–1350.

Barkalow, F. S., Jr. 1962. Latitude related to reproduction in the cottontail rabbit, J. Wildl. Mgt. 26:32–37.

Barkalow, F. S., Jr., and Monica Shorten. 1973. The world of the gray squirrel. Lippincott, Philadelphia.

Barkalow, F. S., Jr., and R. F. Soots, Jr. 1975. Life span and reproductive longevity of the gray squirrel, *Sciurus c. carolinensis* Gmelin. J. Mamm. 56:522–524.

Barker, M. S., Jr., and T. L. Best. 1976. The wolverine (*Gulo luscus*) in Nevada. Southwest. Nat. 27:133.

Barnett, R. J. 1977. Bergmann's rule and variation in structures related to feeding in the gray squirrel. Evolution 31:538–545.

Barnett, S. A. 1963. The rat: a study in behaviour. Aldine, Chicago.

———. 1967 Rats. Sci. Am. 216(1):78ff.

Barr, T. R. B. 1963. Infectious diseases in the opossum: a review. J. Wildl. Mgt. 27:53–71.

Barrett, G. W., G. N. Cameron, and S. L. Herren. 1979. Educational and employment trends in mammalogy. Am. Soc. Mamm. Rep. Available from Dr. G. W. Barrett, Dept. Zool., Miami Univ., Oxford, Ohio 45056.

Barry, W. J. 1976. Environmental effects on food hoarding in deermice *Peromyscus*. J. Mamm. 57:731–746.

Barthelemy, R. E. 1971. Our university's museum of natural history. Minn. Volun. 34(199):30–38.

Bartholomew, G. A., Jr., and G. R. Cary. 1954. Locomo-

tion in pocket mice. J. Mamm. 35:386–392.
Basrur, P. K. 1968. The karyotype of the long-tailed weasel *Mustela frenata noveboracensis* Emmons. Can. J. Genet. Cytol. 10:390–394.
Batten, Elizabeth. 1979. Small mammals of the regional copper-nickel study area. Minn. Env. Qual. Bd. 40 pp. Mimeo.
Batzli, G. O., L. L. Getz, and S. S. Hurley. 1977. Suppression of growth and reproduction of microtine rodents by social factors. J. Mamm. 58:583–591.
Bayless, S. R. 1969. Winter food habits, range use, and home range of antelope in Montana. J. Wildl. Mgt. 33:538–551.
Beach, F. A., and R. W. Gilmore. 1949. Response of male dogs to urine from females in heat. J. Mamm. 30:391–392.
Beasom, S. L., and R. A. Moore. 1977. Bobcat food habit response to a change in prey abundance. Southwest. Nat. 21:451–457.
Beck, A. M. 1973. The ecology of stray dogs. York Press, Baltimore.
———. 1975. The ecology of "feral" and free-roving dogs in Baltimore. Pp. 380–390 *in* M. W. Fox, ed. The wild canids. Van Nostrand Reinhold, New York.
Beck, M. L., and J. T. Mahan. 1978. The chromosomes of *Microtus pinetorum*. J. Hered. 69:343–344.
Becker, C., A. Gier, and C. Long. 1976. New records of rare shrews from the Lake Michigan drainage basin, Wisconsin. Rep. Fauna Flora Wisc. (Univ. Wisc., Stevens Pt.) 11:2.
Bee, J. W., and E. R. Hall. 1956. Mammals of northern Alaska on the Arctic Slope. Misc. Publ. Univ. Kan., Mus. Nat. Hist. 8:1–309.
Beer, J. R. 1950a. The least weasel in Wisconsin. J. Mamm. 31:146–149.
———. 1950b. The reproductive cycle of the muskrat in Wisconsin. J. Wildl. Mgt. 14:151–156.
———. 1953a. The screech owl as a predator on the big brown bat. J. Mamm. 34:384.
———. 1953b. Two new locality records for mammals in Minnesota. J. Mamm. 34:384–385.
———. 1954. A record of a hoary bat from a cave. J. Mamm. 35:116.
———. 1955a. Survival and movements of banded big brown bats. J. Mamm. 36:242–248.
———. 1955b. Movements of tagged beaver. J. Wildl. Mgt. 19:492–493.
———. 1956. A record of a silver-haired bat in a cave. J. Mamm. 37:282.
———. 1961a. Hibernation in *Perognathus flavescens*. J. Mamm. 42:103.
———. 1961b. Winter home ranges of the red-backed mouse and white-footed mouse. J. Mamm. 42:174–180.
———. 1961c. Seasonal reproduction in the meadow vole. J. Mamm. 42:483–489.
———. 1962. Emergence of thirteen-lined ground squirrels from hibernation. J. Mamm. 43:109.
Beer, J. R., Paul Lukens,and David Olson. 1954. Small mammal populations on the islands of Basswood Lake, Minnesota. Ecology 35:437–445.
Beer, J. R., and C. F. MacLeod. 1955. The harvest mouse in Dakota County. Flicker 27:176.
———. 1966. Seasonal population changes in the prairie deer mouse. Am. Midl. Nat. 76:277–289.
Beer, J. R., C. F. MacLeod, and L. D. Frenzel. 1957. Prenatal survival and loss in some cricetid rodents. J. Mamm. 38:392–402.
Beer, J. R., and R. K. Meyer. 1951. Seasonal changes in the endocrine organs and behavior patterns of the muskrat. J. Mamm. 32:173–191.
Beer, J. R., and A. G. Richards. 1956. Hibernation of the big brown bat. J. Mamm. 37:31–41.
Beer, J. R., and W. Truax. 1950. Sex and age ratios in Wisconsin muskrats. J. Wildl. Mgt. 14:323–331.
Bekoff, Marc. 1975. Predation and aversive conditioning in coyotes. Science 187:1096.
———. 1977a. *Canis latrans* Say. Mamm. Species 79:1–9.
———. 1977b. Social communication in canids: evidence for the evolution of a stereotyped mammalian display. Science 197:1097–1099.
———. 1977c. Mammalian dispersal and the ontogeny of individual behavioral phenotypes. Am. Nat. 111:715–732.
———, ed. 1978. Coyotes: biology, behavior, and management. Academic Press, New York.
———. 1979. Ground scratching by male domestic dogs: a composite signal. J. Mamm. 60:847–848.
Bekoff, Marc, and J. Diamond. 1976. Precopulatory and copulatory behavior in coyotes. J. Mamm. 57:372–375.
Bekoff, Marc, H. L. Hill, and J. B. Mitton. 1975. Behavioral taxonomy in canids by discriminant function analyses. Science 190:1223–1225.
Bekoff, Marc, and R. Jamieson. 1975. Physical development in coyotes (*Canis latrans*), with a comparison to other canids. J. Mamm. 56:685–692.
Bekoff, Marc, and M. C. Wells. 1980. The social ecology of coyotes. Sci. Am. 242(4):130ff.
Belan, Ingrid, P. N. Lehner, and Tim Clark. 1978. Vocalization of the American pine marten, *Martes americana*. J. Mamm. 59:871–874.
Bellamy, D., et al. 1973. Ageing in an island population of the house mouse. Age and Ageing 2:235–250.
Bellrose, F. C. 1950. The relationship of muskrat populations to various marsh and aquatic plants. J. Wildl. Mgt. 14:299–315.
Belwood, J. J., and M. B. Fenton. 1976. Variation in the diet of *Myotis lucifugus* (Chiroptera: Vespertilionidae). Can. J. Zool. 54:1674–1678.

Benton, A. H. 1955. Observations on the life history of the eastern pine mouse. J. Mamm. 36:52–62.

Berg, W. E. 1971. Habitat use, movements, and activity patterns of moose in northwestern Minnesota. MS thesis, Univ. Minn.

———. 1980. Ecology of bobcats in northern Minnesota. Proc. Bobcat Res. Conf., Natl. Wildl. Fed. 55–61.

Berg, W. E., and R. A. Chesness. 1978. Ecology of coyotes in northern Minnesota. Pp. 229–246 *in* M. Bekoff, ed. Coyotes: biology, behavior, and management. Academic Press, New York.

Berger, J. 1977. Organizational systems and dominance in feral horses in the Grand Canyon. Behav. Ecol. Sociobiol. 2:131–146.

Bergerud, A. T. 1967. Management of Labrador caribou. J. Wildl. Mgt. 31:621–642.

———. 1971a. Abundance of forage on the winter range of Newfoundland caribou. Can. Field-Nat. 85:39–52.

———. 1971b. The population dynamics of Newfoundland caribou. Wildl. Monogr. 25:1–55.

———. 1972. Food habits of Newfoundland caribou. J. Wildl. Mgt. 36:913–923.

———. 1973. Movement and rutting behaviour of caribou (*Rangifer tarandus*) at Mount Albert, Quebec. Can. Field-Nat. 87:357–369.

———. 1974a. Rutting behaviour of Newfoundland caribou. IUCN Publ. 24(1):395–435.

———. 1974b. The role of the environment in the aggregation, movement, and disturbance behaviour of caribou. IUCN Publ. 24(2):552–584.

———. 1974c. Decline of caribou in North America following settlement. J. Wildl. Mgt. 38:757–770.

———. 1975. The reproductive season of Newfoundland caribou. Can. J. Zool. 53:1213–1221.

Bergman, R. D., P. Swain, and M. W. Weller. 1970. A comparative study of nesting Forster's and black terns. Wilson Bull. 82:435–444.

Berner, Alfred. 1980. Reproductively speaking, a rat by any other name is still a rat. Proc. Symp. Mammal. Ecol. Hab. Mgt. Minn., Bemidji State Univ. 35–37.

Berner, Alfred, and L. W. Gysel. 1967. Raccoon use of large tree cavities and ground burrows. J. Wildl. Mgt. 31:706–714.

Berry, R. J. 1978. Genetic variation in wild house mice: where natural selection and history meet. Am. Sci. 66:52–60.

Berry, R. J., and M. E. Jakobson. 1975a. Ecological genetics of an island population of the house mouse (*Mus musculus*). J. Zool., London 175:523–540.

———. 1975b. Adaptation and adaptability in wild-living house mice (*Mus musculus*). J. Zool., London 176:391–402.

Berryman, J. H. 1972. The principles of predator control. J. Wildl. Mgt. 36:395–400.

Best, T. L. 1973. Ecological separation of three genera of pocket gophers (Geomyidae). Ecology 54:1311–1319.

Bickham, J. W. 1979. Chromosomal variation and evolutionary relationships of vespertilionid bats. J. Mamm. 60:350–363.

Bider, J. R. 1961. An ecological study of the snowshoe hare *Lepus americanus*. Can. J. Zool. 39:81–103.

———. 1968. Animal activity in uncontrolled terrestrial communities as determined by a sand transect technique. Ecol. Monogr. 38:269–308.

Bider, J. R., P. Thibault, and R. Sarrazin. 1968. Schemes dynamiques spatio-temporels de l'activité de *Procyon lotor* en relation avec le comportement. Mammalia 32:137–163.

Birdsall, D. A. 1974. An analysis of selection at two loci in fluctuating populations of *Microtus*. Can. J. Zool. 52:1457–1462.

Birney, E. C. 1974. Twentieth century records of wolverine in Minnesota. Loon 46:78–81.

Birney, E. C., and E. D. Fleharty. 1968. Comparative success in the application of aging techniques to a population of winter-trapped mink. Southwest. Nat. 13:275–282.

Birney, E. C., R. Jenness, and I. D. Hume. 1980. Evolution of an enzyme system: ascorbic acid biosynthesis in monotremes and marsupials. Evolution 34:230–239.

Bishop, S. C. 1947. Curious behavior of a hoary bat. J. Mamm. 28:293–294 and 409.

Black, C. C. 1963. A review of North American Tertiary Sciuridae. Bull. Mus. Comp. Zool. 130:113–248.

———. 1972. Holarctic evolution and dispersal of squirrels (Rodentia: Sciuridae). Evol. Biol. 6:305–322.

Black, H. L. 1972. Differential exploitation of moths by the bats *Eptesicus fuscus* and *Lasiurus cinereus*. J. Mamm. 53:598–601.

Black, J. D. 1935. Vitality of the Virginia opossum as exhibited in the skeleton. J. Mamm. 16:223.

Blair, W. F. 1940. A study of prairie deer-mouse populations in southern Michigan. Am. Midl. Nat. 24:273–305.

———. 1941. Some data on the home ranges and general life history of the short-tailed shrew, red-backed vole, and woodland jumping mouse in northern Michigan. Am. Midl. Nat. 25:681–685.

———. 1942. Size of home range and notes on the life history of the woodland deer mouse and the eastern chipmunk in northern Michigan. J. Mamm. 23:27–36.

———. 1948. Population density, life span, and mortality rates of small mammals in the blue-grass meadow and blue-grass field associations of southern Michigan. Am. Midl. Nat. 40:395–419.

Blanchard, Harold. 1964. Weight of a large fisher. J. Mamm. 45:487–488.

Blaustein, A. R., and S. I. Rothstein. 1978. Multiple captures of *Reithrodontomys megalotis*: social bonding in a mouse? Am. Midl. Nat. 100:376–383.

Bloom, J. C., J. G. Rogers, Jr., and Owen Maller. 1973. Taste responses of the North American porcupine (*Er-*

ethizon dorsatum). Physiol. Behav. 11:95–98.
Bogan, M. A. 1972. Observations on parturition and development in the hoary bat, *Lasiurus cinereus*. J. Mamm. 53:611–614.
Boggess, E. K., R. D. Andrews, and R. A. Bishop. 1978. Domestic animal losses to coyotes and dogs in Iowa. J. Wildl. Mgt. 42:362–372.
Boice, R. 1972. Some behavioral tests of domestication in Norway rats. Behaviour 42:198–231.
Boise, C. M. 1975. Skull measurements as criteria for aging fishers. N.Y. Fish Game J. 22(1):32–37.
Boles, B. K. 1977. Predation by wolves on wolverines. Can. Field-Nat. 91:68–69.
Booth, E. S. 1946. Notes on the life history of the flying squirrel. J. Mamm. 27:28–30.
Boreman, J. 1972. Social dominance in wild and domestic Norway rats (*Rattus norvegicus*). Anim. Behav. 20:534–542.
Bourdelle, E. 1939. American mammals introduced into France in the contemporary period, especially *Myocaster* [*sic*] and *Ondatra*. J. Mamm. 20:287–291.
Bowker, L. S., and P. G. Pearson. 1975. Habitat orientation and interspecific interaction of *Microtus pennsylvanicus* and *Peromyscus leucopus*. Am. Midl. Nat. 94:491–496.
Bowles, J. B. 1975. Distribution and biogeography of mammals of Iowa. Mus. Texas Tech Univ., Spec. Publ. 9:1–184.
Boxall, P. C. 1979. Interaction between a long-tailed weasel and a snowy owl. Can. Field-Nat. 93:67–68.
Boyce, M. S. 1978. Climatic variability and body size variation in the muskrats (*Ondatra zibethicus*) of North America. Oecologia 36:1–19.
Bozeman, F. M., et al. 1975. Epidemic typhus rickettsiae isolated from flying squirrels. Nature 255:545–547.
Bradley, W. G., and M. K. Yousef. 1975. Thermoregulatory responses in the plains pocket gopher, *Geomys bursarius*. Comp. Biochem. Physiol. 52A:35–38.
Bradt, G. W. 1938. A study of beaver colonies in Michigan. J. Mamm. 19:139–162.
Bramble, D. M. 1978. Origin of the mammalian feeding complex: models and mechanisms. Paleobiology 4:271–301.
Brand, C. J., and L. B. Keith. 1979. Lynx demography during a snowshoe hare decline in Alberta. J. Wildl. Mgt. 43:827–849.
Brand, C. J., L. B. Keith, and C. A. Fischer. 1976. Lynx responses to changing snowshoe hare densities in central Alberta. J. Wildl. Mgt. 40:416–428.
Brander, R. B. 1971. Longevity of wild porcupines. J. Mamm. 52:835.
———. 1973. Life-history notes on the porcupine in a hardwood-hemlock forest in Upper Michigan. Mich. Acad. 5:425–433.
Breakey, D. R. 1963. The breeding season and age structure of feral house mouse populations near San Francisco Bay, California. J. Mamm. 44:153–168.
Breckenridge, W. J. 1929. Actions of the pocket gopher (*Geomys bursarius*). J. Mamm. 10:336–339.
———. 1935. Status of the Minnesota caribou. J. Mamm. 16:327–328.
———. 1946. Weights of a Minnesota moose. J. Mamm. 27:90–91.
———. 1947. An unusual melanistic squirrel. J. Mamm. 28:403–404.
Brenner, F. J. 1964. Reproduction of the beaver in Crawford County, Pennsylvania. J. Wildl. Mgt. 28:743–747.
———. 1968. A three year study of two breeding colonies of the big brown bat, *Eptesicus fuscus*. J. Mamm. 49:775–778.
———. 1973. Hereditary basis of fat and hibernation cycle in the eastern chipmunk. Yearb. Am. Phil. Soc. 1972:347–349.
Brenner, F. J., and P. D. Lyle. 1974. Comparison of photoperiodic influences on the seasonal changes in body weight and energy intake in the eastern chipmunk and the 13-lined ground squirrel. Trans. N.Y. Acad. Sci., Ser. 11.
———. 1975. Effect of previous photoperiodic conditions and visual stimulation on food storage in the eastern chipmunk (*Tamias striatus*). Am. Midl. Nat. 93:227–234.
Brickner, Joe. 1953. Red fox preying on muskrats. J. Mamm. 34:389.
Bridgewater, D. D., and D. F. Penny. 1966. Predation by *Citellus tridecemlineatus* on other vertebrates. J. Mamm. 47:345–346.
Briese, L. A., and M. H. Smith. 1973. Competition between *Mus musculus* and *Peromyscus polionotus*. J. Mamm. 54:968–969.
———. 1974. Seasonal abundance and movement of nine species of small mammals. J. Mamm. 55:615–629.
Brink, C. H., and F. C. Dean. 1966. Spruce seed as food of red squirrels and flying squirrels in interior Alaska. J. Wildl. Mgt. 30:503–512.
Brisbin, I. L. 1966. Energy-utilization in a captive hoary bat. J. Mamm. 47:719–720.
Broadbooks, H. E. 1952. Nest and behavior of a short-tailed shrew, *Cryptotis parva*. J. Mamm. 33:241–243.
———. 1974. Tree nests of chipmunks with comments on associated behavior and ecology. J. Mamm. 55:630–639.
———. 1977. Tree nesting and maternal behavior of chipmunks. Southwest. Nat. 22:154–155.
Bromley, P. T. 1969. Territoriality in pronghorn bucks on the National Bison Range, Moiese, Montana. J. Mamm. 50:81–89.
Bronson, F. H. 1962. Daily and seasonal activity patterns of woodchucks. J. Mamm. 43:425–427.
———. 1963. Some correlates of interaction rate in natural populations of woodchucks. Ecology 44:637–643.
———. 1964. Agonistic behaviour in woodchucks. Anim.

Behav. 12:470–478.

Brooks, J. E., and A. M. Bowerman. 1973. Anticoagulant resistance in wild Norway rats in New York. J. Hyg. 71:217–222.

Brothers, D. R. 1972. A case of anticoagulant rodenticide resistance in an Idaho Norway rat (*Rattus norvegicus*) population. Calif. Vector Views 19(6):41–44

Brower, J. E., and T. J. Cade. 1966. Ecology and physiology of *Napaeozapus insignis* (Miller) and other woodland mice. Ecology 47:46–63.

Brown, E. L. 1889–1901. Diaries of E. L. Brown 1889–1901. Brown Papers A.878E BX.2, Archives/Manuscripts Div., Minn. Histor. Soc.

Brown, J. H., and R. C. Lasiewski. 1972. Metabolism of weasels: the cost of being long and thin. Ecology 53:939–943.

Brown, L. G., and L. E. Yeager. 1945. Fox squirrels and gray squirrels in Illinois. Ill. Nat. Hist. Surv., Bull. 23:449–536.

Brown, L. N. 1964. Reproduction of the brush mouse and white-footed mouse in the central United States. Am. Midl. Nat. 72:226–240.

———. 1967a. Ecological distribution of six species of shrews and comparison of sampling methods in the central Rocky Mountains. J. Mamm. 48:617–623.

———. 1967b. Ecological distribution of mice in the Medicine Bow Mountains of Wyoming. Ecology 48:677–680.

———. 1972. Unique features of tunnel-systems of the eastern mole in Florida. J. Mamm. 53:394–395.

Brown. L. N., and D. Metz. 1966. First record of *Perognathus flavescens* in Wyoming. J. Mamm. 47:118.

Browne, R. A. 1977. Genetic variation in island and mainland populations of *Peromyscus leucopus*. Am. Midl. Nat. 97:1–9.

Brumwell, M. J. 1951. An ecological survey of the Fort Leavenworth Military Reservation. Am. Midl. Nat. 45:187–231.

Bruns, E. H. 1977. Winter behavior of pronghorns in relation to habitat. J. Wildl. Mgt. 41:560–571.

Bryant, M. D. 1945. Phylogeny of Nearctic Sciuridae. Am. Midl. Nat. 33:257–390.

Bryson, R. A., D. A. Baerreis, and W. M. Wendland. 1970. The character of late glacial and post glacial climatic changes. Pp. 53–74 *in* W. Dort, Jr., and J. K. Jones, Jr., eds. Pleistocene and Recent environments of the Central Great Plains. Univ. Press Kan., Spec. Publ. 3.

Buchler, E. R. 1976. Prey selection by *Myotis lucifugus* (Chiroptera: Vespertilionidae). Am. Nat. 110:619–628.

Buchwald, Karen. 1978. Wildcat! Minn. Volun. 41(236):8–15.

Buckner, C. H. 1957. Population studies on small mammals of sotheastern Manitoba. J. Mamm. 38:87–97.

———. 1964. Metabolism, food capacity, and feeding behavior of four species of shrews. Can. J. Zool. 42:259–279.

———. 1966. Populations and ecological relationships of shrews in tamarack bogs of southeastern Manitoba. J. Mamm. 47:181–194.

Bue, G. T., and M. H. Stenlund. 1953. Recent records of the mountain lion, *Felis concolor*, in Minnesota. J. Mamm. 34:390–391.

Buech, R. R., K. Siderits, et al. 1977. Small mammal populations after a wildfire in northeast Minnesota. USDA For. Serv. Res. Pap. NC-151:1–8.

Buech, R. R., R. M. Timm, and K. Siderits. 1977. A second population of rock voles, *Microtus chrotorrhinus*, in Minnesota with comments on habitat. Can. Field-Nat. 91:413–414.

Buechner, H. K. 1950. Life history, ecology, and range use of the pronghorn antelope in Texas. Am. Midl. Nat. 43:257–354.

Buell, M. F., and W. A. Niering. 1957. Fir-spruce-birch forest in northern Minnesota. Ecology 38: 602–610.

Burgdorfer, W., J. C. Cooney, and L. A. Thomas. 1974. Zoonotic potential (Rocky Mountain spotted fever and tularemia) in the Tennessee Valley region. Am. J. Trop. Med. Hyg. 23:109–117.

Burghardt, G. M., R. O. Hietala, and M. R. Pelton. 1972. Knowledge and attitudes concerning black bears by users of the Great Smoky Mountains National Park. IUCN Publ. 23:255–273.

Burkholder, B. L. 1962. Observations concerning wolverine. J. Mamm. 43:263–264.

Burt, W. H. 1940. Territorial behavior and populations of some small mammals in southern Michigan. Misc. Publ. Mus. Zool., Univ. Mich. 45:1–58.

———. 1943. Territoriality and home range concepts as applied to mammals. J. Mamm. 24:346–352.

———. 1957. Mammals of the Great Lakes region. Univ. Mich. Press, Ann Arbor.

———. 1958. The history and affinities of the Recent land mammals of western North America. Pp. 131–154 *in* C. L. Hubbs, ed. Zoogeography. AAAS Publ. 51.

———. 1960. Bacula of North American mammals. Misc. Publ. Mus. Zool., Univ. Mich. 113:1–75.

———, and R. P. Grossenheider. 1976. A field guide to the mammals. 3d ed. Houghton Mifflin, Boston.

Cahn, A. R. 1921. The mammals of Itasca County, Minnesota. J. Mamm. 2:68–74.

———. 1937. The mammals of the Quetico Provincial Park of Ontario. J. Mamm. 18:19–30.

Caldwell, L. D., and J. B. Gentry. 1965. Interactions of *Peromyscus* and *Mus* in a one-acre field enclosure. Ecology 46:189–192.

Calef, G. W., and G. M. Lortie. 1975. A mineral lick of the barren-ground caribou. J. Mamm. 56:240–242.

Calhoun, J. B. 1945. Diel activity rhythms of the rodents, *Microtus ochrogaster* and *Sigmodon hispidus hispidus*. Ecology 26:251–273.

———. 1962. The ecology and sociology of the Norway rat. U.S. Pub. Health Serv. Publ. 1008:viii + 288.

Cameron, A. W. 1964. Competitive exclusion between the rodent genera *Microtus* and *Clethrionomys*. Evolution 18:630–634.

Cameron, D. M., Jr. 1976. Distribution of the southern flying squirrel (*Glaucomys volans*) in Maine. Can. Field-Nat. 90:173–174.

Cameron, G. N., and G. W. Barrett. 1979. Graduate training in mammalogy. Am. Soc. Mamm. Rep. Available from Dr. G. N. Cameron, Dept. Biol. Univ. Houston, Houston, TX. 77004.

Campbell, B. 1939. The shoulder anatomy of the moles: a study of phylogeny and adaptation. Am. J. Anat. 64:1–39.

Canby, T. Y. 1977. The rat, lapdog of the devil. Nat. Geogr. 152:60–87.

Canham, R. P., and D. G. Cameron. 1972. Variation in the serum proteins of the red-backed mice *Clethrionomys rutilus* and *C. gapperi* and its taxonomic significance. Can. J. Zool. 50:217–227.

Carleton, M. D., and R. E. Eshelman. 1979. A synopsis of fossil grasshopper mice, genus *Onychomys*, and their relationships to Recent species. Univ. Mich. Mus. Paleont., Pap. Paleont. 21:1–63.

Carleton, M. D., and P. Myers. 1979. Karyotypes of some harvest mice, genus *Reithrodontomys*. J. Mamm. 60:307–313.

Carlock, J. R. 1974. Collections—environmental hypocrisy? Asn. Midw. Coll. Biol. Teach. News (May):2.

Carlsen, J. C., and R. E. Farmes. 1957. Movement of white-tailed deer tagged in Minnesota. J. Wildl. Mgt. 21:397–401.

Carroll, Diane, and L. L. Getz. 1976. Runway use and population density in *Microtus ochrogaster*. J. Mamm. 57:772–776.

Carter, D. C. 1970. Chiropteran reproduction. Pp. 233–246 *in* B. H. Slaughter and D. W. Walton, eds. About bats. South. Meth. Univ. Press, Dallas.

Cary, J. R., and L. B. Keith. 1979. Reproductive changes in the 10-year population cycle of snowshoe hares. Can. J. Zool. 57:375–390.

Case, R. M. 1978. Interstate highway road-killed animals: a data source for biologists. Wildl. Soc. Bull. 6:8–13.

Case, T. J. 1978. Endothermy and parental care in the terrestrial vertebrates. Am. Nat. 112:861–874.

Casey, G. A., and W. A. Webster. 1975. Age and sex determination of striped skunks (*Mephitis mephitis*) from Ontario, Manitoba, and Quebec. Can. J. Zool. 53:223–226.

Casteel, D. A. 1966. Nest building, parturition, and copulation in the cottontail rabbit. Am. Midl. Nat. 75:160–167.

Cauble, C. 1977. The great grizzly grapple. Nat. Hist. 86(7):74–81.

Cengel, D. J., J. E. Estep, and R. L. Kirkpatrick. 1978. Pine vole reproduction in relation to food habits and body fat. J. Wildl. Mgt. 42:822–833.

Chamberlain, E. B. 1929. Behavior of the least shrew. J. Mamm. 10:250–251.

Chambers, G. D. 1978. Little fox on the prairie. Audubon 80(4):62–71.

Chapman, J. A., Amy L. Harman, and D. E. Samuel. 1977. Reproductive and physiological cycles in the cottontail complex in western Maryland and nearby West Virginia. Wildl. Monogr. 56:1–73.

Chapman, J. A., J. G. Hockman, and M. M. Ojeda C. 1980. *Sylvilagus floridanus*. Mamm. Spp. 136.

Chapman, J. A., and R. P. Morgan II. 1973. Systematic status of the cottontail complex in western Maryland and nearby West Virginia. Wildl. Monogr. 36:1–54.

Chapman, R. C. 1978. Rabies: decimation of a wolf pack in Arctic Alaska. Science 201:365–367.

Cheatum, E. L., and C. W. Severinghaus. 1950. Variations in fertility of the white-tailed deer related to range conditions. Trans. N. Am. Wildl. Conf. 15:170–189.

Chelberg, D. A. 1972. The science museum of Minnesota. Minn. Volun. 35(205):36–42.

Chesness, R. A. 1974. The coyote—our lone ranger. Minn. Volun. 37(217):45–52.

Chesness, R. A., M. M. Nelson, and W. A. Longley. 1968. The effect of predator removal on pheasant reproductive success. J. Wildl. Mgt. 32:683–697.

Chiasson, R. B. 1954. The phylogenetic significance of rodent cheek pouches. J. Mamm. 35:425–427.

Choate, J. R. 1972. Variation within and among populations of the short-tailed shrew in Connecticut. J. Mamm. 53:116–128.

Choate, J. R., and E. D. Fleharty. 1973. Habitat preferences and spatial relations of shrews in a mixed grassland in Kansas. Southwest. Nat. 18:110–112.

Choate, J. R., E. D. Fleharty, and R. J. Little. 1973. Status of the spotted skunk, *Spilogale putorius*, in Kansas. Trans. Kan. Acad. Sci. 76:226–233.

Choate, J. R., and H. H. Genoways. 1975. Collections of Recent mammals in North America. J. Mamm. 56:452–502.

Choate, J. R., and Janet E. Krause. 1976. Historical biogeography of the gray fox (*Urocyon cinereoargenteus*) in Kansas. Trans. Kan. Acad. Sci. 77:231–235.

Choate, J. R., and D. M. Terry. 1974. Observations on habitat preference of *Onychomys leucogaster* (Rodentia: Muridae) on the Central Great Plains. Trans. Kan. Acad. Sci. 76:263–265.

Choate, J. S. 1972. Effects of stream channeling on wetlands in a Minnesota watershed. J. Wildl. Mgt. 36:940–944.

Choquette, L. P. E., et al. 1972. Parasites and diseases of bison in Canada. III. Anthrax outbreaks in the last decade in northern Canada and control measures. Can. Field-Nat. 86:127–132.

Chorn, John, and R. S. Hoffman. 1978. *Ailuropoda melanoleuca*. Mamm. Species 110:1–6.

Christian, D. P. 1975. Vulnerability of meadow voles, *Microtus pennsylvanicus*, to predation by domestic cats. Am. Midl. Nat. 93:498–502.

Christian, J. J. 1950. Behavior of the mole (*Scalopus*) and the shrew (*Blarina*). J. Mamm. 31:281–287.

———. 1956. The natural history of a summer aggregation of the big brown bat, *Eptesicus fuscus fuscus*. Am. Midl. Nat. 55:66–95.

———. 1964. Selective feeding on dandelion roots by muskrats. J. Mamm. 45:147.

———. 1970. Social subordination, population density, and mammalian evolution. Science 168:84–90.

Christian, J. J., and D. E. Davis. 1956. The relationship between adrenal weight and population status of urban Norway rats. J. Mamm. 37:475–486.

Christian, J. J., E. Steinberger, and T. D. McKinney. 1972. Annual cycle of spermatogenesis and testis morphology in woodchucks. J. Mamm. 53:708–716.

Christianson, Lee. 1977. Winter movements of *Peromyscus* across a lake in northern Minnesota. J. Mamm. 58:244.

Church, D. C., and W. H. Hines. 1978. Ruminoreticular characteristics of elk. J. Wildl. Mgt. 42:654–659.

Churcher, C. S. 1959. The specific status of the New World red fox. J. Mamm. 40:513–520.

Clark, D. R., Jr., T. H. Kunz, and T. E. Kaiser. 1978. Insecticides applied to a nursery colony of little brown bats (*Myotis lucifugus*): lethal concentrations in brain tissues. J. Mamm. 59:84–91.

Clark, D. R., Jr., and T. G. Lamont. 1976. Organochlorine residues and reproduction in the big brown bat. J. Wildl. Mgt. 40:249–254.

Clark, J. M. 1975. The effects of selection and human preference on coat colour gene frequencies in urban cats. Heredity 35:195–210.

Clark, S. P. 1970. Field experience of feral mink in Yorkshire and Lancashire. Mamm. Rev. 1:41–47.

Clark, T. W. 1970. Early growth, development, and behavior of the Richardson ground squirrel (*Spermophilus richardsoni elegans*). Am. Midl. Nat. 83:197–205.

Clem, M. K. 1977. Interspecific relations of fishers and martens in Ontario during winter. Proc. 1975 Pred. Symp., Univ. Mont., Missoula:165–182.

Clough, G. C. 1959. Extension of range of the woodland jumping mouse. J. Mamm. 40:449.

———. 1963. Biology of the arctic shrew, *Sorex arcticus*. Am. Midl. Nat. 69:69–81.

———. 1964. Local distribution of two voles: evidence for interspecific interaction. Can. Field-Nat. 78:80–89.

Coblentz, B. E. 1970. Food habits of George Reserve deer. J. Wildl. Mgt. 34:535–540.

Cockrum, E. L. 1952. Mammals of Kansas. Univ. Kan. Publ., Mus. Nat. Hist. 7:1–303.

Cockrum, E. L., and S. P. Cross. 1964. Time of bat activity over water holes. J. Mamm. 45:635–636.

Cole, F. R., and G. O. Batzli. 1978. Influence of supplemental feeding on a vole population. J. Mamm. 59:809–819.

Cole, G. F. 1972. Grizzly bear-elk relationships in Yellowstone National Park. J. Wildl. Mgt. 36:556–561.

Cole, Glen. 1979. Mission-oriented research in Voyageurs National Park. Second Conf. Sci. Res. Natl. Parks. In press.

Cole, L. C. 1954. The population consequences of life history phenomena. Quart. Rev. Biol. 29:103–137.

Coles, R. W. 1970. Pharyngeal and lingual adaptations in the beaver. J. Mamm. 51:424–425.

Colinvaux, P. A., and B. D. Barnett. 1979. Lindeman and the ecological efficiency of wolves. Am. Nat. 114:707–718.

Collins, L. R. 1973. Monotremes and marsupials: a reference for zoological institutions. Smithsonian Inst., Publ. 4888.

Coman, B. J., and Hans Brunner. 1972. Food habits of the feral house cat in Victoria. J. Wildl. Mgt. 36:848–853.

Conaway, C. H. 1952. Life history of the water shrew (*Sorex palustris navigator*). Am. Midl. Nat. 48:219–248.

———. 1959. The reproductive cycle of the eastern mole. J. Mamm. 40:180–194.

Conaway, C. H., K. C. Sadler, and D. H. Hazelwood. 1974. Geographic variation in litter size and onset of breeding in cottontails. J. Wildl. Mgt. 38:473–481.

Conaway, C. H., and H. M. Wight. 1962. Onset of reproductive season and first pregnancy of the season in cottontails. J. Wildl. Mgt. 26:278–290.

Conaway, C. H., H. M. Wight, and K. C. Sadler. 1963. Annual production by a cottontail population. J. Wildl. Mgt. 27:171–175.

Connolly, Guy E., et al. 1976. Sheep killing behavior of captive coyotes. J. Wildl. Mgt. 40:400–407.

Connolly, M. S. 1979. Time tables in home range usage by gray squirrels (*Sciurus carolinensis*). J. Mamm. 60:814–817.

Connor, P. F. 1959. The bog lemming, *Synaptomys cooperi*, in southern New Jersey. Publ. Mus. Mich. State Univ., Biol. Ser. 1:161–248.

Conover, M. R., J. G. Francik, and D. E. Miller. 1977. An experimental evaluation of aversive conditioning for controlling coyote predation. J. Wildl. Mgt. 41:775–779.

———. 1979. Aversive conditioning in coyotes: a reply. J. Wildl. Mgt. 43:209–211.

Constantine, D. G. 1958. Ecological observations on lasiurine bats in Georgia. J. Mamm. 39:64–70.

———. 1959. Ecological observations on lasiurine bats in the North Bay area of California. J. Mamm. 40:13–15.

———. 1966. Ecological observations on lasiurine bats in Iowa. J. Mamm. 47:34–41.

———. 1970. Bats in relation to the health, welfare, and economy of man. Pp. 319–449 *in* W. Wimsatt, ed. Biology of bats. Vol. 2. Academic Press, New York.

Cook, D. B., and W. J. Hamilton, Jr. 1957. The forest, the fisher, and the porcupine. J. For. 55:719–722.

Cook, J. A., and C. R. Terman. 1977. Influence of displacement distance and vision on homing behavior of the white-footed mouse (*Peromyscus leucopus noveboracensis*). J. Mamm. 58:58–66.

Cornish, Lynn M., and W. N. Bradshaw. 1978. Patterns in twelve reproductive parameters for the white-footed mouse (*Peromyscus leucopus*). J. Mamm. 59:731–739.

Cory, C. B. 1912. The mammals of Illinois and Wisconsin. Field Mus. Nat. Hist. Publ. 153, Zool. Ser. 11:1–505 (cited in Haugen 1961).

Costello, D. F. 1966. The world of the porcupine. Lippincott, Philadelphia.

Cothran, E. G., E. G. Zimmerman, and C. F. Nadler, 1977. Genic differentiation and evolution in the ground squirrel subgenus *Ictidomys* (genus *Spermophilus*). J. Mamm. 58:610–622.

Coulter, J. C. 1961. An investigation of possible commensalism in red squirrels, yellow-bellied sapsuckers, and hummingbirds at the Lake Itasca Biological Station. Proc. Minn. Acad. Sci. 29:272–274.

Coulter, M. W. 1960. The status and distribution of fisher in Maine. J. Mamm. 41:1–9.

Coutts, R. A., M. B. Fenton, and E. Glen. 1973. Food intake by captive *Myotis lucifugus* and *Eptesicus fuscus* (Chiroptera: Vespertilionidae). J. Mamm. 54:985–990.

Coventry, A. F. 1932. Notes on the Mearns flying squirrel. Can. Field-Nat. 46:75–78.

Cowan, I. M. 1956. What and where are the mule and black-tailed deer? Pp. 334–359 *in* W. P. Taylor, ed. The deer of North America. Stackpole, Harrisburg, Pa.

Cowan, I. M., and C. J. Guiget. 1975. The mammals of British Columbia. B. C. Prov. Mus. Hbk. 11.

Cowan, I. M., and R. H. Mackay. 1950. Food habits of the marten (*Martes americana*) in the Rocky Mountain region of Canada. Can. Field-Nat. 64:100–104.

Cowardin, L. M., and D. H. Johnson. 1973. A preliminary classification of wetland plant communities in north-central Minnesota. USFWS Spec. Sci. Rep., Wildl. 168:1–33.

Cowles, C. J., R. L. Kirkpatrick, and J. O. Newell. 1977. Ovarian follicular changes in gray squirrels as affected by season, age, and reproductive state. J. Mamm. 58:67–73.

Cox, W. T. 1936. Snowshoe rabbit migration, tick infestation, and weather cycles. J. Mamm. 17:216–221.

Crabb, W. D. 1941. Food habits of the prairie spotted skunk in southeastern Iowa. J. Mamm. 22:349–364.

———. 1948. The ecology and management of the prairie spotted skunk in Iowa. Ecol. Monogr. 18:201–232.

Craighead, F. C., Jr., and J. J. Craighead. 1972. Grizzly bear prehibernation and denning activities as determined by radiotracking. Wildl. Monogr. 32:1–35.

Craighead, J. J., G. Atwell, and B. W. O'Gara. 1972. Elk migrations in and near Yellowstone National Park. Wildl. Monogr. 29:1–48.

Craighead, J. J., et al. 1973. Home ranges and activity patterns of nonmigratory elk of the Madison drainage herd as determined by biotelemetry. Wildl. Monogr. 33:1–48.

Crandall, L. S. 1964. The management of wild mammals in captivity. Univ. Chicago Press, Chicago.

Crête, Michel. 1976a. Estimation of winter mortality of deer in Quebec. Can. Field-Nat. 90:397–403.

———. 1976b. Importance of winter climate in the decline of deer harvest in Quebec. Can. Field-Nat. 90:404–409.

Criddle, Stuart. 1943. The little northern chipmunk in southern Manitoba. Can. Field-Nat. 57:81–86.

———. 1950. The *Peromyscus maniculatus bairdi* complex in Manitoba. Can. Field-Nat. 64:169–177.

Crompton, A. W. 1963. On the lower jaw of *Diarthrognathus* and the origin of the mammalian lower jaw. Proc. Zool. Soc. London 140:697–750.

———. 1971. The origin of the tribosphenic molar. J. Linn. Soc. (zool.) 50(Suppl. 1):65–87.

Crompton, A. W., and F. A. Jenkins. 1973. Mammals: a review of mammalian origins. Annu. Rev. Earth Plan. Sci. 1:131–153.

Crompton, A. W., and Pamela Parker. 1978. Evolution of the mammalian masticatory apparatus. Am. Sci. 66:192–201.

Crowcroft, Peter. 1966. Mice all over. Dufour editions, Chester, Pa.

Crowcroft, Peter, and F. P. Rowe. 1963. Social organization and territorial behaviour in the wild house mouse (*Mus musculus* L.). Proc. Zool. Soc. London 140:517–531.

Crowe, D. M. 1975a. Aspects of ageing, growth, and reproduction of bobcats from Wyoming. J. Mamm. 56:117–198.

———. 1975b. A model for exploited bobcat populations in Wyoming. J. Wildl. Mgt. 39:408–415.

Crowe, D. M., and M. D. Strickland. 1975. Dental annulation in the American badger. J. Mamm. 56:269–272.

Crowell, K. L. 1973. Experimental zoogeography: introductions of mice to small islands. Am. Nat. 107:535–558.

Crowner, A. W., and G. W. Barrett. 1979. Effects of fire on the small mammal component of an experimental grassland community. J. Mamm. 60:803–813.

Cumbie, P. M. 1975. Mercury in hair of bobcats and raccoons. J. Wildl. Mgt. 39:419–425.

Curtis, J. D., and E. L. Kozicky. 1944. Observation on the eastern porcupine. J. Mamm. 25:137–146.

Cuthbert, J. H. 1973. The origin and distribution of feral mink in Scotland. Mamm. Rev. 3:97–103.

Cutter, W. L. 1958a. Denning of the swift fox in northern Texas. J. Mamm. 39:70–74.

———. 1958b. Food habits of the swift fox in northern Texas. J. Mamm. 39:527–532.

Dagg, A. I., and A. Taub. 1970. Flehmen. Mammalia 34:686–695.

Dahlberg, B. L., and R. C. Guettinger. 1956. The white-tailed deer in Wisconsin. Wis. Cons. Dept. Tech. Wildl. Bull. 14:1–282.

Dalquest, W. W., and W. Kilpatrick. 1973. Dynamics of pocket gopher distribution on the Edwards plateau of Texas. Southwest. Nat. 18:1–9.

Danell, Kjell. 1978. Intra- and interannual changes in habitat selection by the muskrat. J. Wildl. Mgt. 42:540–549.

Dapson, R. W. 1968. Reproduction and age structure in a population of short-tailed shrews, *Blarina brevicauda*. J. Mamm. 49:205–214.

Dapson, R. W., E. H. Studier, Mary J. Buckingham, and Ann L. Studier. 1977. Histochemistry of odoriferous secretions from integumentary glands in three species of bats. J. Mamm. 58:531–535.

Davis, D. D. 1964. The giant panda: a morphological study of evolutionary mechanisms. Fieldiana: Zool. Mem. 3:1–339.

Davis, D. E. 1966. The moult of woodchucks (*Marmota monax*). Mammalia 30:640–644.

———. 1967a. The annual rhythm of fat deposition in woodchucks (*Marmota monax*). Physiol. Zool. 40:391–402.

———. 1967b. The role of environmental factors in hibernation of woodchucks (*Marmota monax*). Ecology 48:683–689.

———. 1976. Hibernation and circannual rhythms of food consumption in marmots and ground squirrels. Quart. Rev. Biol. 51:477–514.

Davis, D. E., J. J. Christian, and F. Bronson. 1964. Effect of exploitation on birth, mortality, and movement rates in a woodchuck population. J. Wildl. Mgt. 28:1–9.

Davis, D. E., and E. P. Finnie. 1975. Entrainment of circannual rhythm in weight of woodchucks. J. Mamm. 56:199–203.

Davis, J. G., and R. K. Meyer. 1973a. FSH and LH in the snowshoe hare during the increasing phase of the ten-year cycle. Gen. Comp. Endocrinol. 20:53–60.

———. 1973b. Seasonal variation in LH and FSH of bilaterally castrated snowshoe hare. Gen. Comp. Endocrinol. 20:61–68.

Davis, J. W., L. H. Karstad, and D. O. Trainer. 1970. Infectious diseases of wild mammals. Iowa State Univ. Press, Ames.

Davis, W. B. 1942. The moles of Texas. Am. Midl. Nat. 27:380–386.

Davis, W. H. 1964. Winter awakening patterns in the bats *Myotis lucifugus* and *Pipistrellus subflavus*. J. Mamm. 45:645–647.

———. 1966. Population dynamics of the bat *Pipistrellus subflavus*. J. Mamm. 47:383–396.

———. 1967. A *Myotis lucifugus* with two young. Bat Res. News 8:3.

———. 1970. Hibernation: ecology and physiological ecology. Pp. 265–300 *in* W. A. Wimsatt, ed. Biology of bats. Vol. 1. Academic Press, New York.

Davis, W. H., and R. W. Barbour. 1965. The use of vision in flight by the bat, *Myotis sodalis*. Am. Midl. Nat. 74:497–499.

Davis, W. H., R. W. Barbour, and M. D. Hassell. 1968. Colonial behavior of *Eptesicus fuscus*. J. Mamm. 49:44–50.

Davis, W. H., and J. R. Beer. 1959. Winter activity of the eastern chipmunk in Minnesota. J. Mamm. 40:444–445.

Davis, W. H., and C. H. Ernst. 1971. The taxonomic status of *Zapus* in northwestern Minnesota. Am. Midl. Nat. 85:265–267.

Davis, W. H., and J. W. Hardin. 1967. Homing in *Lasionycteris noctivagans*. J. Mamm. 48:323.

Davis, W. H., M. D. Hassell, and C. L. Rippy. 1965. Maternity colonies in the bat *Myotis l. lucifugus* in Kentucky. Am. Midl. Nat. 73:161–165.

Davis, W. H., and H. B. Hitchcock. 1964. Notes on sex ratios of hibernating bats. J. Mamm. 45:475–476.

———. 1965. Biology and migration of the bat, *Myotis lucifugus*, in New England. J. Mamm. 46:296–313.

Davis, W. H., and W. Z. Lidicker, Jr. 1956. Winter range of the red bat, *Lasiurus borealis*. J. Mamm. 37:280–281.

Davis, W. H., and R. E. Mumford. 1962. Ecological notes on the bat *Pipistrellus subflavus*. Am. Midl. Nat. 68:394.

Davis, W. H., and Ola B. Wright. 1967. Responses of bats from temperate regions to changes in ambient temperature. Biol. Bull. 132:320–328.

Davison, R. P., et al. 1978. The efficiency of food utilization and energy requirements of captive female fishers. J. Wildl. Mgt. 42:811–821.

Dawson, M. R. 1967. Fossil history of the families of Recent mammals. Pp. 12–53 *in* S. Anderson and J. K. Jones, Jr. Recent mammals of the world. Ronald Press, New York.

Day, M. G. 1968. Food habits of British stoats (*Mustela erminea*) and weasels (*Mustela nivalis*). J. Zool., London 155:485–497.

Dear, L. S. 1955. Cougar or mountain lion reported in northwestern Ontario. Can. Field-Nat. 69:26.

DeBlase, A. F., and R. E. Martin. 1974. A manual of mammalogy. Wm. C. Brown, Dubuque, Iowa.

De Coursey, Patricia J. 1960a. Daily light sensitivity rhythm in a rodent. Science 131:33–35.

———. 1960b. Phase control of activity in a rodent. Cold Spr. Harbor Symp. Quant. Biol. 25:49–55.

———. 1972. LD ratios and the entrainment of circadian activity in a nocturnal and diurnal rodent. J. Comp. Physiol. 78:221–235.

Delong, K. T. 1966. Population ecology of feral house mice: interference by *Microtus*. Ecology 47:481–484.

DeMar, R., and H. R. Barghusen. 1972. Mechanics and evolution of the synapsid jaw. Evolution 26:622–637.

DeMeules, D. H. 1954. Possible anti-adrenalin action of shrew venom. J. Mamm. 35:425.

Denny, R. N. 1974. Impact of uncontrolled dogs on wildlife and livestock. Trans. N. Am. Wildl. Nat. Resour. Conf. 39:257–291.

de Vos, A. 1951. Recent findings in fisher and marten ecology and management. Trans. N. Amer. Wildl. Conf. 16:498–507.

———. 1953. Bobcat preying on porcupine? J. Mamm. 34:129–130.

———. 1958. Summer observations of moose behavior in Ontario. J. Mamm. 39:128–139.

———. 1964. Range changes of mammals in the Great Lakes region. Am. Midl. Nat. 71:210–231.

de Vos, A., P. Brokx, and V. Geist. 1967. A review of social behavior of the North American cervids during the reproductive period. Am. Midl. Nat. 77:390–417.

deVos, A., and D. I. Gillespie. 1960. A study of woodchucks on an Ontario farm. Can. Field-Nat. 74:130–145.

Dewsbury, D. A. 1975. Copulatory behavior of white-footed mice (*Peromyscus leucopus*). J. Mamm. 56:420–428.

———. 1976. Copulatory behavior of pine voles (*Microtus pinetorum*). Percep. Motor Skills 43:91–94.

Dewsbury, D. A., D. Q. Estep, and D. L. Lanier. 1977. Estrous cycles of nine species of muroid rodents. J. Mamm. 58:89–92.

Dice, L. R., and P. J. Clark. 1962. Variation in measures of behavior among three races of *Peromyscus*. Contr. Lab. Vert. Biol. Univ. Mich. 76:1–28.

Dickerman, R. W., and J. R. Tester. 1957. *Onychomys leucogaster* in Kittson County, Minnesota. J. Mamm. 38:269.

Diersing, V. E. 1980. Systematics and evolution of the pygmy shrews (subgenus *Microsorex*) of North America. J. Mamm. 61:76–101.

Dirschl, H. J. 1962. Sieve mesh size related to analysis of antelope rumen contents. J. Wildl. Mgt. 26:327–328.

Dodds, D. G. 1965. Reproduction and productivity of snowshoe hares in Newfoundland. J. Wildl. Mgt. 29:303–315.

Dodds, D. G., and A. M. Martell. 1971a. The recent status of the marten, *Martes americana americana* (Turton) in Nova Scotia. Can. Field-Nat. 85:61–62.

———. 1971b. The recent status of the fisher, *Martes pennanti* (Erxleben), in Nova Scotia. Can. Field-Nat. 85:62–65.

Dolan, Patricia G., and Dilford C. Carter. 1977. *Glaucomys volans*. Mamm. Species 78:1–6.

Dolbeer, R. A. 1973. Reproduction in the red squirrel (*Tamiasciurus hudsonicus*) in Colorado. J. Mamm. 54:536–540.

Dolbeer, R. A., and W. R. Clark. 1975. Population ecology of snowshoe hares in the central Rocky Mountains. J. Wildl. Mgt. 39:535–549.

Doremus, H. M. 1965. Heart rate, temperature, and respiration rate of the short-tailed shrew in captivity. J. Mamm. 46:424–425.

Dorn, R. D. 1971. White-tailed deer in southeastern Minnesota: winter observations. J. Minn. Acad. Sci. 37:16–18.

Dorney, R. S. 1954. Ecology of marsh raccoons. J. Wildl. Mgt. 18:217–225.

Dorrance, M. J., P. J. Savage, and D. E. Huff. 1975. Effects of snowmobiles on white-tailed deer. J. Wildl. Mgt. 39:563–569.

Doucet, G. J. 1973. House cat as predator of snowshoe hare. J. Wildl. Mgt. 37:591.

Doucet, G. J., and J. R. Bider. 1969. Activity of *Microtus pennsylvanicus* related to moon phase and moonlight revealed by the sand transect technique. Can. J. Zool. 47:1183–1186.

———. 1974. The effects of weather on the activity of the masked shrew. J. Mamm. 55:348–363.

Doucet, G. J., J.-P. R. Sarrazin, and J. R. Bider. 1974. Use of highway overpass embankments by the woodchuck, *Marmota monax*. Can. Field-Nat. 88:187–190.

Doutt, J. K., C. A. Heppenstall, and J. E. Guilday. 1973. Mammals of Pennsylvania. 3d ed. Pa. Game Comm., Harrisburg.

Dow, S. A., Jr., and P. L. Wright. 1962. Changes in mandibular dentition associated with age in pronghorn antelope. J. Wildl. Mgt. 26:1–18.

Downing, S. C., and D. H. Baldwin. 1961. Sharp-shinned hawk preys on red bat. J. Mamm. 42:540.

Drickamer, L. C. 1970. Seed preferences of wild caught *Peromyscus maniculatus bairdii* and *P. leucopus noveboracensis*. J. Mamm. 51:191–194.

———. 1972. Experience and selection behavior in the food habits of *Peromyscus*: use of olfaction. Behaviour 41:269–287.

———. 1974. Sexual maturation of female house mice: social inhibition. Dev. Psychobiol. 7:257–265.

———. 1976. Hypotheses linking food habits and habitat selection in *Peromyscus*. J. Mamm. 57:763–766.

Drolet, C.-A. 1978. Use of forest clear-cuts by white-tailed deer in southern New Brunswick and central Nova Scotia. Can. Field-Nat. 92:275–282.

Duby, R. T., and H. F. Travis. 1972. Photoperiodic control

of fur growth and reproduction in the mink (*Mustela vison*). J. Exp. Zool. 182:217–225.

Duke, K. L. 1957. Reproduction in *Perognathus*. J. Mamm. 38:207–210.

Dunford, C. 1970. Behaviorial aspects of spatial organization in the chipmunk, *Tamias striatus*. Behaviour 36:215–231.

———. 1972. Summer activity of eastern chipmunks. J. Mamm. 53:176–180.

———. 1977. Kin selection for ground squirrel alarm calls. Am. Nat. 111:782–785.

Dunmire, W. W. 1955. Sex dimorphism in the pelvis of rodents. J. Mamm. 36:356–361.

Dunstone, N., and W. Sinclair. 1978a. Comparative aerial and underwater visual acuity of the mink, *Mustela vison* Schreber, as a function of discrimination distance and stimulus luminance. Anim. Behav. 26:6–13.

———. 1978b. Orienting behaviour during aerial and underwater visual discrimination by the mink (*Mustela vison* Schreber). Anim. Behav. 26:14–21.

Dusek, G. L. 1975. Range relations of mule deer and cattle in prairie habitat. J. Wildl. Mgt. 39:605–616.

Dwyer, P. D. 1971. Temperature regulation and cave-dwelling bats: an evolutionary perspective. Mammalia 35:424–455.

Eadie, W. R. 1944. The short-tailed shrew and field mouse predation. J. Mamm. 25:359–364.

———. 1948. Shrew-mouse predation during low mouse abundance. J. Mamm. 29:35–37.

———. 1952. Shrew predation and vole populations on a localized area. J. Mamm. 33:185–189.

———. 1953. Response of *Microtus* to vegetative cover. J. Mamm. 34:263–264.

———. 1954. Skin gland activity and pelage descriptions in moles. J. Mamm. 35:186–196.

Eadie, W. R., and W. J. Hamilton, Jr. 1956. Notes on reproduction in star-nosed mole. J. Mamm. 37:223–231.

Easterla, D. A. 1967. First specimens of plains pocket mouse from Missouri. J. Mamm. 48:479–480.

Easterla, D. A., and L. C. Watkins. 1967. Silver-haired bat in southwestern Iowa. J. Mamm. 48:327.

———. 1970. Breeding of *Lasionycteris noctivagans* and *Nycticeius humeralis* in southwestern Iowa. Am. Midl. Nat. 84:254–255.

Eberhardt, L. L., T. J. Peterle, and R. Schofield. 1963. Problems in a rabbit population study. Wildl. Monogr. 10:1–51.

Ebert, Patricia D., and Janet S. Hyde. 1976. Selection for agonistic behavior in wild female *Mus musculus*. Behav. Genet. 6:291–304.

Ecke, D. H. 1955. The reproductive cycle of the Mearns cottontail in Illinois. Am. Midl. Nat. 53:294–311.

Eddy, Samuel, and A. C. Hodson. 1961. Taxonomic keys to the common animals of the north central states. Burgess, Minneapolis.

Edwards, W. R., and L. Eberhardt. 1967. Estimating cottontail abundance from live-trapping data. J. Wildl. Mgt. 31:87–96.

Egoscue, H. J. 1960. Laboratory and field studies of the northern grasshopper mouse. J. Mamm. 41:99–110.

———. 1979. *Vulpes velox*. Mamm. Spp. 122.

Egoscue, H. J., J. G. Bittmenn, and J. A. Petrovich. 1970. Some fecundity and longevity records for captive small mammals. J. Mamm. 51:622–623.

Ehrlich, P. R., and L. C. Birch. 1967. The "balance of nature" and "population control." Am. Nat. 101:97–107.

Einarsen, A. S. 1948. The pronghorn antelope. Stackpole, Harrisburg, Pa.

———. 1956. Life of the mule deer. Pp. 363–390 *in* W. P. Taylor, ed. The deer of North America. Stackpole, Harrisburg, Pa.

Eisenberg, J. F. 1963. The behavior of heteromyid rodents. Univ. Calif. Publ. Zool.

Eisenberg, J. F., and D. G. Kleiman. 1972. Olfactory communication in mammals. Annu. Rev. Ecol. Syst. 3:1–32.

Elder, W. H., and C. M. Hayden. 1977. Use of discriminant function in taxonomic determination of canids from Missouri. J. Mamm. 58:17–24.

Ellins, S. M., S. R. Catalano, and S. A. Schechinger. 1977. Conditioned taste aversion: a field application to coyote predation on sheep. Behav. Biol. 20:91–95.

Elliott, L. 1978. Social behavior and foraging ecology of the eastern chipmunk (*Tamias striatus*) in the Adirondack Mountains. Smithson. Contr. Zool. 265.

Ellis, J. E., and Michael Travis. 1975. Comparative aspects of foraging behaviour of pronghorn antelope and cattle. J. Appl. Ecol. 12:411–420.

Ellis, L. S., V. E. Diersing, and D. F. Hoffmeister. 1978. Taxonomic status of short-tailed shrews (*Blarina*) in Illinois. J. Mamm. 59:305–311.

Ellis, L. S., and L. R. Maxson. 1979. Evolution of the chipmunk genera *Eutamias* and *Tamias*. J. Mamm. 60:331–334.

Elton, C., and M. Nicholson. 1942. Fluctuations in numbers of the muskrat (*Ondatra zibethica*) in Canada. J. Anim. Ecol. 11:96–126.

Elton, C. S. 1942. Voles, mice, and lemmings: problems in population dynamics. Oxford Univ. Press, London.

———. 1958. The ecology of invasions by animals and plants. Methuen, London.

Elwell, A. S. 1962. Blue jay preys on young bats. J. Mamm. 43:434.

Elwell, A. S., C. S. Holt, and C. H. Fuchsman. 1973. Environmental biological aspects of water management alternatives in the Red Lake River subbasin, Minnesota. Bemidji State College, Center for Envir. Studies: ix + 155.

Enders, R. K. 1952. Reproduction in the mink (*Mustela

vison). Proc. Am. Philos. Soc. 96:691–755.

Engel, R. L., and T. A. Vaughan. 1966. A coyote-golden eagle association. J. Mamm. 47:143.

Engstrom, M. D., and J. R. Choate. 1979. Systematics of the northern grasshopper mouse (*Onychomys leucogaster*) on the central Great Plains. J. Mamm. 60:723–739.

Erickson, A. B. 1939. Beaver populations in Pine County, Minnesota. J. Mamm. 20:195–201.

———. 1946. Incidence of worm parasites in Minnesota Mustelidae and host lists and keys to North American species. Am. Midl. Nat. 36:494–509.

Erickson, A. B., and G. T. Bue. 1954. Additional mule deer records for Minnesota. J. Mamm. 35:457–458.

Erickson, A. B., et al. 1961. The white-tailed deer of Minnesota. Minn. Dept. Cons. Tech. Bull. 5:1–64.

Erickson, Bruce. 1962. A description of *Castoroides ohioensis* from Minnesota. Proc. Minn. Acad. Sci. 30(1):6–13.

Ernst, C. H. 1971. Skull key to adult Minnesota mammals. J. Minn. Acad. Sci. 37:32–35.

Ernst, C. H., and E. M. Ernst. 1972. The eastern chipmunk, *Tamias striatus*, in southwestern Minnesota, U.S.A. Can. Field-Nat. 86:377.

Ernst, C. H., and Lee French. 1977. Mammals of southwestern Minnesota. J. Minn. Acad. Sci. 43:28–31.

Errington, P. L. 1933. Bobwhite winter survival in an area heavily populated with gray foxes. Iowa State Coll. J. Sci. 8:127–130.

———. 1935. Food habits of mid-west foxes. J. Mamm. 16:192–200.

———. 1937a. Summer food habits of the badger in northwestern Iowa. J. Mamm. 18:213–216.

———. 1937b. The breeding season of the muskrat in northwest Iowa. J. Mamm. 18:333–337.

———. 1937c. Drowning as a cause of mortality in muskrats. J. Mamm. 18:497–500.

———. 1938. Observations on muskrat damage to corn and other crops in central Iowa. J. Agr. Res. 57:415–421.

———. 1939a. Reactions of muskrat populations to drought. Ecology 20:168–186.

———. 1939b. Observations on young muskrats in Iowa. J. Mamm. 20:465–478.

———. 1940. Natural restocking of muskrat-vacant habitats. J. Wildl. Mgt. 4:173–185.

———. 1941. Versatility in feeding and population maintenance of the muskrat. J. Wildl. Mgt. 5:68–89.

———. 1943. An analysis of mink predation upon muskrats in north-central United States. Iowa Exp. Sta. Res. Bull. 320:799–924.

———. 1946. Predation and vertebrate populations. Q. Rev. Biol. 21:144–177, 221–245.

———. 1948. Environmental control for increasing muskrat populations. Trans. N. Am. Wildl. Conf. 13:596–607.

———. 1951. Concerning fluctuations in populations of the prolific and widely distributed muskrat. Am. Nat. 85:273–292.

———. 1954a. On the hazards of overemphasizing numerical fluctuations in studies of "cyclic" phenomena in muskrat populations. J. Wildl. Mgt. 18:66–90.

———. 1954b. The special responsiveness of minks to epizootics in muskrat populations. Ecol. Monogr. 24:377–393.

———. 1956. Factors limiting higher vertebrate populations. Science 124:304–307.

———. 1957a. Of population cycles and unknowns. Cold Spring Harbor Symp. Quant. Biol. 22:287–300.

———. 1957b. Of men and marshes. Iowa State Univ. Press, Ames.

———. 1963a. Muskrat populations. Iowa State Univ. Press, Ames.

———. 1963b. The phenomenon of predation. Amer. Sci. 51:180–192.

———. 1967. Of predation and life. Iowa State Univ. Press, Ames.

Errington, P. L., Frances Hamerstrom, and F. N. Hamerstrom, Jr. 1940. Great horned owl and its prey in north central United States. Agr. Exp. Sta., Iowa State Coll. Agr. Mech. Arts, Res. Bull. 277:757–850.

Euler, D. L., B. Snider, and H. R. Timmerman. 1976. Woodland caribou and plant communities on the Slate Islands, Lake Superior. Can. Field-Nat. 90:17–21.

Evans, C. A. 1934. Hibernating bats in Minnesota. J. Mamm. 15:240–241.

Evans, F. C. 1951. Notes on a population of the striped ground squirrel (*Citellus tridecemlineatus*) in an abandoned field in southeastern Michigan. J. Mamm. 32:437–449.

Ewer, R. F. 1973. The carnivores. Cornell Univ. Press, Ithaca, N.Y.

Fall, M. W. 1971. Seasonal variations in the food consumption of woodchucks (*Marmota monax*). J. Mamm. 52:370–375.

Fashingbauer, B. A. 1961. Deer studies at Camp Ripley, Minnesota. Proc. Minn. Acad. Sci. 29:275–279.

———. 1965a. The mule deer in Minnesota. Pp. 49–56 *in* J. B. Moyle, ed. Big game in Minnesota. Minn. Dept. Cons. Tech. Bull. 9.

———. 1965b. The elk in Minnesota. Pp. 99–132 *in* J. B. Moyle, ed. Big game in Minnesota. Minn. Dept. Cons. Tech. Bull. 9.

———. 1965c. The woodland caribou in Minnesota. Pp. 133–166 *in* J. B. Moyle, ed. Big game in Minnesota. Minn. Dept. Cons. Tech. Bull. 9.

———. 1965d. The bison in Minnesota. Pp. 167–173 *in* J. B. Moyle, ed. Big game in Minnesota. Minn. Dept. Cons. Tech. Bull. 9.

———. 1965e. The pronghorn antelope in Minnesota. Pp. 175–178 *in* J. B. Moyle, ed. Big game in Minnesota.

Minn. Dept. Cons. Tech. Bull. 9.

Fashingbauer, B. A., and J. B. Moyle. 1963. Nutritive value of red-osier dogwood and mountain maple as deer browse. Proc. Minn. Acad. Sci. 31:73–77.

Fassler, D. J., and R. D. Leavitt. 1975. Terrestrial activity of the northern pocket gopher (Geomyidae) as indicated by owl predation. Southwest. Nat. 19:452–453.

Faulkner, C. E., and W. E. Dodge. 1962. Control of the porcupine in New England. J. Forestry 60:36–37.

Feng, A. S., J. A. Simmons, and S. A. Kick. 1978. Echo detection and target-ranging neurons in the auditory system of the bat *Eptesicus fuscus*. Science 202:645–648.

Ferguson, M. A. D., and H. G. Merriam. 1978. A winter feeding relationship between snowshoe hares and porcupines. J. Mamm. 59:878–880.

Ferron, Jean. 1975. Solitary play of the red squirrel (*Tamiasciurus hudsonicus*). Can. J. Zool. 53:1495–1499.

———. 1976. Cycle annuel d'activité de l'écureil roux (*Tamiasciurus hudsonicus*), adultes et jeunes en semi-liberté au Québec. Naturaliste Can. 103:1–10.

Ferron, Jean, and Jacques Prescott. 1977. Gestation, litter size, and number of litters of the red squirrel (*Tamiasciurus hudsonicus*) in Quebec. Can. Field-Nat. 91:83–85.

Fertig, D. S., and V. W. Edmonds. 1969. The physiology of the house mouse. Sci. Am. 221(4):103–110.

Fichter, E., G. Schildman, and J. H. Sather. 1955. Some feeding patterns of coyotes in Nebraska. Ecol. Monogr. 25:1–37.

Findley, J. S. 1956a. Comments on the winter food of red foxes in eastern South Dakota. J. Wildl. Mgt. 20:216–217.

———. 1956b. Mammals of Clay County, South Dakota. Univ. S. D. Publ. Biol. 1:1–45.

———. 1967. Insectivores and dermopterans. Pp. 87–108 *in* S. Anderson and J. K. Jones, Jr., eds. Recent mammals of the world. Ronald Press, New York.

———. 1972. Phenetic relationships among bats of the genus *Myotis*. Syst. Zool. 21:31–52.

Findley, J. S., and C. Jones. 1964. Seasonal distribution of the hoary bat. J. Mamm. 45:461–470.

Finerty, J. P. 1979. Cycles in Canadian lynx. Am. Nat. 114:453–455.

Fish, P. G. 1974. Notes on the feeding habits of *Microtus ochrogaster* and *M. pennsylvanicus*. Am. Midl. Nat. 92:460–461.

Fish, P. G., and J. O. Whitaker. 1971. *Microtus pinetorum* with grooved incisors. J. Mamm. 52:827.

Fisler, G. F. 1965. Adaptations and speciation in harvest mice of the marshes of San Fransisco Bay. Univ. Calif. Publ. Zool. 77:1–108.

———. 1969. Mammalian organizational systems. Contr. Sci., Los Angeles Co. Mus. 167:1–32.

———. 1970. Communication systems and organizational systems in three species of rodents. So. Calif. Acad. Sci., Bull. 69:43–51.

———. 1971. Age structure and sex ratio in populations of *Reithrodontomys*. J. Mamm. 52:653–662.

Fitch, H. S. 1957. Aspects of reproduction and development in the prairie vole (*Microtus ochrogaster*) in northeastern Kansas. Univ. Kan. Publ., Mus. Nat. Hist. 10:129–161.

Fitch, H. S., and L. L. Sandidge. 1953. Ecology of the opossum on a natural area in northeastern Kansas. Univ. Kan. Publ., Mus. Nat. Hist. 7:305–338.

Fitch, H. S., and H. W. Shirer. 1970. A radiotelemetric study of spatial relationships in the the opossum. Am. Midl. Nat. 84:170–186.

Fitch, J. H. 1979. Patterns of habitat selection and occurrence in the deermouse *Peromyscus maniculatus gracilis*. Publ. Mus. Mich. State. Univ., Biol. Ser. 5:443–484.

Fitch, J. H., and K. A. Shump, Jr. 1979. *Myotis keenii*. Mamm. Species 121:1–3.

Flake, L. D. 1973. Food habits of four species of rodents on a shortgrass prairie in Colorado. J. Mamm. 54:636–647.

———. 1974. Reproduction of four rodent species in a shortgrass prairie of Colorado. J. Mamm. 55:213–216.

Flattum, Roger. 1962. Distribution of mice populations in the Winona area. Proc. Minn. Acad. Sci. 30:14–17.

Fleming, T. H., and R. J. Rauscher. 1978. On the evolution of litter size in *Peromyscus leucopus*. Evolution 32:45–55.

Flinders, J. T., and R. H. M. Hansen. 1973. Abundance and dispersion of leporids within a shortgrass ecosystem. J. Mamm. 54:287–290.

Floyd, T. J., L. D. Mech, and P. A. Jordan. 1978. Relating wolf scat content to prey consumed. J. Wildl. Mgt. 42:528–532.

Floyd, T. J., L. D. Mech, and M. E. Nelson. 1979. An improved method of censusing deer in deciduous-coniferous forests. J. Wildl. Mgt. 43:258–261.

Fluharty, S. L., D. H. Taylor, and G. W. Barrett. 1976. Sun-compass orientation in the meadow vole, *Microtus pennsylvanicus*. J. Mamm. 57:1–9.

Flyger, V. F. 1960. Movements and home range of the gray squirrel *Sciurus carolinensis*, in two Maryland woodlots. Ecology 41:365–369.

Fogl, J. G., and H. S. Mosby. 1978. Aging gray squirrels by cementum annuli in razor-sectioned teeth. J. Wildl. Mgt. 42:444–448.

Folk, G. E., Jr. 1940. Shift of population among hibernating bats. J. Mamm. 21:306–315.

Follmann, E. H. 1974. Comparative ecology and behavior of red and gray foxes. Diss. Abstr. Int. B Sci. Eng. 34:(9):4332–4333.

Forbes, R. B. 1966a. Notes on a litter of least chipmunks. J. Mamm. 47:159–161.

———. 1966b. Studies of the biology of Minnesotan chipmunks. Am. Midl. Nat. 76:290–308.

———. 1966c. Fall accumulation of fat in chipmunks. J. Mamm. 47:715–716.

———. 1967. Some aspects of the water economics of two species of chipmunks. J. Mamm. 48:466–468.

Ford, S. D. 1977. Range, distribution, and habitat of the western harvest mouse, *Reithrodontomys megalotis*, in Indiana. Am. Midl. Nat. 98:422–432.

Foreman, D. 1974. Structural and functional homologies of the accessory reproductive glands of two species of sciurids, *Cynomys ludovicianus* and *Citellus tridecemlineatus*. Anat. Rec. 180:331–339.

Foresman, K. R., and R. A. Mead. 1973. Duration of post-implantation in a western subspecies of the spotted skunk (*Spilogale putorius*). J. Mamm. 54:521–523.

———. 1974. Pattern of luteinizing hormone secretion during delayed implantation in the spotted skunk (*Spilogale putorius latifrons*). Biol. Reprod. 11:475–480.

Foresman, K. R., J. J. Reeves, and R. A. Mead. 1974. Pituitary responsiveness to luteinizing hormone-releasing hormone during delayed implantation in the spotted skunk (*Spilogale putorius latifrons*): validation of LH radioimmunoassay. Biol. Reprod. 11:102–107.

Forsyth, D. J. 1976. A field study of growth and development of nestling masked shrews (*Sorex cinereus*). J. Mamm. 57:708–721.

Forsyth, D. J., and D. A. Smith. 1973. Temporal variability in home ranges of eastern chipmunks (*Tamias striatus*) in a southeastern Ontario woodlot. Am. Midl. Nat. 90:107–117.

Foster, J. B. 1961. Life history of the phenacomys vole. J. Mamm. 42:181–198.

Foster, J. B., and R. L. Peterson. 1961. Age variation in phenacomys. J. Mamm. 42:44–53.

Fox, J. F. 1978. Forest fires and the snowshow hare-Canada lynx cycle. Oecologica 31:349–374.

Fox, M. W. 1969a. The anatomy of aggression and its ritualization in Canidae: a developmental and comparative study. Behaviour 35:242–258.

———. 1969b. Ontogeny of prey-killing in Canidae. Behaviour 35:259–272.

———. 1970. A comparative study of the development of facial expressions in canids, wolf, coyote, and foxes. Behaviour 36:49–73.

———. 1971a. Behaviour of wolves, dogs, and related canids. Harper and Row, New York.

———. 1971b. Possible examples of high order behaviour in wolves. J. Mamm. 52:640–641.

———. 1972a. Socio-ecological implications of individual differences in wolf litters: a developmental and evolutionary perspective. Behaviour 41:298–313.

———. 1972b. The social significance of genital licking in the wolf, *Canis lupus*. J. Mamm. 53:637–640.

———. 1973. Social dynamics of three captive wolf packs. Behaviour 47:290–301.

———. 1975. Evolution of social behavior in canids. Pp. 429–460 *in* M. W. Fox, ed. The wild canids. Van Nostrand Reinhold, New York.

———. 1978. The dog: its domestication and behavior. Garland STPM Press, New York.

Fox, M. W., and R. V. Andrews. 1973. Physiological and biochemical correlates of individual differences in behavior of wolf cubs. Behaviour 46:129–140.

Francq, E. N. 1969. Behavioral aspects of feigned death in the opossum, *Didelphis marsupialis*. Am. Midl. Nat. 81:556–568.

———. 1970. Electrocardiograms of the opossum, *Didelphis marsupialis*, during feigned death. J. Mamm. 51:395.

Franey, E. M. 1953. Are there wolverines in Minnesota? Cons. Volun. 16(91):18–20.

Franzmann, A. W., and P. D. Arneson. 1976. Marrow fat in Alaskan moose femurs in relation to mortality factors. J. Wildl. Mgt. 40:336–339.

Franzmann, A. W., and R. E. LeResche. 1978. Alaskan moose blood studies with emphasis on condition evaluation. J. Wildl. Mgt. 42:334–351.

Frazzetta, T. H. 1975. Complex adaptations in evolving populations. Sinauer, Sunderland, Mass.

Frederickson, L. H. 1970. Breeding biology of American coots in Iowa. Wilson Bull. 82:445–457.

Freeland, W. J. 1974. Vole cycles: another hypothesis. Am. Nat. 108:238–245.

Freeland, W. J., and D. H. Janzen. 1974. Strategies of herbivory in mammals: the role of plant secondary compounds. Am. Nat. 108:269–289.

French, N. R., W. E. Grant, W. Grodzinski, and D. M. Swift. 1976. Small mammal energetics in grassland ecosystems. Ecol. Monogr. 46:201–220.

Frenzel, L. D., Jr. 1965. An estimate of winter utilization of the Cedar Creek Natural History area by the white-tailed deer. J. Minn. Acad. Sci. 32:98–100.

———. 1974. Occurrence of moose in food of wolves as revealed by scat analyses: a review of North American studies. Naturaliste Can. 101:467–479.

Frenzel, L. D., Jr., and L. D. Mech. 1967. Wolf-deer relations in northeastern, Minnesota: a preliminary study. Minn. Naturalist 18(1):8–9.

Frick, C. 1937. Horned ruminants of North America. Bull. Am. Mus. Nat. Hist. 69:1–699.

Friley, C. E., Jr. 1949. Age determination, by use of the baculum, in the river otter, *Lutra c. canadensis* Schreber. J. Mamm. 30:102–110.

Fritts, S. H. 1979. Dynamics, movements, and feeding ecology of a newly-protected wolf (*Canis lupus*) population in northwestern Minnesota. Ph.D. thesis, Univ. Minn., Minneapolis.

———. 1980. The timber wolf in Minnesota. Proc. Symp. Mammal. Ecol. Hab. Mgt. Minn., Bemidji State Univ. 70.

Fritts, S. H., and J. A. Sealander. 1978a. Reproductive biology and population characteristics of bobcats (*Lynx rufus*) in Arkansas. J. Mamm. 59:347–353.

———. 1978b. Diets of bobcats in Arkansas with special reference to age and sex differences. J. Wildl. Mgt. 42:533–539.

Fritzell, E. K. 1977. Dissolution of raccoon sibling bonds. J. Mamm. 58:427–428.

———. 1978a. Habitat use by prairie raccoons during the waterfowl breeding season. J. Wildl. Mgt. 42:118–127.

———. 1978b. Aspects of raccoon (*Procyon lotor*) social organization. Can. J. Zool. 56:260–271.

Frydendall, M. J. 1969. Rodent populations on four habitats in central Kansas. Trans. Kan. Acad. Sci. 72:213–222.

Fryxell, F. M. 1926. Squirrels migrate from Wisconsin to Iowa. J. Mamm. 7:60.

Fulk, G. W. 1972. The effect of shrews on the space utilization of voles. J. Mamm. 53:461–478.

Fuller, T. K. 1978. Variable home-range sizes of female gray foxes. J. Mamm. 59:446–449.

Fuller, W. A. 1977a. Demography of a subarctic population of *Clethrionomys gapperi*: size and growth. Can. J. Zool. 55:415–425.

———. 1977b. Demography of a subarctic population of *Clethrionomys gapperi*: numbers and survival. Can. J. Zool. 55:42–51.

Furlong, E. L. 1935. Pliocene antelopes of the pronghorn type. Science 82:250–251.

Gabrielson, J. N. 1928. Notes on the habits and behavior of the porcupine in Oregon. J. Mamm. 9:33–38.

Gadgil, M., and W. H. Bossert. 1970. Life history consequences of natural selection. Am. Nat. 104:1–24.

Gaines, M. S., L. R. McClenaghan, Jr., and R. K. Rose. 1978. Temporal patterns of allozymic variation in fluctuating populations of *Microtus ochrogaster*. Evolution 32:723–739.

Gaines, M. S., R. K. Rose, and L. M. McClenaghan. 1977. The demography of *Synaptomys cooperi* populations in eastern Kansas. Can. J. Zool. 55:1584–1594.

Galbavy, E. J., et al. 1972. Blood gases of woodchucks at reduced body temperatures. J. Mamm. 53:919–921.

Gardner, A. L. 1973. The systematics of the genus *Didelphis* (Marsupialia: Didelphidae) in North and Middle America. Spec. Publ. Mus., Texas Tech Univ. 4:3–81.

Garretson, M. S. 1938. The American bison; the story of its extermination as a wild species and its restoration under federal protection. N.Y. Zool. Soc., New York.

Garrison, M. V., and B. E. Johns. 1975. Antifertility effects of SC-20775 in Norway and Polynesian rats. J. Wildl. Mgt. 39:26–29.

Gashwiler, J. S., W. L. Robinette, and O. W. Morris. 1960. Foods of bobcats in Utah and eastern Nevada. J. Wildl. Mgt. 24:226–229.

———. 1961. Breeding habits of bobcats in Utah. J. Mamm. 42:76–84.

Gates, J. E., and D. M. Gates. 1975. Fox squirrel use of cocklebur seeds. J. Mamm. 56:239–240.

Gaughran, G. R. L. 1954. A comparative study of the osteology and myology of the cranial and cervical regions of the shrew, *Blarina brevicanda*, and the mole, *Scalopus aquaticus*. Misc. Publ. Mus. Zool. Univ. Mich. 80:1–82.

Geier, A. R., and L. B. Best. 1980. Habitat selection by small mammals of riparian communities: evaluating effects of habitat alterations. J. Wildl. Mgt. 44:16–24.

Geist, V. 1966. The evolution of horn-like organs. Behaviour 27:175–214.

———. 1974. On the relationship of ecology and behaviour in the evolution of ungulates: theoretical considerations. IUCN Publ. 24(1):235–246.

———. 1981. Chapter V. Pp. 157–224 *in* O. C. Wallmo, ed. Mule and blacktailed deer of North America. Univ. Nebraska Press, Lincoln.

Geluso, K. N., J. S. Altenbach, and D. E. Wilson. 1976. Bat mortality: pesticide poisoning and migratory stress. Science 194:184–186.

Genoways, H. H., and J. R. Choate. 1972. A multivariate analysis of systematic relationships among populations of the short-tailed shrew (genus *Blarina*) in Nebraska. Syst. Zool. 21:106–116.

Genoways, H. H., and J. K. Jones, Jr. 1972. Mammals from southwestern North Dakota. Occ. Pap. Mus., Texas Tech Univ. 6:1–36.

Genoways, H. H., et al. 1976. Systematists, other users, and uses of North American collections of Recent mammals. Museology 3:1–87.

George, W. G. 1974. Domestic cats as predators and factors in winter shortages of raptor prey. Wilson Bull. 86:384–396.

Georges, Stan. 1976. A range extension of the coyote in Quebec. Can. Field-Nat. 90:78–79.

Gerber, J. D., and E. C. Birney. 1968. Immunological comparisons of four subgenera of ground squirrels. Syst. Zool. 17:413–416.

Getty, Thomas. 1979. An observation on *Tamias striatus* reproducing in a tree nest. J. Mamm. 60:636.

Getz, L. L. 1960. A population study of the vole, *Microtus pennsylvanicus*. Am. Midl. Nat. 64:392–405.

———. 1961a. Factors influencing the local distribution of *Microtus* and *Synaptomys* in southern Michigan. Ecology 42:110–119.

———. 1961b. Home ranges, territoriality, and movement of the meadow vole. J. Mamm. 42:24–36.

———. 1961c. Factors influencing the local distribution of

shrews. Am. Midl. Nat. 65:67–88.

———. 1961d. Notes on the local distribution of *Peromyscus leucopus* and *Zapus hudsonius*. Am. Midl. Nat. 65:486–500.

———. 1962a. Notes on the water balance of the redback vole. Ecology 43:565–566.

———. 1962b. A local concentration of the least shrew. J. Mamm. 43:281–282.

———. 1962c. Aggressive behavior of the meadow and prairie voles. J. Mamm. 43:351–358.

———. 1963. A comparison of the water balance of the prairie and meadow voles. Ecology 44:202–207.

———. 1968a. Influences of water balance and microclimate on the local distribution of the redback vole and the white-footed mouse. Ecology 49:276–286.

———. 1968b. Influence of weather on the activity of the redbacked vole. J. Mamm. 49:565–570.

———. 1970. Influence of vegetation on the local distribution of the meadow vole in southern Wisconsin. Univ. Conn. Occ. Pap. 1:213–241.

Giacalone-Madden, J. R. 1977. The behavioral ecology of the southern flying squirrel, *Glaucomys volans*, on Long Island, New York. Diss. Abstr. Int. B Sci. Eng. 37(8):3722.

Gier, H. T. 1948. Rabies in the wild. J. Wildl. Mgt. 12:142–153.

———. 1968. Coyotes in Kansas. Kan. Agr. Exp. Sta., Bull. 393.

———. 1975. Ecology and behavior of the coyote (*Canis latrans*). Pp. 247–262 *in* M. W. Fox, ed. The wild canids. Van Nostrand Reinhold, New York.

Gilbert, B. 1970. The weasels. Random House, Toronto.

Gilbert, B. K. 1973. Scent marking and territoriality in pronghorn (*Antilocapra americana*) in Yellowstone National park. Mammalia 37:24–33.

Gilbert, F. F. 1969. Physiological effects of natural DDT residues and metabolites on ranch mink. J. Wildl. Mgt. 33:933–943.

Gill, Don. 1975. The feral house cat as a predator of varying hares. Can. Field-Nat. 89:78–79.

Gill, Don, and L. D. Cordes. 1972. Winter habitat preference of porcupines in the southern Alberta foothills. Can. Field-Nat. 86:349–355.

Gilpin, M. E. 1973. Do hares eat lynx? Am. Nat. 107:727–730.

Gingerich, P. D., and D. A. Winkler. 1979. Patterns of variation and correlation in the dentition of the red fox, *Vulpes vulpes*. J. Mamm 60:691–704.

Gipson, P. S., I. K. Gipson, and J. A. Sealander. 1975. Reproductive biology of wild *Canis* (Canidae) in Arkansas. J. Mamm. 56:605–612.

Glass, B. P. 1951. A key to the skulls of North American mammals. Okla. State Univ. Bookstore, Stillwater.

Godfrey, Gillian, and Peter Crowcroft. 1960. The life of the mole (*Talpa europaea* Linnaeus). Museum Press, London.

Godin, A. J. 1977. Wild mammals of New England. John Hopkins Univ. Press, Baltimore.

Goehring, H. H. 1954. *Pipistrellus subflavus obscurus, Myotis keenii*, and *Eptesicus fuscus fuscus* hibernating in a storm sewer in central Minnesota. J. Mamm. 35:434–435.

———. 1955. Observations on hoary bats in a storm. J. Mamm. 36:130–131.

———. 1958. A six year study of big brown bat survival. Proc. Minn. Acad. Sci. 26:222–224.

———. 1971a. Big brown bat survives sub-zero temperatures. J. Mamm. 52:832–833.

———. 1971b. Two rhino mice (*Mus musculus*) from Minnesota. J. Mamm. 52:834–835.

———. 1972. Twenty-year study of *Eptesicus fuscus* in Minnesota. J. Mamm. 53:201–207.

Goertz, J. W. 1971. An ecological study of *Microtus pinetorum* in Oklahoma. Am. Midl. Nat. 86:1–12.

Goertz, J. W., R. M. Dawson, and Elmer Mowbray. 1975. Response to nest boxes and reproduction by *Glaucomys volans* in northern Louisiana. J. Mamm. 56:933–939.

Goldberg, Jennie S., and Wendy Haas. 1978. Interactions between mule deer dams and their radio-collared and unmarked fawns. J. Wildl. Mgt. 42:422–425.

Goldman, E. A. 1950. Raccoons of North and Middle America. N. Am. Fauna 60.

Gonyea, W., and R. Ashworth. 1975. The form and function of retractile claws in the Felidae and other representative carnivorans. J. Morph. 145:229–238.

Goodpaster, W., and D. F. Hoffmeister. 1950. Bats as prey for mink in Kentucky cave. J. Mamm. 31:457.

Goodwin, M. K. 1979. Notes on caravan and play behavior in young captive *Sorex cinereus*. J. Mamm. 60:411–413.

Gordon, K. R. 1977. Molar measurements as a taxonomic tool in *Ursus*. J. Mamm. 58:247–248.

Goudie, R. I. 1978. Red squirrels, *Tamiasciurus hudsonicus*, in the Salmonier River Valley, Newfoundland. Can. Field-Nat. 92:193–194.

Gould, E. 1955. The feeding efficiency of insectivorous bats. J. Mamm. 36:399–407.

Graham, R. W. 1976. Late Wisconisn mammalian faunas and environmental gradients of the eastern United States. Paleobiology 2:343–350.

Graham, R. W., and H. A. Semken, Jr. 1976. Paleoecological significance of the short-tailed shrew (genus: *Blarina*) with a systematic discussion of *Blarina ozarkensis*. J. Mamm. 57:433–449.

Grant, P. R. 1969. Experimental studies of competitive interaction in a two-species system. I. *Microtus* and *Clethrionomys* species in enclosures. Can. J. Zool. 47:1059–1082.

———. 1971. The habitat preference of *Microtus pennsyl-*

vanicus, and its relevance to the distribution of this species on islands. J. Mamm. 52:351–361.

———. 1974. Reproductive compatibility of voles from separate continents (Mammalia: *Clethrionomys*). J. Zool., London. 174:245–254.

———. 1975. Population performance of *Microtus pennsylvanicus* confined to woodland habitat, and a model of habitat occupancy. Can. J. Zool. 53:1447–1465.

Grant, W. E., and E. C. Birney. 1979. Small mammal community structure in North American grasslands. J. Mamm. 60:23–36.

Grant, W. E., N. R. French, and D. M. Swift. 1977. Response of a small mammal community to water and nitrogen treatments in a shortgrass prairie ecosystem. J. Mamm. 58:637–652.

Gray, G. D., and D. A. Dewsbury. 1973. A quantitative description of copulatory behavior in prairie voles, *Microtus ochrogaster*. Brain Behav. Evol. 8:437–452.

———. 1975. A quantitative description of the copulatory behavior of meadow voles (*Microtus pennsylvanicus*). Anim. Behav. 23:261–267.

Grayson, D. K. 1977. On the Holocene history of some northern Great Basin lagomorphs. J. Mamm. 58:507–513.

Greaves, J. H., and P. B. Ayres. 1977. Unifactoral inheritance of warfarin resistance in *Rattus norvegicus* from Denmark. Genet. Res. 29:215–222.

Greaves, W. S. 1978. The jaw lever system in ungulates: a new model. J. Zool., London 184:271–285.

———. 1980. The mammalian jaw mechanism—the high glenoid cavity. Am. Nat. 116:432–440.

Greeley, F., and J. R. Beer. 1949. The pipistrel (*Pipistrellus subflavus*) in northern Wisconsin. J. Mamm. 30:198.

Green, J. C., and R. B. Janssen. 1975. Minnesota birds: where, when, and how many. Univ. Minn. Press, Minneapolis.

Green, R. G., and C. A. Evans. 1940. Studies on a population cycle of snowshow hares on the Lake Alexander area. J. Wildl. Mgt. 4:220–238, 267–278, 347–358.

Greenbaum, Ira F., and R. J. Baker. 1978. Determination of the primitive karyotype for *Peromyscus*. J. Mamm. 59:820–834.

Greenbaum, Ira F., R. J. Baker, and P. R. Ramsey. 1978. Chromosomal evolution and the mode of speciation in three species of *Peromyscus*. Evolution 32:646–654.

Greenewalt, C. H. 1962. Dimensional relationships for flying animals. Smithson. Misc. Coll. 144(2):1–46.

Greensides, R. D., and R. A. Mead. 1973. Ovulation in the spotted skunk (*Spilogale putorius latifrons*). Biol. Reprod. 8:576–584.

Greer, K. R. 1955. Yearly food habits of the river otter in the Thompson Lakes region, Northwestern Montana, as indicated by scat analysis. Am. Midl. Nat. 54:299–313.

Griffin, D. R. 1940. Migrations of New England bats. Bull. Mus. Comp. Zool. 86:217–246.

———. 1958. Listening in the dark. Yale Univ. Press, New Haven, Conn.

———. 1970. Migrations and homing of bats. Pp. 233–264 *in* W. A. Wimsatt, ed. Biology of bats. Vol. 1. Academic Press, New York.

———. 1976. The question of animal awareness: evolutionary continuity of mental experience. Rockefeller Univ. Press, New York.

Griffin, D. R., and H. B. Hitchcock. 1965. Probably 24-year longevity record for *Myotis lucifugus*. J. Mamm. 46:332.

Griffiths, David. 1975. Prey availability and the food of predators. Ecology 56:1209–1214.

Grizzel, R. A. 1955. A study of the southern woodchuck *Marmota monax monax*. Am. Midl. Nat. 53:257–293.

Grout, F. F., R. P. Sharp, and G. M. Schwartz. 1959. The geology of Cook County, Minnesota. Bull. Minn. Geol. Surv. 39.

Guggisberg, C. A. W. 1975. Wild cats of the world. Taplinger, New York.

Guilday, J. E. 1958. The prehistoric distribution of the opossum. J. Mamm. 39:39–43.

———. 1968. Grizzly bears from eastern North America. Am. Midl. Nat. 79:247–250.

———. 1972. Archaeological evidence of *Scalopus aquaticus* in the upper Ohio Valley. J. Mamm. 53:905–907.

Guilday, J. E., and H. W. Hamilton. 1973. The Late Pleistocene small mammals of Eagle Cave, Pendleton County, West Virginia. Ann. Carnegie Mus. 44:45–48.

———. 1978. Ecological significance of displaced boreal mammals in West Virginia caves. J. Mamm. 59:176–181.

Guilday, J. E., H. W. Hamilton, and A. D. McCrady. 1971. The Welsh Cave peccaries (*Platygonus*) and associated fauna, Kentucky, Pleistocene. Ann. Carnegie Mus. 43:249–320.

Guilday, J. E., and P. W. Parmalee. 1972. Quaternary periglacial records of voles of the genus *Phenacomys* Merriam (Cricetidae: Rodentia). Quat. Res. 2:170–175.

Guilday, J. E., P. W. Parmalee, and H. W. Hamilton. 1977. The Clark's Cave bone deposit and the Late Pleistocene paleoecology of the central Appalachian Mountains of Virginia. Bull. Carnegie Mus. Nat. Hist. 2:1–87.

Gunderson, H. L. 1944. Notes on a heavy Norway rat population. J. Mamm. 25:307–308.

———. 1948. Mule deer record for Minnesota. J. Mamm. 29:184.

———. 1950. A study of some small mammal populations at Cedar Creek Forest, Anoka County, Minnesota. Minn. Mus. Nat. Hist. Occ. Pap. 4.

———. 1955. Nutria, *Myocaster coypus*, in Minnesota. J. Mamm. 36:465.

———. 1959. Red-backed vole habitat studies in central Minnesota. J. Mamm. 40:405–412.

———. 1961. Opossum records in Minnesota. Minn. J.

Sci. 5(1):48–50.

———. 1965. Marten records for Minnesota. J. Mamm. 46:688.

———. 1976. Mammalogy. McGraw-Hill, New York.

———. 1978. A mid-continent irruption of Canada lynx, 1962–1963. Prairie Nat. 10:71–80.

Gunderson, H. L., and J. R. Beer. 1953. The mammals of Minnesota. Minn. Mus. Nat. Hist., Occ. Pap. 6.

Gunderson, H. L., and B. R. Mahan. 1980. Analysis of sonograms of American bison (*Bison bison*). J. Mamm. 61:379–381.

Gunson, J. R., and R. R. Bjorge. 1979. Winter denning of the striped skunk in Alberta. Can. Field-Nat. 93:252–258.

Gupta, B. B. 1966a. Notes on the gliding mechanism of the flying squirrel. Occ. Pap. Mus. Zool. Univ. Mich. 645:1–7.

———. 1966b. Skeleton of *Erethizon* and *Coendou*. Mammalia 30:495–497.

———. 1966c. The anatomy of the posterior axial region and the hind limb of the eastern American mole. Mammalia 30:667–680.

Gustavson, C. R. 1979. An experimental evaluation of aversive conditioning for controlling coyote predation: a critique. J. Wildl. Mgt. 43:208–209.

Gustavson, C. R., J. Garcia, W. G. Hankins, and K. W. Rusiniak. 1974. Coyote predation control by aversive conditioning. Science 184:581–583.

Gustavson, C. R., D. J. Kelly, and J. Garcia. 1975. Predation and aversive conditioning in coyotes. Science 187:1096.

Gustavson, C. R., et al. 1976. Prey-lithium aversions. I: Coyotes and wolves. Behav. Biol. 17:61–72.

Guthrie, Mary J. 1933a. Notes on the seasonal movements and habits of some cave bats. J. Mamm. 14:1–19.

———. 1933b. The reproductive cycles of some cave bats. J. Mamm. 14:199–216.

Guthrie, R. D. 1965. Variability in characters undergoing rapid evolution, an analysis of *Microtus* molars. Evolution 19:214–233.

———. 1966. Pelage of fossil bison—a new osteological index. J. Mamm. 47:725–727.

———. 1970. Bison evolution and zoogeography in North America during the Pleistocene. Q. Rev. Biol. 45:1–15.

Gwynn, G. W. 1972a. Effects of a chemosterilant on fecundity of wild Norway rats. J. Wildl. Mgt. 36:550–556.

———. 1972b. Field trial of a chemosterilant in wild Norway rats. J. Wildl. Mgt. 36:823–828.

Gwynn, G. W., and S. M. Kurtz. 1970. Acceptability and efficacy of an antifertility agent in wild Norway rats. J. Wildl. Mgt. 34:514–519.

Haag, W. G. 1962. The Bering Strait land bridge. Sci. Am. 206(1):112–123.

Haberman, C. G., and E. D. Fleharty. 1971a. Natural history notes on Franklin's ground squirrel in Boone County, Nebraska. Trans. Kan. Acad. Sci. 74:76–80.

———. 1971b. Energy flow in *Spermophilus franklinii*. J. Mamm. 52:710–716.

Hafez, E. S. E. 1969. The behaviour of domestic animals. Williams and Wilkins, Baltimore.

Hagmeier, E. M. 1956. Distribution of marten and fisher in North America. Can. Field-Nat. 70:149–168.

———. 1958. Inapplicability of the subspecies concept to the North American marten. Syst. Zool. 7:1–7.

———. 1961. Variation and relationships in North American marten. Can. Field-Nat. 75:122–138.

Hagmeier, E. M., and C. D. Stults. 1964. A numerical analysis of the distribution patterns of North American mammals. Syst. Zool. 13:125–155.

Haigh, Gale R. 1979. Sun-compass orientation in the thirteen-lined ground squirrel, *Spermophilus tridecemlineatus*. J. Mamm. 60:629–632.

Hale, J. G. 1966. Influence of beaver on some trout streams along the Minnesota north shore of Lake Superior. Minn. Fish. Investig. 4:5–29.

Hall, E. R. 1945. A revised classification of the American ermines with a description of a new subspecies from the western Great Lakes region. J. Mamm. 26:175–182.

———. 1946. Mammals of Nevada. Univ. Calif. Press, Berkeley.

———. 1951a. A synopsis of the North American Lagomorpha. Univ. Kan. Publ., Mus. Nat. Hist. 5:119–202.

———. 1951b. American weasels. Univ. Kan. Publ., Mus. Nat. Hist. 4:1–466.

———. 1958. Introduction, Part II. Pp. 371–373 *in* C. L. Hubbs, ed. Zoogeography. AAAS, Publ. 51.

———. 1962. Collecting and preparing study specimens of vertebrates. Misc. Publ. Mus. Nat. Hist., Univ. Kan. 30:1–46.

Hall, E. R., and E. L. Cockrum. 1953. A synopsis of the North American microtine rodents. Univ. Kan. Publ., Mus. Nat. Hist. 5:373–498.

Hall, E. R., and W. W. Dalquest. 1950. A synopsis of the American bats of the genus *Pipistrellus*. Univ. Kan. Publ., Mus. Nat. Hist. 1:591–602.

Hall, E. R., and K. R. Kelson. 1959. The mammals of North America. 2 vols. Ronald Press, New York.

Hall, J. S., R. J. Cloutier, and D. R. Griffin. 1957. Longevity records and notes on tooth wear of bats. J. Mamm. 38:407–409.

Hall, S. A. 1972. Holocene *Bison occidentalis* from Iowa. J. Mamm. 53:604–606.

Halls, L. K., and R. Alcaniz. 1968. Browse plants yield best in forest openings. J. Wildl. Mgt. 32:185–186.

Hamerstrom, F. N., Jr., and Frances Hamerstrom. 1951. Food of young raptors on the Edwin S. George Reserve. Wilson Bull. 63:16–25.

Hamilton, R. B., and D. T. Stalling. 1972. *Lasiurus borealis* with five young. J. Mamm. 53:190.

Hamilton, W. D. 1963. The evolution of altruistic behavior. Am. Nat. 97:354–356.
———. 1971. Geometry for the selfish herd. J. Theor. Biol. 31:295–311.
Hamilton, W. J., Jr. 1929. Breeding habits of the short-tailed shrew, *Blarina brevicauda*. J. Mamm. 10:125–134.
———. 1930. The food of the Soricidae. J. Mamm. 11:26–39.
———. 1931. Habits of the starnosed mole, *Condylura cristata*. J. Mamm. 12:345–355.
———. 1933a. The insect food of the big brown bat. J. Mamm. 14:155–156.
———. 1933b. The weasels of New York: their natural history and economic status. Am. Midl. Nat. 14:289–344.
———. 1934. The life history of the rufescent woodchuck *Marmota monax rufescens* Howell. Ann. Carnegie Mus. 23:85–178.
———. 1935. Habits of jumping mice. Am. Midl. Nat. 16:187–200.
———. 1936. Seasonal food of skunks in New York. J. Mamm. 17:240–246.
———. 1937. Winter activity of the skunk. Ecology 18:326–327.
———. 1938. Life history notes on the northern pine mouse. J. Mamm. 19:163–170.
———. 1939. American mammals. McGraw-Hill, New York.
———. 1941a. The reproduction of the field mouse *Microtus pennsylvanicus* (Ord). Cornell Univ. Agr. Exp. Sta. Mem. 273:1–23.
———. 1941b. The food of small forest mammals in eastern United States. J. Mamm. 22:250–263.
———. 1943a. Caterpillars as food of the gray squirrel. J. Mamm. 24:104.
———. 1943b. The mammals of eastern United States. Comstock, Ithaca, N.Y.
———. 1944. The biology of the little short-tailed shrew, *Cryptotis parva*. J. Mamm. 25:1–7.
———. 1958. Life history and economic relations of the opossum (*Didelphis marsupialis virginiana*) in New York State. Mem. Cornell Univ. Agr. Exp. Sta. 354:1–48.
———. 1963. Reproduction of the striped skunk in New York. J. Mamm. 44:123–124.
———. 1974. Food habits of the coyote in the Adirondacks. N.Y. Fish Game J. 21:177–181.
Hamilton, W. J., Jr., and A. H. Cook. 1955. The biology and management of the fisher in New York. N.Y. Fish Game J. 2:13–35.
Hamilton, W. J., Jr., and W. R. Eadie. 1964. Reproduction in the otter *Lutra canadensis*. J. Mamm. 45:242–252.
Hamilton, W. J., Jr., and W. J. Hamilton, III. 1954. The food of some small mammals from the Gaspé Peninsula, P. Q. Can. Field-Nat. 68:108–109.
Hamilton, W. J., Jr., and R. P. Hunter. 1939. Fall and winter food habits of Vermont bobcats. J. Wildl. Mgt. 3:99–103.
Hamilton, W. J., Jr., and J. O. Whitaker, Jr. 1979. Mammals of the eastern United States. Cornell Univ. Press, Ithaca, N.Y.
Hancock, J. A., W. E. Mercer, and T. H. Northcott. 1976. Lynx attack on man carrying hares in Newfoundland. Can. Field-Nat. 90:46–47.
Handley, C. O., Jr. 1954. *Phenacomys* in Minnesota. J. Mamm. 35:260.
Hansen, H. L., and R. M. Brown. 1950. Deer, brush, and the Itasca Park Forest. Cons. Volun. 13(78):1–5.
Hansen, H. L., L. W. Krefting, and V. Kurmis. 1973. The forest of Isle Royale in relation to fire history and wildlife. U. Minn. Agr. Exp. Sta. Tech. Bull. 294, For. Ser. 13:1–43.
Hansen, L. P., and G. O. Batzli. 1979. Influence of supplemental food on local populations of *Peromyscus leucopus*. J. Mamm. 60:335–342.
Hansen, R. M. 1962a. Dispersal of Richardson ground squirrel in Colorado. Am. Midl. Nat. 68:58–66.
———. 1962b. Movements and survival of *Thomomys talpoides* in a mima-mound habitat. Ecology 43:151–154.
———. 1975. Plant matter in the diet of *Onychomys*. J. Mamm. 56:530–531.
Hansen, R. M., and G. D. Bear. 1963. Winter coats of white-tailed jackrabbits in southwestern Colorado. J. Mamm. 44:420–422.
Hansen, R. M., and R. C. Clark. 1977. Foods of elk and other ungulates at low elevations in northwestern Colorado. J. Wildl. Mgt. 41:76–80.
Hansen, R. M., and M. K. Johnson. 1976. Stomach content weight and food selection by Richardson ground squirrels. J. Mamm. 57:749–751.
Hansen, R. M., and R. S. Miller. 1959. Observation on the plural occupancy of pocket gopher burrow systems. J. Mamm. 40:577–584.
Hansen, R. M., and M. J. Morris. 1968. Movement of rocks by northern pocket gophers. J. Mamm. 49:391–399.
Hansen, R. M., and L. D. Reed. 1975. Diet overlap of deer, elk, and cattle in southern Colorado. J. Range Mgt. 28:43–47.
Hansen, R. M., and D. N. Ueckert. 1970. Dietary similarity of some primary consumers. Ecology 51:640–648.
Harder, L. D. 1979. Winter feeding by porcupines in montane forests of southwestern Alberta. Can. Field-Nat. 93:405–410.
Harlow, R. F., et al. 1975. Deer foods during years of oak mast abundance and scarcity. J. Wildl. Mgt. 39:330–336.

Harper, Francis. 1929. Notes on mammals of the Adirondacks. N.Y. State Mus. Hbk. 8:51–118.

———. 1955. The barren ground caribou of Keewatin. Univ. Kan. Mus. Nat. Hist., Misc. Publ. 6:1–164.

Harriman, A. E. 1973. Self-selection of diet in northern grasshopper mice (*Onychomys leucogaster*). Am. Midl. Nat. 90:97–106.

Harrington, F. H., and L. D. Mech. 1978a. Howling at two Minnesota wolf pack summer homesites. Can. J. Zool. 56:2024–2028.

———. 1978b. Wolf vocalization. Pp. 109–132 *in* R. L. Hall and H. S. Sharp, eds. Wolf and man: evolution in parallel. Academic Press, New York.

———. 1979. Wolf howling and its role in territory maintenance. Behaviour 68:207–249.

Harris, V. T. 1952. An experimental study of habitat selection by prairie and forest races of the deermouse, *Peromyscus maniculatus*. Contr. Lab. Vert. Biol., Univ. Mich. 56:1–53.

———. 1954. Experimental evidence of reproductive isolation between two subspecies of *Peromyscus maniculatus*. Contr. Lab. Vert. Biol., Univ. Mich. 70:1–13.

Hart, F. M., and J. A. King. 1966. Distress vocalizations of young in two subspecies of *Peromyscus maniculatus*. J. Mamm. 47:287–293.

Hart, J. S. 1971. Rodents. Pp. 1–149 *in* C. Whittow, ed. Comparative physiology of thermoregulation. Vol. 2. Academic Press, New York.

Hartgrove, R. W., and R. E. Webb. 1973. The development of benzypyrene hydroxylase activity in endrin susceptible and resistant pine mice. Biochem. Physiol. 3:61–65.

Hartman, C. G. 1952. Possums. Univ. Texas Press, Austin.

Hartung, T. G., and D. A. Dewsbury. 1978. A comparative analysis of copulatory plugs in muroid rodents and their relationship to copulatory behavior. J. Mamm. 59:717–723.

Harvey, M. J. 1976. Home range, movements, and diet activity of the eastern mole, *Scalopus aquaticus*. Am. Midl. Nat. 95:436–445.

Hatfield, D. M. 1937. Notes on Minnesota squirrels. J. Mamm. 18:242–243.

———. 1939a. Checklist of the mammals of Minnesota with keys for identification. Edward Bros., Ann Arbor, Mich.

———. 1939b. Winter food habits of foxes in Minnesota. J. Mamm. 20:202–206.

———. 1939c. Northern pine mouse in Minnesota. J. Mamm. 20:376.

Hatt, R. T. 1929. The red squirrel: its life history and habits. Bull. N.Y. State Coll. For., Roosevelt Wild Life Annals 2:1–140.

———. 1930. The biology of the voles of New York. Roosevelt Wildl. Bull. 5:513–623.

Haugen, A. O. 1942. Life history studies of the cottontail rabbit in southwestern Michigan. Am. Midl. Nat. 28:204–244.

———. 1961. Wolverine in Iowa. J. Mamm. 42:174–175.

———. 1974. Reproduction in the plains bison. Iowa State J. Res. 49:1–8.

———. 1975. Reproductive performance of white-tailed deer in Iowa. J. Mamm. 56:151–159.

Havera, S. P. 1979. Temperature variation in a fox squirrel nest box. J. Wildl. Mgt. 43:251–253.

Havera, S. P., and C. M. Nixon. 1980. Winter feeding of fox and gray squirrel populations. J. Wildl. Mgt. 44:41–55.

Havera, S. P., C. M. Nixon, and F. I. Collins. 1976. Fox squirrels feeding on buckeye pith. Am. Midl. Nat. 95:462–464.

Hawley, David. 1974. Strange saga of the swift fox. Minn. Volun. 37(217):41–44.

Hawley, V. D., and F. E. Newby. 1957. Marten home ranges and population fluctuations. J. Mamm. 38:174–184.

Hayes, S. R. 1976. Daily activity and body temperature of the southern woodchuck, *Marmota monax monax*, in northwestern Arkansas. J. Mamm. 57:291–299.

Hayward, B. J., and R. Davis. 1964. Flight speeds in western bats. J. Mamm. 45:236–242.

Hazard, E. B. 1960. A field study of activity among squirrels (Sciuridae) in southern Michigan. Ph.D. thesis, Univ. Mich., Ann Arbor.

———. 1963. Records of the opossum in northern Minnesota. J. Mamm. 44:118.

———. 1973. Environmental impact on terrestrial vertebrates. Pp. E-III-1 to E-III-13 *in* C. H. Fuchsman, ed. Environmental review of the Headwaters of the Mississippi reservoir projects. Center for Envir. Studies, Bemidji State Univ.

———. 1978a. Minnesota. Pp. 112–114 *in* M. K. Brown, ed. The pine marten. N.Y. State Dept. Env. Cons., Div. Fish Wildl., unnum. report.

———. 1978b. Spotted skunk in northern Hubbard County. Loon 50:207.

———. 1980. The distribution of mammals in Minnesota. Proc. Symp. Mammal. Ecol. Hab. Mgt. Minn., Bemidji State Univ. 6–13.

Heaney, L. R., and E. C. Birney. 1975. Comments on the distribution and natural history of some mammals in Minnesota. Can. Field-Nat. 89:29–34.

Heidt, G. A. 1970. The least weasel *Mustela nivalis* Linnaeus. Developmental biology in comparison with other North American *Mustela*. Mich. State Univ. Publ., Mus. Nat. Hist. 4:227–282.

———. 1972. Anatomical and behavioral aspects of killing and feeding by the least weasel, *Mustela nivalis* L. Ark. Acad. Sci. Proc. 26:53–54.

Heidt, G. A., et al. 1976. A bibliography of mustelids. IV: Weasels. Mich. Agric. Exp. Sta., J. Art. 7662:47pp.

Heidt. G. A., and J. Hargraves. 1974. Blood chemistry and hematology of the spotted skunk, *Spilogale putorius*. J. Mamm. 55:206–208.

Heidt, G. A., and Jane N. Huff. 1970. Ontogeny of vocalization in the least weasel. J. Mamm. 51:385–386.

Heidt, G. A., M. K. Petersen, and F. L. Kirkland, Jr. 1968. Mating behavior and development of least weasels (*Mustela nivalis*) in captivity. J. Mamm. 49:413–419.

Heinselman, M. L. 1973. Fire in the virgin forests of the Boundary Waters Canoe Area, Minnesota. Quaternary Res. 3:329–382.

Hemmer, Helmut. 1978. The evolutionary systematics of living Felidae: present status and current problems. Carnivore 1:71–79.

Henderson, Carrol. 1978a. State of Minnesota nongame wildlife program-1977. Div. Fish Wildl., Minn. Dept. Nat. Resour.:1–7.

———. 1978b. 182 Program: summary of uncommon wildlife observations through December 1, 1977. Div. Fish Wildl., Minn. Dept. Nat. Resour.:14 unnum. pp.

———. 1978c. River otter status report, 1977. Div. Fish Wildl., Minn. Dept. Nat. Resour.:1–10.

———. 1978d. Minnesota 1977 Canada lynx status report. Div. Fish Wildl., Minn. Dept. Nat. Resour.:1–23.

———. 1978e. Minnesota bobcat status report. Div. Fish Wildl., Minn. Dept. Nat. Resour.:1–7.

———. 1978f. The lynx link. Minn. Volun. 41(236):16–21.

———. 1978g. Wildlife watch. Minn. Volun. 41(237):56–57.

———. 1979. A preliminary review of the taxonomy, distribution, legal status, and utilization of nongame mammals in Minnesota. Minn. D.N.R. report:1–27.

———. 1980. Nongame Wildlife Program, 1979 Summary. Proc. Symp. Mammal. Ecol. Hab. Mgt. Minn., Bemidji State Univ. 14–19.

Henderson, Carrol, and Julie Reitter. 1979a-j. Guide to the nongame mammals of: northwest Minnesota—Region 1N. Minn. Dept. Nat. Resour.:1–7; northwest Minnesota—Region 1S. MDNR:1–6; northeast Minnesota—Region 2. MDNR:1–10; east central Minnesota—Region 3E. MDNR:1–7; central Minnesota—Region 3W. MDNR:1–6; south central Minnesota—Region 4E. MDNR:1–7; southwest Minnesota—Region 4S. MDNR:1–6; southwest Minnesota—Region 4W. MDNR:1–6; southeast Minnesota—Region 5. MDNR:1–7; metropolitan Minnesota—Region 6. MDNR:1–7.

Henderson, F. R. 1960. Beaver in Kansas. Univ. Kan. Mus. Nat. Hist., Misc. Publ. 26:1–85.

Henderson, J. A., and F. F. Gilbert. 1978. Distribution and density of woodchuck burrow systems in relation to land-use practices. Can. Field-Nat. 92:128–136.

Hendricksen, R. L. 1973. Variation in the plains pocket gopher (*Geomys bursarius*) along a transect across Kansas and eastern Colorado. Trans. Kan. Acad. Sci. 75:322–368.

Henning, W. L. 1952. Method for keeping the eastern mole in captivity. J. Mamm. 33:392–395.

———. 1957. The eastern prairie mole (*Scalopus aquaticus machrinus* Rafinesque). Chap. 44 *in* The U.F.A.W. Handbook, Univ. Fed. Anim. Welfare, London.

Henry, J. D. 1977. The use of urine marking in the scavenging behaviour of the red fox (*Vulpes vulpes*). Behaviour 61:82–106.

Herman, Thomas, and Kathleen Fuller. 1974. Observations of the marten, *Martes americana*, in the Mackenzie District, Northwest Territories. Can. Field-Nat. 88:501–503.

Herrero, S. 1972. Aspects of evolution and adaptation in American black bears (*Ursus americanus* Pallas) and brown and grizzly bears (*U. arctos* Linné) of North America. IUCN Publ. 23:221–231.

Herrero, S., and D. Hamer. 1977. Courtship and copulation of a pair of grizzly bears, with comments on reproductive plasticity and strategy. J. Mamm. 58:441–444.

Herrick, C. L. 1892. The mammals of Minnesota . . . Bull. Geo. Nat. Hist. Surv. Minn. 7:1–299.

Hershkovitz, Philip. 1949. Status of names credited to Oken, 1816. J. Mamm. 30:289–301.

Hiaasen, B., et al. 1978. A bibliography of the genus *Pitymys* (Rodentia, Mammalia) and homonyms *Microtus pinetorum* and *Microtus subterraneus*. Dept. Nat. Resour., Cornell Univ. Agr. Exp. Sta., Ser. 10:1–43.

Hibbard, C. W. 1970. Pleistocene mammalian local faunas from the Great Plains and Central Lowland provinces of the United States. Pp. 395–433 *in* W. Dort, Jr., and J. Knox Jones, Jr., eds. Pleistocene and Recent environments of the central Great Plains. Univ. Pr. Kan., Spec. Publ. 3.

Hibbard, C. W., et al. 1965. Quaternary mammals of North America. Pp. 509–525 *in* H. E. Wright, Jr., and D. G. Frey, eds. The Quarternary of the United States. Princeton Univ. Press, Princeton, N.J.

Hibbard, E. A. 1954. Fox and gray squirrels. N.D. Outdoors 17:4–5.

———. 1955. Study of the fox and gray squirrel in North Dakota. N.D. Game Fish Dept. Pittman-Robertson Div., Proj. W36 R-1 and 2. 27 p. Mimeo.

———. 1956a. Range and spread of the fox and gray squirrels in North Dakota. J. Mamm. 37:525–531.

———. 1956b. History of the bounty system in North Dakota and its effect on predator populations. N.D. Outdoors 18(8):10–13.

———. 1957. Age ratios in wild mink populations. J. Mamm. 38:412–413.

———. 1958. Movements of beaver transplanted in North

Dakota. J. Wildl. Mgt. 22:209–211.
———. 1965. Northwestward extensions of the known range of the opossum. Loon 37:89–90.
———. 1970. Additional Minnesota opossum records. Loon 42:77–78.
Hibbard, E. A., and J. R. Beer. 1960. The plains pocket mouse in Minnesota. Flicker 32:89–94.
Hicks, E. A. 1949. Ecological factors affecting the activity of the western fox squirrel, *Sciurus niger rufiventer* (Geoffroy). Ecol. Monogr. 19:287–302.
Highby, P. R. 1940. The story of Minnesota beaver. Cons. Volun. 1(2):41–46.
———. 1941. A management program for Minnesota muskrat. Proc. Minn. Acad. Sci. 9:30–34.
Hildebrand, M. 1952. The integument in Canidae. J. Mamm. 33:419–428.
Hill, R. W. 1972. The amount of maternal care in *Peromyscus leucopus* and its thermal significance for the young. J. Mamm. 53:774–790.
Hiner, L. E. 1938. Observations on the foraging habits of beavers. J. Mamm. 19:317–319.
Hirth, D. H. 1977. Social behavior of white-tailed deer in relation to habitat. Wildl. Monogr. 53:1–55.
Hirth, D. H., and D. R. McCullough. 1977. Evolution of alarm signals in ungulates with special reference to white-tailed deer. Am. Nat. 111:31–42.
Hirth, H. F. 1959. Small mammals in an old field succession. Ecology 40:417–425.
Hisaw, F. L. 1923a. Feeding habits of moles. J. Mamm. 4:9–20.
———. 1923b. Observations on the burrowing habits of moles (*Scalopus aquaticus machrinoides*). J. Mamm. 4:79–88.
Hitchock, H. B. 1949. Hibernation of bats in southeastern Ontario and adjacent Quebec. Can. Field-Nat. 63:47–59.
———. 1965. Twenty-three years of bat banding in Ontario and Quebec. Can. Field-Nat. 79:4–14.
Hodgdon, H. E., and J. S. Larson. 1973. Some sexual differences in behaviour within a colony of marked beavers (*Castor canadensis*). Anim. Behav. 21:147–152.
Hodgson, J. R. 1972. Local distribution of *Microtus montanus* and *M. pennsylvanicus* in southwestern Montana. J. Mamm. 53:487–499.
Hoffman, C. O., and J. L. Gottschang. 1977. Numbers, distribution, and movements of a raccoon population in a surburban residential community. J. Mamm. 58:623–636.
Hoffman, R. S. 1958. The role of reproduction and mortality in population fluctuations of voles (*Microtus*). Ecol. Monogr. 28:79–109.
Hoffman, R. S., and J. K. Jones, Jr. 1970. Influence of late-glacial and post-glacial events on the distribution of Recent mammals on the northern Great Plains. Pp. 335–394 *in* W. Dort, Jr., and J. Knox Jones, Jr., eds. Pleistocene and Recent environments of the central Great Plains. Dept. Geol., Univ. Kan. Spec. Pub. 3.
Hoffman, R. S., and R. S. Peterson. 1967. Systematics and zoogeography of *Sorex* in the Bering Strait area. Syst. Zool. 16:127–136.
Hoffmeister, D. F., and W. L. Downes. 1964. Blue jays as predators of red bats. Southwest. Nat. 9:102–109.
Hoffmeister, D. F., and L. L. Getz. 1968. Growth and age-class in the prairie vole, *Microtus ochrogaster*. Growth 32:57–69.
Hohn, Bonnie M., and W. H. Marshall. 1966. Annual and seasonal weight changes in a thirteen-lined ground squirrel population, Itasca State Park, Minnesota. J. Minn. Acad. Sci. 33:102–106.
Holleman, D. F., and R. A. Dieterich. 1973. Body water content and turnover in several species of rodents as evaluated by the tritiated water method. J. Mamm. 54:456–465.
Holleman, D. F., J. R. Luick, and R. G. White. 1979. Lichen intake estimates for reindeer and caribou during winter. J. Wildl. Mgt. 43:192–201.
Hollister, N. 1914. A systematic account of the grasshopper mice. Proc. U.S. Nat. Mus. 47:427–489.
Holter, J. B., G. Tyler, and T. W. Walski. 1974. Nutrition of the snowshoe hare (*Lepus americanus*). Can. J. Zool. 52:1553–1558.
Homan, J. A., and H. H. Genoways. 1978. An analysis of hair structure and its phylogenetic implications among heteromyid rodents. J. Mamm. 59:740–760.
Hooper, E. T. 1942. An effect on the *Peromyscus maniculatus* Rassenkreis of land utilization in Michigan. J. Mamm. 23:193–196.
———. 1958. The male phallus in mice of the genus *Peromyscus*. Misc. Publ. Mus. Zool., Univ. Mich. 105:1–24.
———. 1968. Anatomy of the middle-ear walls and cavities in nine species of microtine rodents. Occ. Pap. Mus. Zool., Univ. Mich. 657:1–28.
Hooper, E. T., and G. G. Musser. 1964. The glans penis in neotropical cricetines (family Muridae) with comments on classification of muroid rodents. Misc. Publ. Mus. Zool., Univ. Mich. 123:1–57.
Hooper, E. T., and B. T. Ostenson. 1949. Age groups in Michigan otter. Occ. Pap. Mus. Zool., Univ. Mich. 518:1–22.
Hopson, J. A. 1966. The origin of the mammalian ear. Am. Zool. 6:437–450.
———. 1969. The origin and adaptive radiation of mammal-like reptiles and non-therian mammals. Ann. N.Y. Acad. Sci. 167(art. 1):199–216.
———. 1970. The classification of non-therian mammals. J. Mamm. 51:1–9.
———. 1973. Endothermy, small size, and the origin of

mammalian reproduction. Am. Nat. 107:446–452.

Hopson, J. A., and W. W. Crompton. 1969. Origin of mammals. Evol. Biol. 3:15–72.

Horner, B. E. 1947. Paternal care of young mice of the genus *Peromyscus*. J. Mamm. 28:31–36.

———. 1954. Arboreal adaptations of *Peromyscus* with special reference to use of the tail. Contr. Lab. Vert. Biol., Univ. Mich. 61:1–85.

Hornicke, H., and F. Batsch. 1977. Coecotrophy in rabbits—a circadian function. J. Mamm. 58:240–242.

Hornocker, M. G. 1969. Winter territoriality in mountain lions. J. Wildl. Mgt. 33:457–464.

———. 1970a. The American lion. Nat. Hist. 79(9):40–49, 68–71.

———. 1970b. An analysis of mountain lion predation upon mule deer and elk in the Idaho Primitive Area. Wildl. Monogr. 21:1–39.

———. 1972. Predator ecology and management—what now? J. Wildl. Mgt. 36:401–404.

Hoskinson, R. L., and L. D. Mech. 1976. Whitetailed deer migration and its role in wolf predation. J. Wildl. Mgt. 40:429–441.

Hoskinson, R. L., and J. R. Tester. 1980. Migration behavior of pronghorn in southeastern Idaho. J. Wildl. Mgt. 44:132–144.

Hough, F. 1957. Set hunting pattern of a little brown bat. J. Mamm. 38:121–122.

Houseknecht, C. R. 1969. Denning habits of the striped skunk and the exposure potential for disease. Wildl. Dis. Ass. Bull. 5:302–306.

Houseknecht, C. R., and R. A. Huempfner. 1970. A red fox-striped skunk encounter. Am. Midl. Nat. 83:304–306.

Houseknecht, C. R., and J. R. Tester. 1978. Denning habits of striped skunks (*Mephitis mephitis*). Am. Midl. Nat. 100:424–430.

Houston, D. B. 1974. Aspects of the social organization of moose. IUCN Publ. 24(2):690–696.

Houtcooper, W. C. 1978. Food habits of rodents in a cultivated ecosystem. J. Mamm. 59:427–430.

Howard, V. W., Jr. 1966. An observation of parturition in the pronghorn antelope. J. Mamm. 47:708.

Howard, W. E. 1949a. A means to distinguish skulls of coyotes and domestic dogs. J. Mamm. 30:169–171.

———. 1949b. Dispersal, amount of inbreeding, and longevity in a local population of prairie deermice on the George Reserve, southern Michigan. Contr. Lab. Vert. Biol., Univ. Mich. 43:1–50.

———. 1965. Interaction of behavior, ecology, and genetics of introduced mammals. Pp. 461–480 *in* H. G. Baker and G. L. Stebbins, eds. The genetics of colonizing species. Academic Press, New York.

Howard, W. E., and F. C. Evans. 1961. Seeds stored by prairie deer mice. J. Mamm. 42:260–263.

Howell, A. H. 1938. Revision of the North American ground squirrels. N. Am. Fauna 56.

Hoyt, S. Y., and S. F. Hoyt. 1950. Gestation period of the woodchuck, *Marmota monax*. J. Mamm. 31:454.

Hudson, J. W. 1973. Torpidity in mammals. Pp. 97–165 *in* G. C. Whittow, ed. Comparative physiology of thermoregulation. Vol. 3. Academic Press, New York.

Hudson, R. J. 1976. Resource division within a community of large herbivores. Naturaliste Can. 103:153–167.

Huenecke, H. A., A. B. Erickson, and W. H. Marshall. 1958. Marsh gases in muskrat houses in winter. J. Wildl. Mgt. 22:240–245.

Humphrey, S. R., and J. B. Cope. 1976. Population ecology of the little brown bat, *Myotis lucifugus*, in Indiana and north-central Kentucky. Am. Soc. Mamm., Spec. Publ. 4:1–81.

Hunsaker, Don II. 1977. Ecology of New World marsupials. Pp. 95–156 *in* Don Hunsaker II, ed. The biology of the marsupials. Academic Press, New York.

Hunsaker, Don II, and Donald Shupe. 1977. Behavior of New World marsupials. Pp. 279–347 *in* Don Hunsaker II, ed. The biology of the marsupials. Academic Press, New York.

Hunt, R. N., and L. M. Mangus. 1954. Deer management study: Mud Lake National Wildlife Refuge, Holt, Minnesota. J. Wildl. Mgt. 18:482–495.

Husband, T. P. 1976. Energy metabolism and body composition of the fox squirrel. J. Wildl. Mgt. 40:255–263.

Idstrom, J. M. 1965. The moose in Minnesota. Pp. 57–98 *in* J. B. Moyle, ed. Big game in Minnesota. Minn. Dept. Cons. Tech. Bull. 9.

Ingram, W. M. 1942. Snail associates of *Blarina brevicauda talpoides* (Say). J. Mamm. 23:255–258.

Innes, D. G. L. 1978. A reexamination of litter size in some North American microtines. Can. J. Zool. 56:1488–1496.

Irwin, L. L. 1975. Deer-moose relationships on a burn in northeastern Minnesota. J. Wildl. Mgt. 39:653–662.

Isley, T. E., and L. W. Gysel. 1975. Sound-source localization by the red fox. J. Mamm. 56:397–404.

Iversen, J. A. 1972. Basal energy metabolism of mustelids. J. Comp. Physiol. 81:341–344.

Iverson, S. L., R. W. Seabloom, and J. M. Hnatiuk. 1967. Small mammal distribution across the prairie-forest transition of Minnesota and North Dakota. Am. Midl. Nat. 78:188–197.

Iverson, S. L., and B. N. Turner. 1972a. Natural history of a Manitoba population of Franklin's ground squirrel. Can. Field-Nat. 86:145–149.

———. 1972b. Winter coexistence of *Clethrionomys gapperi* and *Microtus pennsylvanicus* in a grassland habitat. Am. Midl. Nat. 88:440–445.

———. 1974. Winter weight dynamics in *Microtus pennsylvanicus*. Ecology 55:1030–1041.

Jackson, H. H. T. 1915. A review of the American moles. N. Am. Fauna 38:1–100.

———. 1921. A recent migration of the gray squirrel in Wisconsin. J. Mamm. 2:113–114.

———. 1922a. Wolverene in Itasca County, Minnesota. J. Mamm. 3:53.

———. 1922b. Some habits of the prairie mole, *Scalopus aquaticus machrinus*. J. Mamm. 3:115.

———. 1926. An unrecognized water shrew from Wisconsin. J. Mamm. 7:57–58.

———. 1928. A taxonomic review of the American long-tailed shrews (genera *Sorex* and *Microsorex*). N. Am. Fauna 51.

———. 1955. The Wisconsin puma. Proc. Biol. Soc. Wash. 68:149–150.

———. 1961. Mammals of Wisconsin. Univ. Wisconsin Press, Madison.

Jacobsen, N. K. 1979. Alarm-bradycardia in white-tailed deer fawns (*Odocoileus virginianus*). J. Mamm. 60:343–349.

Jacobson, H. A., R. L. Kirkpatrick, and B. S. McGinnes. 1978. Disease and physiologic characteristics of two cottontail populations in Virginia. Wildl. Monogr. 60:1–53.

Jahoda, J. C. 1970. Effects of rainfall on activity of *Onychomys leucogaster breviauritus*. Am. Zool. 10:326.

———. 1973. The effect of the lunar cycle of the activity pattern of *Onychomys leucogaster breviauritus*. J. Mamm. 54:544–549.

James, T. R., and R. W. Seabloom. 1969. Reproductive biology of the white-tailed jackrabbit in North Dakota. J. Wildl. Mgt. 33:558–568.

Jameson, E. W., Jr. 1947. Natural history of the prairie vole. Univ. Kan. Publ., Mus. Nat. Hist. 1:125–151.

———. 1949. Some factors influencing the local distribution and abundance of woodland small mammals in central New York. J. Mamm. 30:221–235.

Janes, D. W. 1959. Home range and movements of the eastern cottontail in Kansas. Univ. Kan. Publ., Mus. Nat. Hist. 10:553–572.

Jenkins, F. A., Jr. 1971. Limb posture and locomotion in the Virginia opossum (*Didelphis marsupialis*) and in other non-cursorial mammals. J. Zool. 165:303–315.

Jenkins, S. H., and P. E. Busher. 1979. *Castor canadensis*. Mamm. Spp. 120:1–8.

Johnson, A. M. 1922. An observation on the carnivorous propensities of the gray gopher. J. Mamm. 3:187.

———. 1926. Tree-climbing woodchucks again. J. Mamm. 7:132–133.

Johnson, C. E. 1916. A brief descriptive list of Minnesota mammals. Fins, Feathers and Fur 8:1–8.

———. 1921. The "hand-stand" habit of the spotted skunk. J. Mamm. 2:87–89 .

———. 1922. Notes on the mammals of northern Lake County, Minnesota. J. Mamm. 3:33–39.

———. 1923a. A recent report of the wolverene in Minnesota. J. Mamm. 4:54–55.

———. 1923b. Aquatic habits of the woodchuck. J. Mamm. 4:105–107.

———. 1930. Recollections of the mammals of northwestern Minnesota. J. Mamm. 11:435–452.

Johnson, D. R. 1961. The food habits of rodents on rangelands of southern Idaho. Ecology 42:407–410.

———. 1969. Returns of the American Fur Company, 1835–1839. J. Mamm. 50:836–839.

Johnson, E. A., and J. S. Rowe. 1975. Fire in the subarctic wintering ground of the Beverly caribou herd. Am. Midl. Nat. 94:1–14.

Johnson, H. N. 1959. Rabies. Pp. 405–431 *in* T. M. Rivers and F. L. Horsfall, Jr., eds. Viral and rickettsial infections of man. 3d ed. Lippincott, Philadelphia and Montreal.

Johnson, M. L. 1967. An external identification character of the heather vole, *Phenacomys intermedius*: the orange aural hairs. Murrelet 48:17.

———. 1973. Characters of the heather vole, *Phenacomys*, and the red tree vole, *Arborimus*. J. Mamm. 54:239–244.

Johnson, M. S. 1930. Common injurious mammals of Minnesota. Univ. Minn. Agr. Exp. Sta., Bull. 259.

Johnson, R. P. 1973. Scent marking in mammals. Anim. Behav. 21:521–535.

Johnson-Murray, J. L. 1977. Myology of the gliding membranes of some petauristine rodents (genera: *Glaucomys*, *Pteromys*, *Petinomys*, and *Petaurista*). J. Mamm. 58:374–384.

Jolicoeur, Pierre. 1959. Multivariate geographical variation in the wolf *Canis lupus* L. Evolution 13:283–299.

———. 1975. Sexual dimorphism and geographical distance as factors of skull variation in the wolf, *Canis lupus* L. Pp. 54–61 *in* M. W. Fox, ed. The wild canids. Van Nostrand Reinhold, New York.

Jones, C., and R. D. Suttkus. 1973. Colony structure and organization of *Pipistrellus subflavus* in southern Louisiana. J. Mamm. 54:962–968.

Jones, J. K. Jr. 1960. The pronghorn, *Antilocapra americana*, in western Iowa. Am. Midl. Nat. 63:249.

———. 1964. Distribution and taxonomy of mammals of Nebraska. Univ. Kan. Publ., Mus. Nat. Hist. 16:1–356.

Jones, J. K., Jr., D. C. Carter, and H. H. Genoways. 1975. Revised checklist of North American mammals north of Mexico. Occ. Papers, Mus. Texas Tech Univ. 28:1–14.

Jones, J. K., Jr., J. R. Choate, and R. B. Wilhelm. 1978. Notes on distribution of three species of mammals in South Dakota. Prairie Nat. 10:65–70.

Jones, J. K., Jr., E. D. Fleharty, and P. B. Dunnigan. 1967. The distribution status of bats in Kansas. Univ. Kan.

Mus. Nat. Hist., Misc. Publ. 46:1–33.

Jones, J. K., Jr., and B. Mursaloğlu. 1961. Geographic variation in the harvest mouse, *Reithrodontomys megalotis*, on the central Great Plains and in adjacent regions. Univ. Kan. Publ., Mus. Nat. Hist. 14:9–27.

Jonkel, C. J., and I. M. Cowan. 1971. The black bear in the spruce-fir forest. Wildl. Monogr. 27:1–57.

Jonkel, C. J., and R. P. Weckwerth. 1963. Sexual maturity and implantation of blastocysts in the wild pine marten. J. Wildl. Mgt. 27:93–98.

Jope, E. M. 1976. The evolution of plants and animals under domestication: the contribution of studies at the molecular level. Phil. Trans. Roy. Soc. London, B (Biol. Sci.) 275:99–116.

Jordan, J. S. 1948. A midsummer study of the southern flying squirrel. J. Mamm. 29:44–48.

———. 1956. Notes on a population of eastern flying squirrels. J. Mamm. 37:294–295.

Jordan, P. A., P. C. Shelton, and D. L. Allen. 1967. Numbers, turnover, and social structure of the Isle Royale wolf population. Am. Zool. 7:233–252.

Jordan, R. H. 1974. Threat behavior of the black bear. IUCN Publ., new ser. 40:57–63.

Jorgensen, S. E., and L. D. Mech, eds. 1971. Proceedings of a symposium on the native cats of North America, their status and management. USFWS, Bur. Sport Fish. Wildl., Region 3.

Jorgenson, J. W., M. Novotny, M. Carmack, G. B. Copland, S. R. Wilson, S. Katona, and W. K. Whitten. 1978. Chemical scent constituents in the urine of the red fox (*Vulpes vulpes* L.) during the winter season. Science 199:796–798.

Joyal, R. 1976. Winter foods of moose in La Vérendrye Park, Quebec: an evaluation of two browse survey methods. Can. J. Zool. 54:1765–1770.

Joyal, R., and B. Scherrer. 1978. Summer movements and feeding by moose in western Quebec. Can. Field-Nat. 92:252–258.

Judd, S., J. Idstrom, and C. Kinsey. 1971. The black bear in Minnesota. Minn, Volun. 34(198):5–9.

Julander, O., W. L. Robinette, and D. A. Jones. 1961. Relation of summer range condition to mule deer herd productivity. J. Wildl. Mgt. 25:54–60.

Juniper, I. 1978. Morphology, diet, and parasitism in Quebec black bears. Can. Field-Nat. 92:186–189.

Justice, K. E. 1961. A new method for measuring home ranges of small mammals. J. Mamm. 42:462–470.

Kalin, Oscar. 1964. Soils and other factors affecting the distribution of two species of pocket gophers (*Geomys bursarius* and *Thomomys talpoides*) in northeastern North Dakota, and a taxonomic study of two *Geomys* populations on opposite sides of the Red River. M. S. Thesis, Univ. N. D., Grand Forks.

Kaplan, M. K., and H. Koprowski. 1980. Rabies. Sci. Am. 242(1):120ff.

Karns, P. D. 1967a. *Pneumostrongylus tenuis* in deer in Minnesota and implications for moose. J. Wildl. Mgt. 31:299–303.

———. 1967b. The moose in northeastern Minnesota. J. Minn. Acad. Sci. 34:114–116.

———. 1973. Minnesota's 1971 moose hunt: a preliminary report on the biological collections II. N. Am. Moose Conf. & Workshop 9:96–100.

———. 1979a. Managing our northwoods monarch. Minn. Volun. 42(242):32–40.

———. 1979b. Help for starving deer. Minn. Volun. 42(243):46–51.

———. 1980. *Parelaphostrongylus* infections in moose, caribou, and white-tailed deer. Proc. Symp. Mammal. Ecol. Hab. Mgt. Minn., Bemidji State Univ. 46–47.

Karns, P. D., and V. F. J. Crichton. 1978. Effects of handling and physical restraint on blood parameters in woodland caribou. J. Wildl. Mgt. 42:904–908.

Karns, P. D., H. Haswell, F. F. Gilbert, and A. E. Patton. 1974. Moose management in the coniferous-deciduous ecotone of North America. Naturaliste Can. 101:643–656.

Kasper, T. C., and J. F. Parrish III. 1974. The porcupine, *Erethizon dorsatum* Linnaeus, (Rodentia), in Texas. Southwest. Nat. 19:214–215.

Kaufman, D. W., and E. D. Fleharty. 1974. Habitat selection by nine species of rodents in north-central Kansas. Southwest. Nat. 18:443–452.

Kavanau, J. Lee. 1969. Influences of light on activity of small mammals. Ecology 50:549–557.

Kavanau, J. Lee, and Judith Ramos. 1975. Influences of light on activity and phasing of carnivores. Am. Nat. 109:391–418.

Keith, J. O., R. M. Hansen, and A. L. Ward. 1959. Effects of 2,4–D on abundance and foods of pocket gophers. J. Wildl. Mgt. 23:137–145.

Keith, L. B. 1966. Habitat vacancy during a snowshoe hare decline. J. Wildl Mgt. 30:828–832.

Keith, L. B., E. C. Meslow, and O. J. Rongstad. 1968. Techniques for snowshoe hare population studies. J. Wildl. Mgt. 32:801–812.

Keith, L. B., O. J. Rongstad, and E. C. Meslow. 1966. Regional differences in reproductive traits of the snowshoe hare. Can. J. Zool. 44:953–961.

Keith, L. B., and D. C. Surrendi. 1971. Effects of fire on a snowshoe hare population. J. Wildl. Mgt. 35:16–26.

Keith, L. B., and L. A. Windberg. 1978. A demographic analysis of the snowshoe hare cycle. Wildl. Monogr. 58:6–70.

Keller, B. L., and C. J. Krebs. 1970. *Mictorus* population biology. III. Reproductive changes in fluctuating populations of *M. ochrogaster* and *M. pennsylvanicus* in southern Indiana, 1965–1967. Ecol. Monogr. 40:263–294.

Keller, L. F. 1937. Porcupines killed and eaten by a coyote. J. Mamm. 16:232.

Kellogg, Remington, 1956. What and where are the whitetails? Pp. 31–55 *in* W. P. Taylor, ed. The deer of North America. Stackpole, Harrisburg, Pa., and Wildl. Mgt. Inst., Washington.

Kelly, J. P. 1978. British Columbia. Pp. 26–36 *in* M. K. Brown, ed. The pine marten. N. Y. State Dept. Env. Cons., Div. Fish. Wildl., unnum. report.

Kelsall, J. P. 1968. The migratory barren-ground caribou of Canada. Can. Wildl. Ser., Queen's Printer, Ottawa.

———. 1969. Structural adaptations of moose and deer for snow. J. Mamm. 50:302–310.

Kemp. G. A. 1972. Black bear population dynamics at Cold Lake, Alberta. IUCN Publ. 23:26–31.

———. 1976. The dynamics and regulation of black bear, *Ursus americanus*, populations in northern Alberta. IUCN Publ., new ser. 40:191–197.

Kemp, G. A., and L. B. Keith. 1970. Dynamics and regulation of red squirrel (*Tamiasciurus hudsonicus*) populations. Ecology 51:763–779.

Kennelly, J. J., and B. E. Johns. 1976. The estrous cycle of coyotes. J. Wildl. Mgt. 40:272–277.

Kennelly, J. J., and J. D. Roberts. 1969. Fertility of coyote-dog hybrids. J. Mamm. 50:830–831.

Kennerly, T. E., Jr. 1959. Contact between the ranges of two allopatric species of pocket gophers. Evolution 13:247–263.

———. 1964. Microenvironmental conditions of the pocket gopher burrow. Texas J. Sci. 16:395–441.

Kermack, K. A. 1967. The interrelations of early mammals. J. Linnaean Soc. (Zool.) 47:241–249.

Kile, T. L., and R. L. Marchinton. 1977. White-tailed deer rubs and scrapes: spatial, temporal, and physical characteristics and social role. Am. Midl. Nat. 97:257–266.

Kilgore, D. L., Jr. 1969. An ecological study of the swift fox (*Vulpes velox*) in the Oklahoma Panhandle. Am. Midl. Nat. 81:512–534.

King, Carolyn M. 1975. The sex ratio of trapped weasels (*Mustela nivalis*). Mamm. Rev. 5(1):1–8.

King, J. A. 1958. Maternal behavior and behavioral development in two subspecies of *Peromyscus maniculatus*. J. Mamm. 39:177–190.

———, ed. 1968. Biology of *Peromyscus* (Rodentia). Spec. Publ. Am. Soc. Mamm. 2.

King, J. A., J. C. Deshaies, and R. Webster. 1964. Age of weaning in two subspecies of deermice. Science 139:483–484.

King, J. A., D. Maas, and R. G. Weismann. 1964. Geographic variation in nest size among species of *Peromyscus*. Evolution 18:231–234.

King, J. A., E. O. Price, and P. L. Weber. 1967. Behavioral comparisons within the genus *Peromyscus*. Mich. Acad. Sci. Arts Lett. Pap. 53:113–136.

Kinsella, J. M. 1967. Unusual habitat of the water shrew in western Montana. J. Mamm. 48:475–477.

Kinsey, Charles. 1965. The black bear in Minnesota. Pp. 179–210 *in* J. B. Moyle, ed. Big game in Minnesota. Minn. Dept. Cons., Tech. Bull. 9.

———. 1967. "Brother bear"—the Indians call him Muckwa. Minn. Volun. 30(174):49–53.

Kirkland, G. L., Jr. 1976. Small mammals of a mine waste situation in the central Adirondacks, New York: a case of opportunism by *Peromyscus maniculatus*. Am. Midl. Nat. 95:103–110.

———. 1977a. The rock vole, *Microtus chrotorrhinus* (Miller) (Mammalia: Rodentia) in West Virginia. Ann. Carnegie Mus. 46:45–53.

———. 1977b. Responses of small mammals to the clearcutting of northern Appalachian forests. J. Mamm. 58:600–609.

Kirkland, G. L., Jr., and R. J. Griffin. 1974. Microdistribution of small mammals at the coniferous-deciduous forest ecotone in northern New York. J. Mamm. 55:417–427.

Kirkland, G. L., Jr., and C. J. Kirkland. 1979. Are small mammal hibernators K-selected? J. Mamm. 60:164–168.

Kirkpatrick, R. L., and G. L. Valentine. 1970. Reproduction in captive pine voles, *Microtus pinetorum*. J. Mamm. 51:779–785.

Kirsch, J. A. W. 1968. Prodromus of the comparative serology of marsupials. Nature (London) 217: 418–420.

———. 1977a. The six-percent solution: second thoughts on the adaptedness of the Marsupialia. Am. Sci. 65: 276–288.

———. 1977b. The classification of marsupials. Pp. 1–50 *in* Don Hunsaker II, ed. 1977. The biology of marsupials. Academic Press, New York.

———. 1977c. Biological aspects of the marsupial-placental dichotomy: a reply to Lillegraven. Evolution 31:898–900.

Kitchen, D. W. 1974. Social behavior and ecology of the pronghorn. Wildl. Monogr. 38:1–96.

Kitchen, D. W., and P. T. Bromley. 1974. Agonistic behavior of territorial pronghorn bucks. IUCN Publ. 24(1):365–381.

Kittredge, Joseph. 1928. Can the flying squirrel count? J. Mamm. 9:251–252.

———. 1938. The interrelations of habitat, growth rate, and associated vegetation in the aspen community of Minnesota and Wisconsin. Ecol. Monogr. 8:151–246.

Kleiman, Devra G. 1966. Scent marking in the Canidae. Symp. Zool. Soc. London 18:167–177.

———. 1967. Some aspects of social behavior in the Canidae. Am. Zool. 7:365–372.

———. 1974. Patterns of behaviour in hystricomorph rodents. Symp. Zool. Soc. London 34:171–209.

———. 1977. Monogamy in mammals. Quart, Rev. Biol. 52:39–69.

Kleiman, Devra G., and J. F. Eisenberg. 1973. Comparisons of canid and felid social systems from an evolutionary perspective. Anim. Behav. 21:637–659.

Klein, H. G. 1957. Inducement of torpidity in the woodland jumping mouse. J. Mamm. 38:272–274.

———. 1960. Ecological relationships of *Peromyscus leucopus noveboracensis* and *Peromyscus maniculatus gracilis* in central New York. Ecol. Monogr. 30:387–407.

Kline, P. D. 1959. Additional mule deer records for Iowa. J. Mamm. 40:148–149.

———. 1963. Notes on the biology of the jackrabbit in Iowa. Proc. Iowa Acad. Sci. 70:196–204.

Klingener, D. 1963. Dental evolution of *Zapus*. J. Mamm. 44:248–260.

———. 1964. The comparative myology of four dipodoid rodents (genera *Zapus*, *Napaeozapus*, *Sicista*, and *Jaculus*). Misc. Publ. Mus. Zool., Univ. Mich. 124:1–100.

Klinghammer, E., ed. 1978. The behavior and ecology of wolves. Garland STPM Press, New York.

Knight, R. R. 1970. The Sun River elk herd. Wildl. Monogr. 23:1–66.

Knopf, F. L., and D. F. Balph. 1969. Badgers plug burrows to confine prey. J. Mamm. 50:635–636.

Knowlton, F. F. 1972. Preliminary interpretations of coyote population mechanics with some management implications. J. Wildl. Mgt. 36:369–382.

Knudsen, G. J. 1962. Relationship of beaver to forest, trout, and wildlife in Wisconsin. Wis. Dept. Cons., Tech. Bull. 25.

Knudsen, G. J., and J. B. Hale. 1968. Food habits of otters in the Great Lakes region. J. Wildl. Mgt. 32:89–93.

Koehler, G. M., and M. G. Hornocker. 1977. Fire effects on marten habitat in the Selway-Bitterroot Wilderness. J. Wildl. Mgt. 41:500–505.

Koehler, G. M., M. G. Hornocker, and H. S. Hash. 1979. Lynx movements and habitat use in Montana. Can. Field-Nat. 93:441–442.

Koeppl, J. W., R. S. Hoffman, and C. F. Nadler. 1978. Pattern analysis of acoustical behavior in four species of ground squirrels. J. Mamm. 59:677–696.

Kohn, B. E., and J. J. Mooty. 1971. Summer habitat of white-tailed deer in north-central Minnesota. J. Wildl. Mgt. 35:476–487.

Kohn, P. H., and R. H. Tamarin. 1978. Selection of electrophoretic loci for reproductive parameters in island and mainland voles. Evolution 32:15–28.

Kolenosky, G. B. 1971. Hybridization between wolf and coyote. J. Mamm. 52:446–449.

———. 1972. Wolf predation on wintering deer in east-central Ontario. J. Wildl. Mgt. 36:357–369.

Komarek, E. V. 1932. Distribution of *Microtus chrotorrhinus*, with description of a new subspecies. J. Mamm. 13:155–158.

Koopman, K. F. 1967. Artiodactyls. Pp. 385–406 *in* S. Anderson and J. K. Jones, Jr., eds. Recent mammals of the world. Ronald Press, New York.

Koopman, K. F., and E. L. Cockrum. 1967. Bats. Pp. 109–150 *in* S. Anderson and J. K. Jones, Jr. Recent mammals of the world. Ronald Press, New York.

Koplin, J. R., and R. S. Hoffmann. 1968. Habitat overlap and competitive exclusion in voles (*Microtus*). Am. Midl. Nat. 80:494–507.

Korschgen, L. J. 1959. Food habits of the red fox in Missouri. J. Wildl. Mgt. 23:168–176.

Kramm, K. R. 1975. Entrainment of circadian activity rhythms in squirrels. Am. Nat. 109:379–389.

Kramm, K. R., D. E. Maki, and J. M. Glime. 1975. Variation within and among populations of red squirrel in the Lake Superior region. J. Mamm. 56:258–262.

Krausman, P. R. 1978. Forage relationships between two deer species in Big Bend National Park, Texas. J. Wildl. Mgt. 42:101–107.

Krebs, C. J. 1970. *Microtus* population biology: behavioral changes associated with the population cycle in *M. ochrogaster* and *M. pennsylvanicus*. Ecology 51:34–52.

———. 1977. Competition between *Microtus pennsylvanicus* and *Microtus ochrogaster*. Am. Midl. Nat. 97:42–49.

Krebs, C. J., M. S. Gaines, B. L. Keller, J. H. Myers, and R. H. Tamarin. 1973. Population cycles in small rodents. Science 179:35–41.

Krebs, C. J., B. L. Keller, and R. H. Tamarin. 1969. *Microtus* population biology: demographic changes in fluctuating populations of *M. ochrogaster* and *M. pennsylvanicus* in southern Indiana. Ecology 50:587–607.

Krebs, C. J., and J. H. Myers. 1974. Population cycles in small mammals. Adv. Ecol. Res. 8:267–399.

Kreeger, T. J. 1977. Impact of dog predation on Minnesota whitetail deer. J. Minn. Acad. Sci. 43(2):8–13.

———. 1979. Man's best friend? Minn. Volun. 42(242):2–11.

Krefting, L. W. 1969. The rise and fall of the coyote on Isle Royale. Naturalist (Minn.) 20(3):24–31.

———. 1974. The ecology of the Isle Royale moose with special reference to the habitat. Univ. Minn. Agric. Exp. Sta. Tech. Bull. 297, For. Serv. 15:1–76.

Krefting, L. W., and C. E. Ahlgren. 1974. Small mammals and vegetation changes after fire in a mixed conifer-hardwood forest. Ecology 55:1391–1398.

Krosch, H. F. 1973. Some effects of the bite of the short-tailed shrew, *Blarina brevicauda*. J. Minn. Acad. Sci. 39:21.

Krutzsch, P. H. 1954. North American jumping mice (genus *Zapus*). Univ. Kan. Publ., Mus. Nat. Hist. 7:349–472.

———. 1961. A summer colony of male little brown bats. J. Mamm. 42:529–530.

———. 1966. Remarks on silver-haired and Leib's bats in eastern United States. J. Mamm. 47:121.

Kruuk, H. 1972. Surplus killing by carnivores. J. Zool., London 166:233–244.

Kufeld, R. C. 1973. Foods eaten by the Rocky Mountain elk. J. Range Mgt. 26:106–113.

Kunz, T. H. 1971. Reproduction of some vespertilionid bats in central Iowa. Am. Midl. Nat. 86:477–486.

———. 1973. Resource utilization: temporal and spatial components of bat activity in central Iowa. J. Mamm. 54:14–32.

———. 1974. Reproduction, growth, and mortality of the vespertilionid bat, *Eptesicus fuscus*, in Kansas. J. Mamm. 55:1–13.

Kunz, T. H., E. L. P. Anthony, and W. R. Rumage III. 1977. Mortality of little brown bats following multiple pesticide applications. J. Wildl. Mgt. 41:476–483.

Kurtén, B., and R. Rausch. 1959. A comparison between Alaskan and Fennoscandian wolverine (*Gulo gulo* Linnaeus). Pp. 5–20 *in* Biometric comparisons between North American and European mammals. Acta Arctica 11:44.

Lagler, K. F., and B. T. Ostenson. 1942. Early spring food of the otter in Michigan. J. Wildl. Mgt. 6:244–254.

Lampe, R. P. 1976. Aspects of the predatory strategy of the North American badger, *Taxidea taxus*. Ph.D. Thesis, Univ. Minn. Diss. Abstr. Int. B. Sci. Eng. 37(12):5951.

———. 1980. Food habits and predatory behavior of *Taxidea taxus* in east-central Minnesota. Proc. Symp. Mammal. Ecol. Hab. Mgt. Minn., Bemidji State Univ. 66.

Landers, J. L., R. J. Hamilton, A. S. Johnson, and R. L. Marchinton. 1979. Foods and habitat of black bears in southeastern North Carolina. J. Wildl. Mgt. 43:143–153.

Landwer, M. F. 1935. An outside nest of a flying squirrel. J. Mamm. 16:67.

Lang, E. 1978. Social behavior and foraging ecology of the eastern chipmunk (*Tamias striatus*) in the Adirondack Mountains. Smithson. Contr. Zool. 266:1–107.

Langenau, E. E., Jr. 1979. Nonconsumptive uses of the Michigan deer herd. J. Wildl. Mgt. 43:620–625.

Lanier, D. L., and D. A. Dewsbury. 1977. Studies of copulatory behaviour in northern grasshopper mice (*Onychomys leucogaster*). Anim. Behav. 25:185–192.

Larson, J. S. 1967. Age structure and sexual maturity within a western Maryland beaver (*Castor canadensis*) population. J. Mamm. 48:408–413.

Larson, T. J., O. J. Rongstad, and F. W. Terbilcox. 1978. Movement and habitat use of white-tailed deer in south-central Wisconsin. J. Wildl. Mgt. 42:113–117.

Latham, R. M. 1952. The fox as a factor in the control of weasel populations. J. Wildl. Mgt. 16:516–517.

———. 1953. Simple method for identification of least weasel. J. Mamm. 34:385.

LaVal, R. K., and M. L. LaVal. 1979. Notes on reproduction, behavior and abundance of the red bat, *Lasiurus borealis*. J. Mamm. 60:209–212.

Lavocat, R. 1974. What is an hystricomorph? Symp. Zool. Soc. London 34:7–20.

Lawlor, T. E. 1979. Handbook to the orders and families of living mammals. Mad River Press, Eureka, CA.

Lawrence, Barbara, and W. H. Bossert. 1967. Multiple character analysis of *Canis lupus*, *latrans*, and *familiaris*, with a discussion of the relationships of *Canis niger*. Am. Zool. 7:223–232.

———. 1969. The cranial evidence for hybridization in New England *Canis*. Breviora, Mus. Comp. Zool., Harvard 330:1–13.

———. 1975. Relationships of North American *Canis* shown by a multiple character analysis of selected populations. Pp. 73–86 *in* M. W. Fox, ed. The wild canids. Van Nostrand Reinhold, New York.

Laycock, George. 1979. You'll never see a bobcat. Audubon 81(4):56–63.

Layne, J. N. 1951. The use of the tail by an opossum. J. Mamm. 32:464–465.

———. 1954a. The biology of the red squirrel. *Tamiasciurus hudsonicus loquax* (Bangs), in central New York. Ecol. Monogr. 24:227–267.

———. 1954b. The os clitoridis of some North American Sciuridae. J. Mamm. 35:357–366.

———. 1958a. Reproductive characteristics of the gray fox in southern Illinois. J. Wildl. Mgt. 22:157–163.

———. 1958b. Notes on mammals of southern Illinois. Am. Midl. Nat. 60:219–254.

———. 1967. Lagomorphs. Pp. 192–205 *in* S. Anderson and J. K. Jones, Jr., eds. Recent mammals of the world. Ronald Press, New York.

Layne, J. N., and W. J. Hamilton, Jr. 1954. The young of the woodland jumping mouse, *Napaeozapus insignis insignis* (Miller). Am. Midl. Nat. 52:242–247.

Layne, J. N., and W. H. McKeon. 1956a. Some aspects of red fox and gray fox reproduction in New York. N. Y. Fish Game J. 3:44–74.

———. 1956b. Notes on red fox and gray fox den sites in New York. N. Y. Fish Game J. 3:248–249.

Leach, Doug. 1977a. The forelimb musculature of marten (*Martes americana* Turton) and fisher (*Martes pennanti* Erxleben). Can. J. Zool. 55:31–41.

———. 1977b. The descriptive and comparative postcranial osteology of marten (*Martes americana* Turton) and fisher (*Martes pennanti* Erxleben): the appendicular skeleton. Can. J. Zool. 55:199–214.

Leach, Doug, and Anne Innis Dagg. 1976. The morphology

of the femur in marten and fisher. Can. J. Zool. 54:559–565.

Leach, Doug, and V. S. de Kleer. 1978. The description and comparative postcranial osteology of marten (*Martes americana* Turton) and fisher (*Martes pennanti* Erxleben): the axial skeleton. Can. J. Zool. 56:1180–1191.

Leamy, L. 1977. Genetic and environmental correlations of morphometric traits in randombred house mice. Evolution 31:357–369.

Ledin, Don. 1966. A moose season? Cons. Volun. 29(165):34–35.

Ledin, Don, and Pat Karns. 1963. On Minnesota's moose. Cons. Volun. 26(150):40–48.

Leedy, D. L. 1947. Spermophiles and badgers move eastward in Ohio. J. Mamm. 28:290–292.

Lehner, P. N. 1976. Coyote behavior: implications for management. Wildl. Soc. Bull. 4:120–126.

Lehner, P. N., R. Krumm, and A. T. Cringan. 1976. Tests for olfactory repellents for coyotes and dogs. J. Wildl. Mgt. 40:145–150.

Lent, P. C. 1974. Mother-infant relationships in ungulates. IUCN Publ. 24(1):14–55.

Leopold, Aldo S. 1933. Game management. Scribner's, New York.

———. 1937. Killing technique of the weasel. J. Mamm. 18:98–99.

Leraas, H. J. 1942. Notes on mammals from west-central Minnesota. J. Mamm. 23:343–345.

Levin, E. Y., and V. Flyger. 1971. Uroporphyrinogen III cosynthetase activity in the fox squirrel (*Sciurus niger*). Science 174:59–60.

Leyhausen, Paul. 1965. The communal organization of solitary mammals. Symp. Zool. Soc. Lond. 14:249–263.

———. 1978. Cat behavior: the predatory and social behavior of domestic and wild cats. Garland STPM Press, New York.

Liers, E. E. 1951. Notes on the river otter (*Lutra canadensis*). J. Mamm. 32:1–9.

———. 1953. An otter's story. Viking, New York.

———. 1958. Early breeding in the river otter. J. Mamm. 39:438–439.

Lillegraven. J. A. 1974. Biogeographical considerations of the marsupial-placental dichotomy. Annu. Rev. Ecol. Syst. 5:74–94.

———. 1975. Biological considerations of the marsupial-placental dichotomy. Evolution 29:707–722.

Lillegraven. J. A., Z. Kielen-Jaworowska, and W. A. Clemens, eds. 1979. Mesozoic mammals. Univ. Calif. Press, Berkeley.

Lindzey, F. G. 1978. Movement patterns of badgers in northwestern Utah. J. Wildl. Mgt. 42:418–422.

Lindzey, F. G., and E. C. Meslow. 1976. Winter dormancy in black bears in southwestern Washington. J. Wildl. Mgt. 40:403–415.

———. 1977a. Population characteristics of black bears on an island in Washington. J. Wildl. Mgt. 41:408–412.

———. 1977b. Home range habitat use by black bears in southwestern Washington. J. Wildl. Mgt. 41:413–425.

Lindzey, F. G., S. K. Thompson, and J. I. Hodges. 1977. Scent station index of black bear abundance. J. Wildl. Mgt. 41:151–153.

Linhart, S. B., and F. F. Knowlton. 1967. Determining age of coyotes by tooth cementum layers. J. Wildl. Mgt. 31:362–365.

Linhart, S. B., and W. B. Robinson. 1972. Some relative carnivore densities in areas under sustained coyote control. J. Mamm. 53:880–884.

Linzey, D. W., and A. V. Linzey. 1979. Growth and development of the southern flying squirrel (*Glaucomys volans volans*). J. Mamm. 60:615–620.

Lishak, R. S. 1977. Censusing 13-lined ground squirrels with adult and young alarm calls. J. Wildl. Mgt. 41:755–759.

Lloyd, J. A. 1975. Social structure and reproduction in two freely-growing populations of house mice (*Mus musculus* L.). Anim. Behav. 23:413–424.

LoBue, J., and R. M. Darnell. 1959. Effect of habitat disturbance on a small mammal population. J. Mamm. 40:425–437.

Lockie, J. D. 1966. Territory in small carnivores. Symp. Zool. Soc. London. 18:143–165.

Loken, K. I. 1980. Tularemia. Proc. Symp. Mammal. Ecol. Hab. Mgt. Minn., Bemidji State Univ. 56–59.

Long, C. A. 1962. Records of reproduction for harvest mice. J. Mamm. 43:103–104.

———. 1964. Taxonomic status of the Pleistocene badger, *Taxidea marylandica*. Am. Midl. Nat. 72:176–180.

———. 1965. The mammals of Wyoming. Univ. Kan. Publ. Mus. Nat. Hist. 14:493–758.

———. 1969. Gross morphology of the penis in seven species of the Mustelidae. Mammalia 33:145–160.

———. 1972a. Taxonomic revision of the mammalian genus *Microsorex* Coues. Trans. Kan. Acad. Sci. 74:181–196.

———. 1972b. Notes on habitat preference and reproduction in pigmy shrews, *Microsorex*. Can. Field-Nat. 86:155–160.

———. 1972c. Taxonomic revision of the North American badger, *Taxidea taxus*. J. Mamm. 53:725–759.

———. 1973. Reproduction in the white-footed mouse at the northern limits of its geographical range. Southwest. Nat. 18:11–20.

———. 1974. *Microsorex hoyi* and *Microsorex thompsoni*. Mamm. Spp. 33:1–4.

———. 1975a. Growth and development of the teeth and skull of the wild North American badger, *Taxidea taxus*. Trans. Kan. Acad. Sci. 77:106–120.

———. 1975b. Molt in the North American badger, *Taxidea*

taxus. J. Mamm. 56:921–924.

———. 1976a. Notes on reproduction in pigmy shrews and observed ratios of mammae to body weights. Rep. Fauna Flora Wisc. (Univ. Wisc., Stevens Pt.) 11:5–6.

———. 1976b. Evolution of mammalian cheek pouches and a possibly discontinuous origin of a higher taxon (Geomyoidea). Am. Nat. 110:1093–1097.

Long, C. A., and F. A. Copes. 1968. Note on the rate of dispersion of the opossum in Wisconsin. Am. Midl. Nat. 80:283–284.

Long, C. A., and Thomas Howard. 1976. Intraspecific overt fighting in the wild mink. Rep. Fauna Flora. Wisc. (Univ. Wisc., Stevens Pt. 11:4–5.

Long, C. A., and W. C. Kerfoot. 1963. Mammalian remains from owl pellets in eastern Wyoming. J. Mamm. 44:129–131.

Long, C. A., and R. G. Severson. 1969. Geographical variation in the big brown bat in the north-central United States. J. Mamm. 50:621–624.

Longley, W. H. 1962. Movements of red fox. J. Mamm. 43:107.

———. 1963. Minnesota gray and fox squirrels. Am. Midl. Nat. 69:82–98.

———. 1972. Portrait of the masked ones. Minn. Volun. 35(204):52–56.

Longley, W. H., and J. B. Moyle. 1963. The beaver in Minnesota. Minn. Dept. Cons., Tech. Bull. 6:1–87.

Longley, W. H., and Charles Wechsler. (1977). Minnesota mammals. Minn. Dept. Nat. Resour., Bur. Info. Ed. :1–28.

Lord, R. D., Jr. 1958. The importance of juvenile breeding to the annual cottontail crop. Trans. N. Am. Wildl. Conf. 23:269–275.

———. 1961. A population study of the gray fox. Am. Midl. Nat. 66:87–109.

———. 1963. The cottontail rabbit in Illinois. Ill. Dept. Cons., Tech. Bull. 3.

Lore, R., and K. Flannelly. 1977. Rat societies. Sci. Amer. 236(5):106–111, 113–116.

Lott, D. F. 1974. Sexual and agressive behaviour of American bison *Bison bison*. IUCN Publ. 24(1):382–394.

Lotze, J.-H., and S. Anderson. 1979. *Procyon lotor*. Mamm. Spp. 119:1–8.

Lovejoy, D. A. 1973. Ecology of the woodland jumping mouse (*Napaeozapus insignis*) in New Hampshire. Can. Field-Nat. 87:145–149.

Low, Bobbi S. 1978. Environmental uncertainty and the parental strategies of marsupials and placentals. Am. Nat. 112:197–213.

Lowry, D. A., and K. L. McArthur. 1978. Domestic dogs as predators on deer. Wildl. Soc. Bull. 6(1):38–39.

Luckens, M. M., and W. H. Davis. 1964. Bats: sensitivity to DDT. Science 146:948.

Ludwick, R. L., J. P. Fontenot, and H. S. Mosby. 1969. Energy metabolism of the eastern gray squirrel. J. Wildl. Mgt. 32:569–575.

Lundelius, E. L., Jr. 1974. The last fifteen thousand years of faunal change in North America. Pp. 141–160 *in* C. C. Black, ed. History and prehistory of the Lubbock Lake site. The Mus. J. 15.

Lyman, C. P. 1954. Activity, food consumption, and hoarding in hibernators. J. Mamm. 35:545–552.

———. 1963. Hibernation in mammals and birds. Am. Sci. 51:127–138.

Lynch, Carol B. 1977. Inbreeding effects upon animals derived from a wild population of *Mus musculus*. Evolution 31:526–537.

Lynch, G. M. 1972. Effects of strychnine control on nest predators of dabbling ducks. J. Wildl. Mgt. 36:436–440.

———. 1974. Some densities of Manitoba raccoons. Can. Field-Nat. 88:494–495.

Lynch, G. R., and G. E. Folk, Jr. 1968. Distribution and habitat of the red squirrel, *Tamiasciurus hudsonicus*, in the north central states. Proc. Iowa Acad. Sci. 75:463–466.

Lyon, M. W., Jr. 1932. Franklin's ground squirrel and its distribution in Indiana. Am. Midl. Nat. 13:16–21.

———. 1936. Mammals of Indiana. Am. Midl. Nat 17:1–384.

MacArthur, R. A. 1978. Winter movements and home range of the muskrat. Can Field-Nat. 92:345–349.

MacArthur, R. A., and M. Aleksiuk. 1979. Seasonal microenvironments of the muskrat (*Ondatra zibethicus*) in a northern marsh. J. Mamm. 60:146–154.

MacConnell-Yount, Elizabeth, and Carol Smith. 1978. Mule deer-coyote interactions in north-central Colorado. J. Mamm. 59:422–423.

MacGregor, A. E. 1942. Late fall and winter food of foxes in central Massachusetts. J. Wildl. Mgt. 6:221–224.

Mackiewicz, J., and R. H. Backus. 1956. Oceanic records of *Lasionycteris noctivagans* and *Lasiurus borealis*. J. Mamm. 37:442–443.

Macpherson, A. H. 1965. The origin of diversity in mammals of the Canadian Arctic tundra. Syst. Zool. 14:153–173.

Madden, J. R. 1974. Female territoriality in a Suffolk County, Long Island, population of *Glaucomys volans*. J. Mamm. 55:647–652.

Madison, D. M. 1978. Movement indicators of reproductive events among female meadow voles as revealed by radiotelemetry. J. Mamm. 59:835–843.

Magnus, L. T. 1956. Mountain lion observation in Lake of the Woods country. Flicker 18(1):43–44.

Mahan, B. R., M. P. Munger, and H. L. Gunderson. 1978. Analysis of the Flehmen display in American bison (*Bison bison*). Prairie Nat. 10:33–42.

Mallory, F. F., and F. V. Clulow. 1977. Evidence of pregnancy failure in the wild meadow vole, *Microtus pennsylvanicus*. Can. J. Zool. 55:1–17.

Manley, G. A. 1972. A review of some current concepts of the functional evolution of the ear in terrestrial vertebrates. Evolution 26:608–621.

Manville, R. H. 1949. A study of small mammal populations in northern Michigan. Misc. Publ. Mus. Zool., Univ. Mich. 73:1–83.

Margulis, H. L. 1977. Rat fields, neighborhood sanitation, and rat complaints in Newark, New Jersey. Geogr. Rev. 67:221–231.

Marks, S. A., and A. W. Erickson. 1966. Age determination in the black bear. J. Wildl. Mgt. 30:389–410.

Marschner, F. J. (1930). The original vegetation of Minnesota. Map, 1974, North Central Forest Experiment Station, St. Paul, based on an unpublished map prepared by Marschner in 1930.

Marsden, H. M. and C. H. Conaway. 1963. Behavior and the reproductive cycle in the cottontail. J. Wildl. Mgt. 27:161–170.

Marsden, H. M., and N. R. Holler. 1964. Social behavior in confined populations of the cottontail and the swamp rabbit. Wildl. Monogr. 13:1–39.

Marshall, W. H. 1936. A study of the winter activities of the mink. J. Mamm. 17:382–392.

———. 1937. Muskrat sex-ratios in Utah. J. Mamm. 18:518–519.

———. 1946. Winter food habits of the pine marten in Montana. J. Mamm. 27:83–84.

———. 1951a. Predation on shrews by frogs. J. Mamm. 32:219.

———. 1951b. Accidental death of a porcupine. J. Mamm. 32:221.

———. 1951c. Pine marten as a forest product. J. For. 49:899–905.

———. 1956. Summer weights of raccoons in northern Minnesota. J. Mamm. 37:445.

Marshall, W. H., G. W. Gullion, and R. G. Schwab. 1962. Early summer activities of porcupines as determined by radio-positioning techniques. J. Wildl. Mgt. 26:75–79.

Marshall, W. H., and D. G. Miquelle. 1978. Terrestrial wildlife of Minnesota peatlands: a literature search. Rep. submitted to Minn. Dept. Nat. Resour., May 1978.

Marston, M. A. 1942. Winter relations of bobcats to white-tailed deer in Maine. J. Wildl. Mgt. 6:328–337.

Martell, A. M. 1974. A northern range extension for the northern bog lemming, *Synaptomys borealis borealis* (Richardson). Can. Field-Nat. 88:348.

Martell, A. M., and Andrew Radvanyi. 1977. Changes in small mammal populations after clearcutting of northern Ontario black spruce forest. Can. Field-Nat. 91:41–46.

Martin, E. P. 1956. A population study of the prairie vole (*Microtus ochrogaster*) in northeastern Kansas. Univ. Kan. Publ., Mus. Nat. Hist. 8:361–416.

Martin, K. A., and M. B. Fenton. 1978. A possible defensive function for calls given by bats (*Myotis lucifugus*) arousing from torpor. Can. J. Zool. 56:1430–1432.

Martin, L. D., B. M. Gilbert, and D. B. Adams. 1977. A cheetah-like cat in the North American Pleistocene. Science 195:981–982.

Martin, P. S. 1958. Pleistocene ecology and biogeography of North America. Pp. 375–420 *in* C. L. Hubbs, ed. Zoogeography. Publ. Am. Assoc. Adv. Sci. 51.

Martin, R. A. 1967. Notes on the male reproductive tract of *Nectogale* and other soricid insectivores. J. Mamm. 48:664–666.

Martin, R. L. 1971a. The natural history and taxonomy of the rock vole, *Microtus chrotorrhinus*. Ph.D. thesis, Univ. Conn., Storrs.

———. 1971b. Interspecific associations of rock voles. Beta Kappa Chi Bull. 30:5–7.

———. 1972. Parasites and diseases of the rock vole. Univ. Conn. Occ. Pap., Biol. Sci. Ser. 2:107–113.

———. 1973a. Molting in the rock vole, *Microtus chrotorrhinus*. Mammalia 37:342–357.

———. 1973b. The dentition of *Microtus chrotorrhinus* (Miller) and related forms. Univ. Conn. Occ. Pap., Biol. Sci. Ser. 2:183–201.

Martinka, C. J. 1967. Mortality of northern Montana pronghorns in a severe winter. J. Wildl. Mgt. 31:159–164.

Martinsen, D. L. 1968. Temporal patterns in the home ranges of chipmunks (*Eutamias*). J. Mamm. 49:83–91.

———. 1969. Energetics and activity patterns of shorttailed shrews (*Blarina*) on restricted diets. Ecology 50:505–510.

Mascarello, J. T., A. D. Stock, and Sen Pathak. 1974. Conservatism in the arrangement of genetic material in rodents. J. Mamm. 55:695–704.

Mathisen, J. E. 1972. Wildlife management plan. Chippewa Natl. For. report. 29pp. + 4p. appendix.

Mathwig, H. J. 1973. Food and population characteristics of Iowa coyotes. Iowa State J. Res. 47:167–189.

Matocha, K. G. 1977. The vocal repertoire of *Spermophilus tridecemlineatus*. Am. Midl. Nat. 98:482–487.

Matson, J. R. 1946. Notes on dormancy in the black bear. J. Mamm. 27:203–212.

———. 1954. Observations on the dormant phase of a female black bear. J. Mamm. 35:28–35.

Matthew, W. D. 1934. A phylogenetic chart of the Artiodactyla. J. Mamm. 15:207–209.

Mautz, W. W. 1978. Sledding on a bushy hillside: the fat cycle in deer. Wildl. Soc. Bull. 6:88–90.

Maxell, M. H., and L. N. Brown. 1968. Ecological distribution of rodents on the High Plains of eastern Wyoming. Southwest. Nat. 13:143–158.

Maxham, Glenn. 1970. Return of the marten. Cons. Volun 33(189):8–12.

Maxwell, R. K., L. L. Rogers, and R. B. Brander. 1972.

The energetics of wintering bears (*Ursus americanus*) in northeastern Minnesota. Bull. Ecol. Soc. Am. 53(2):21.

McAlpine, D. F. 1976. First record of the eastern pipistrelle in New Brunswick. Can. Field-Nat. 90:476.

McCabe, R. A., and E. L. Kozicky. 1972. A position on predator management. J. Wildl. Mgt. 36:382–394.

McCaffery, K. R. 1976. Deer trail counts as an index to populations and habitat use. J. Wildl. Mgt. 40:308–316.

McCaffery, K. R., John Tranetzki, and James Piechura. Jr. 1974. Summer foods of deer in northern Wisconsin. J. Wildl. Mgt. 38:215–219.

McCann, L. J. 1944. Notes on growth, sex and age ratios, and suggested management of Minnesota muskrats. J. Mamm. 25:59–63.

McCarley, Howard. 1966. Annual cycle, population dynamics, and adaptive behavior of *Citellus tridecemlineatus*. J. Mamm. 47:294–316.

———. 1975. Long-distance vocalizations of coyotes. J. Mamm. 56:847–856.

McCarty, Richard. 1978. *Onychomys leucogaster*. Mamm. Spp. 87:1–6.

McCloskey, R. J., and K. C. Shaw. 1977. Copulatory behavior of the fox squirrel. J. Mamm. 58:663–665.

McClure, H. E. 1942. Summer activities of bats (genus *Lasiurus*) in Iowa. J. Mamm. 23:430–434.

McCord, C. M. 1974. Selection of winter habitat by bobcats (*Lynx rufus*) on the Quabbin Reservation, Massachusetts. J. Mamm. 55:428–437.

McCullough, Yvette. 1979. Carbohydrate and urea influences on in vitro deer forage digestibility. J. Wildl. Mgt. 43:650–656.

McGee, L. E. 1965. Extension of the range eastward for *Taxidea taxus* (Carnivora: Mustelidae). Southwest. Nat. 10:78.

McHugh, T. C. 1958. Social behavior of American buffalo (*Bison bison bison*). Zoologica 43:1–40.

McInvaille, W. B., Jr., and L. B. Keith. 1974. Predator-prey relations and breeding biology of the great horned owl and red-tailed hawk in central Alberta. Can. Field-Nat. 88:1–20.

McKeever, Sturgis. 1960. Food of the northern flying squirrel in northeastern California. J. Mamm. 41:270–271.

McKenna, M. G. 1969. The origin and early differentiation of therian mammals. Ann. N.Y. Acad. Sci. 167(art. 1):217–240.

McLaughlin, C. A. 1967. Aplodontoid, sciuroid, geomyoid, castoroid, and anomaluroid rodents. Pp. 210–225 *in* S. Anderson and J. K. Jones, Jr., eds. Recent mammals of the world. Ronald Press, New York.

M'Closkey, R. T. 1975a. Habitat dimensions of white-footed mice, *Peromyscus leucopus*. Am. Midl. Nat. 93:158–167.

———. 1975b. Habitat succession and rodent distribution. J. Mamm. 56:950–955.

M'Closkey, R. T., and B. Fieldwick. 1975. Ecological separation of sympatric rodents. J. Mamm. 56:119–129.

M'Closkey, R. T., and D. T. Lajoie. 1975. Determinants of local distribution and abundance in white-footed mice. Ecology 56:467–472.

McManus, J. J. 1967. Observations on sexual behavior in the opossum, *Didelphis marsupialis*. J. Mamm 48:486–487.

———. 1969. Temperature regulation in the opossum, *Didelphis marsupialis virginiana*. J. Mamm. 50:550–558.

———. 1974a. Bioenergetics and water requirements of the redback vole, *Clethrionomys gapperi*. J. Mamm 55:30–44.

———. 1974b. *Didelphis virginiana*. Mamm. Species 40:1–6.

———. 1974c. Activity and thermal preference of the little brown bat, *Myotis lucifugus*, during hibernation. J. Mamm. 55:844–846.

McMiller, P. R. 1947. Principal soil regions of Minnesota. Bull. U. Minn. Agr. Exp. Sta. 392:1–48.

McMillin, J. M., U. S. Seal, and P. D. Karns. 1980. Hormonal correlates of hypophagia in white-tailed deer (*Odocoileus virginianus borealis*). Symp. prot. and fat metab. during mamm. hypophagia and hib. 30th Ann. Fall Mtg., Am. Physiol. Soc.

McMillin, J. M., et. al. 1976. Annual testosterone rhythm in the black bear, *Ursus americanus*. Biol. Reprod. 15:163–167.

McNab, B. K. 1966. The metabolism of fossorial rodents: a study in convergence. Ecology 47:712–733.

———. 1978. The evolution of endothermy in the phylogeny of mammals. Am. Nat. 112:1–21.

McNab, B. K., and P. R. Morrison. 1963. Body temperatures and metabolism in subspecies of *Peromyscus* from arid and mesic environments. Ecol. Monogr. 33:63–82.

Mead, R. A. 1963. Some aspects of parasitism in skunks of the Sacramento Valley of California. Am. Mid. Nat. 70:164–167.

———. 1968a. Reproduction in eastern forms of the spotted skunk (genus *Spilogale*). J. Zool., London 156:119–136.

———. 1968b. Reproduction in western forms of the spotted skunk (genus *Spilogale*). J. Mamm. 49:373–390.

———. 1972. Pineal gland: its role in controlling delayed implantation in the spotted skunk. J. Repr. Fertil. 30:147–150.

———. 1975. Effects of hypophysectomy on blastocyst survival, progesterone secretion, and nidation in the spotted skunk. Biol. Reprod. 12:526–533.

Meagher, M. M. 1973. The bison of Yellowstone National Park. Natl. Park Serv. Sci. Monogr., Ser. 1:1–161.

———. 1974. Yellowstone's bison: a unique wild heritage. Natl. Parks Cons. Mag. 48(5):9–14.

Mech, L. D. 1966a. The wolves of Isle Royale. U. S. Natl. Park Serv., Fauna Ser. 7:1–210.

———. 1966b. Hunting behavior of timber wolves in Minnesota. J. Mamm. 47:347–348.

———. 1967. Telemetry as a technique in the study of predation. J. Wildl. Mgt. 31:492–496.

———. 1970a. Implications of wolf ecology to management. Pp. 39–44 *in* S. E. Jorgensen, C. E. Faulkner, and L. D. Mech, eds. Proceedings of a symposium on wolf management in selected areas of North America. U.S. Fish Wildl. Serv., Bur. Sport Fish Wildl., Region 3.

———. 1970b. The wolf: the ecology and behavior of an endangered species. Doubleday, New York.

———. 1972. Spacing and possible mechanisms of population regulation in wolves. Am. Zool. 12:642.

———. 1973a. Wolf numbers in the Superior National Forest of Minnesota. USDA For. Serv., Res. Pap. NC-97:1–10.

———. 1973b. Canadian lynx invasion of Minnesota. Biol. Cons. 5:151–152.

———. 1974a. A new profile for the wolf. Nat. Hist. 83(4):26–31.

———. 1974b. *Canis lupus*. Mamm. Species 37:1–6.

———. 1975a. Hunting behavior in two similar species of social canids. Pp. 363–368 *in* M. W. Fox, ed. The wild canids. Van Nostrand Reinhold, New York.

———. 1975b. Disproportionate sex ratios of wolf pups. J. Wildl. Mgt. 39:737–740.

———. 1977a. Population trend and winter deer consumption in a wolf pack. Pp. 55–83 *in* Proc. 1975 Pred. Symp., Bull. Mont. For. Cons. Exp. Sta., Univ. Mont.

———. 1977b. Productivity, mortality, and population trends of wolves in northeastern Minnesota. J. Mamm. 58:559–574.

———. 1977c. Wolf-pack buffer zones as prey reservoirs. Science 198:320–321.

———. 1977d. A recovery plan for the eastern timber wolf. Minn. Volun. 40(235):28–39.

———. 1977e. Where can the wolf survive? Nat. Geogr. 152:518–537.

———. 1977f. The eastern timber wolf. Minn. Volun. 40(235):28–33., 36–39.

———. 1977g. Record movement of a Canadian lynx. J. Mamm. 58:676–677.

———. 1979. Why some deer are safe from wolves. Nat. Hist. 88(1)70–77.

———. 1980. Age, sex, reproduction, and spatial organization of lynxes colonizing northeastern Minnesota. J. Mamm. 61:261–267.

Mech, L. D., D. M. Barnes, and J. R. Tester. 1968. Seasonal weight changes, mortality, and population structure of raccoons in Minnesota. J. Mamm. 49:63–73.

Mech, L. D., and L. D. Frenzel, Jr. 1971. An analysis of the age, sex, and condition of deer killed by wolves in northeastern Minnesota. Pp. 35–51 *in* L. D. Mech and L. D. Frenzel, Jr., eds. Ecological studies of the timber wolf in northeastern Minnesota. USDA For. Serv. Res. Pap. NC 52.

Mech, L. D., L. D. Frenzel, Jr., and P. D. Karns. 1971. The effect of snow conditions on the vulnerability of white-tailed deer to wolf predation. Pp. 51–59 *in* L. D. Mech and L. D. Frenzel, Jr., eds. Ecological studies of the timber wolf in northeastern Minnesota. USDA For. Serv. Res. Pap. NC 52.

Mech, L. D., L. D. Frenzel, Jr., R. R. Ream, and J. W. Winship. 1971. Movements, behavior, and ecology of timber wolves in northeastern Minnesota. USDA For. Ser. Res. Pap. NC 52.

Mech, L. D., K. L. Heezen, and D. B. Siniff. 1966. Onset and cessation of activity in cottontail rabbits and snowshoe hares in relation to sunset and sunrise. Anim. Behav. 14:410–413.

Mech, L. D., and P. D. Karns. 1977. Role of the wolf in a deer decline in the Superior National Forest. USDA For. Serv. Res. Pap. NC-148:1–23.

Mech, L. D., and R. P. Peters. 1977. The study of chemical communication in free-ranging mammals. Pp. 321–332 *in* D. Muller-Schwarze and M. M. Mozell, eds. Chemical signals in vetebrates. Plenum, New York.

Mech, L. D., and L. L. Rogers. 1977. Status, distribution, and movements of martens in northeastern Minnesota. USDA For. Serv., Res. Pap. NC-143. N. Cent. For. Exp. Sta., St. Paul.

Mech, L. D., J. R. Tester, and D. W. Warner. 1966. Fall daytime resting habits of raccoons as determined by telemetry. J. Mamm. 47:450–466.

Mech, L. D., and F. J. Turkowski. 1966. Twenty-three raccoons in one winter den. J. Mamm. 47:529–530.

Medin, D. A., and A. E. Anderson. 1979. Modeling the dynamics of a Colorado mule deer population. Wildl. Monogr. 68:1–77.

Medjo, D. C., and L. D. Mech. 1976. Reproductive activity in nine- and ten-month-old wolves. J. Mamm. 57:406–408.

Meierotto, R. R. 1967. The distribution of small mammals across a prairie-forest ecotone. Ph.D. thesis, Univ. Minn.

Mellanby, K. 1971. The mole. Collins, London.

Mengel, R. M. 1971. A study of dog-coyote hybrids and implications concerning hybridization in *Canis*. J. Mamm. 52:316–336.

Menges, R. W., R. T. Habermann, and H. J. Stains. 1955. A distemper-like disease in raccoons and isolation of *Histoplasma capsulatum* and *Haplosporangium parvum*. Trans. Kan. Acad. Sci. 58:58–67.

Meredith, D. H. 1977. Interspecific agonism in two parapatric species of chipmunks (*Eutamias*). Ecology 58:423–430.

Merriam, H. G. 1966. Temporal distribution of woodchuck interburrow movements. J. Mamm. 47:103–110.

———. 1971. Woodchuck burrow distribution and related movement patterns. J. Mamm. 52:732–746.

Merriam, H. G., and A. Merriam. 1965. Vegetation zones around woodchuck burrows. Can. Field-Nat. 79:177–180.

Merrill, Evelyn H. 1978. Bear depredations at backcountry campgrounds in Glacier National Park. Wildl. Soc. Bull. 6:123–127.

Merritt, J. F., and J. M. Merritt. 1978. Population ecology and energy relationships of *Clethrionomys gapperi* in a Colorado subalpine forest. J. Mamm. 59:576–598.

Meserve, P. L. 1971. Population ecology of the prairie vole, *Microtus ochrogaster*, in the western mixed prairie of Nebraska. Am. Midl. Nat. 86:417–433.

———. 1977. Three-dimensional home ranges of cricetid rodents. J. Mamm. 58:549–558.

Meslow, E. C., and L. B. Keith. 1968. Demographic parameters of a snowshoe hare population. J. Wildl. Mgt. 32:812–834.

Metzgar, L. H. 1971. Behavioral population regulation in the woodmouse, *Peromyscus leucopus*. Am. Midl. Nat. 86:434–448.

———. 1973. Home range shape and activity in *Peromyscus leucopus*. J. Mamm. 54:383–390.

Michener, D. R. 1972. Notes on home range and social behavior in adult Richardson's ground squirrels (*Spermophilus richardsonii*). Can. Field-Nat. 86:77–79.

———. 1974. Annual cycle of activity and weight changes in Richardson's ground squirrel, *Spermophilus richardsonii*. Can. Field-Nat. 88:409–413.

Michener, D. R., and G. R. Michener. 1971. Sex ratio and interyear residence in a population of *Spermophilus richardsonii*. J. Mamm. 52:853.

Michener, Gail R. 1971. Maternal behaviour in Richardson's ground squirrel, *Spermophilus richardsonii richardsonii*: retrieval of young by lactating females. Anim. Behav. 19:653–656.

———. 1973a. Field observations on the social relationships between adult female and juvenile Richardson's ground squirrels. Can. J. Zool. 51:33–38.

———. 1973b. Maternal behaviour in Richardson's ground squirrel (*Spermophilus richardsonii richardsonii*): retrieval of young by non-lactating females. Anim. Behav. 21:157–159.

———. 1973c. Climatic conditions and breeding in Richardson's ground squirrel. J. Mamm. 54:499–503.

———. 1973d. Intraspecific aggression and social organization in ground squirrels. J. Mamm. 54:1001–1002.

———. 1974. Development of adult-young identification in Richardson's ground squirrel. Dev. Psychobiol. 7:375–384.

———. 1977a. Effect of climatic conditions on the annual activity and hibernation cycle of Richardson's ground squirrels and Columbian ground squirrels. Can. J. Zool. 55:693–703.

———. 1977b. Gestation period and juvenile age at emergence in Richardson's ground squirrel. Can. Field-Nat. 91:410–413.

———. 1978. Effect of age and parity on weight gain and entry into hibernation in Richardson's ground squirrels. Can. J. Zool. 56:2573–2577.

———. 1979a. The circannual cycle of Richardson's ground squirrels in southern Alberta. J. Mamm. 60:760–768.

———. 1979b. Yearly variations in the population dynamics of Richardson's ground squirrels. Can. Field-Nat. 93:363–370.

———. 1980. Estrous and gestation periods in Richardson's ground squirrels. J. Mamm. 61:531–534.

Michener, Gail R., and D. R. Michener. 1973. Spatial distribution of yearlings in a Richardson's ground squirrel population. Ecology 54:1138–1142.

———. 1977. Population structure and dispersal in Richardson's ground squirrels. Ecology 58:359–368.

Michener, Gail R., and D. H. Sheppard. 1972. Social behavior between adult female Richardson's ground squirrels (*Spermophilus richardsonii*) and their own and alien young. Can. J. Zool. 50:1343–1349.

Mihok, Steve. 1976. Behaviour of subarctic red-backed voles (*Clethrionomys gapperi athabascae*). Can. J. Zool. 54:1932–1945.

Miles, C. H., and D. P. Yaeger. 1979. Minnesota outdoor atlas. The Map Store, West St. Paul.

Millar, J. S. 1970. The breeding season and reproductive cycle of the western red squirrel. Can. J. Zool. 48:471–473.

———. 1975. Tactics of energy partitioning in breeding *Peromyscus*. Can. J. Zool. 53:967–976.

Miller, D. H., and L. L. Getz. 1969. Life-history notes on *Microtus pinetorum* in central Connecticut. J. Mamm. 50:777–784.

———. 1977a. Comparisons of population dynamics of *Peromyscus* and *Clethrionomys* in New England. J. Mamm. 58:1–16.

———. 1977b. Factors influencing local distribution and species diversity of forest small mammals in New England, U.S.A. Can. J. Zool. 55:806–814.

Miller, D. R., and J. D. Robertson. 1967. Results of tagging caribou at Little Duck Lake, Manitoba. J. Wildl. Mgt. 31:150–159.

Miller, F. L. 1974. Age determination of caribou by annulations in dental cementum. J. Wildl. Mgt. 38:47–53.

Miller, G. S., and Remington Kellogg. 1955. List of North American Recent mammals. U. S. Natl. Mus., Bull. 205.

Miller, R. S. 1964. Ecology and distribution of pocket gophers (Geomyidae) in Colorado. Ecology 45:256–272.

Mills, R. S., G. W. Barrett, and M. P. Farrell. 1975. Population dynamics of the big brown bat (*Eptesicus fuscus*) in southwestern Ohio. J. Mamm. 56:591–604.

Miner, J. R. 1966. Rodents on some islands of Pokegama Lake, Minnesota. Masters thesis, Bemidji State Univ.

Minkoff, Eli C. 1976. Mammalian superorders. Zool. J. Linn. Soc. 58:147–158.

Minkoff, Eli C., et al. 1979. The facial musculature of the opossum (*Didelphis virginiana*). J. Mamm. 60:46–57.

Minnesota Volunteer Staff. 1977. DNR's response to the Wolf Recovery Plan: Volunteer Staff report. Minn. Volun. 40(235):40–43.

Mirand, E. A., and A. R. Shadle. 1953. Gross anatomy of the male reproductive system of the porcupine. J. Mamm. 34:210–220.

Mischler, T. W., P. Welaj, and P. Nemith. 1971. Biological evaluation of two estrogenic steroids as possible rodent chemosterilants. J. Wildl. Mgt. 35:449–454.

Mizelle, J. D. 1935. Swimming of the muskrat. J. Mamm. 16:22–25.

Moen, A. N. 1968. Energy exchange of white-tailed deer, western Minnesota. Ecology 49:676–682.

Mohr, Carl O. 1943. Weight and length of white-tailed jackrabbits in Blue Earth County, Minnesota. J. Mamm. 24:504–506.

———. 1965. Home area and comparative biomass of the North American red squirrel. Can. Field-Nat. 79:162–171.

Mohr, E. 1933. The muskrat, *Ondatra zibethica* (Linnaeus), in Europe. J. Mamm. 14:58–62.

Mohr, W. P., and C. O. Mohr. 1936. Recent jackrabbit populations at Rapidan, Minnesota. J. Mamm. 17:112–113.

Moody, P. A., and D. E. Doniger. 1956. Serological light on porcupine relationships. Evolution 10:47–55.

Moore, J. C. 1957. The natural history of the fox squirrel, *Sciurus niger shermani*. Am. Mus. Nat. Hist., Bull. 113:1–71.

———. 1959a. The relationships of the gray squirrel, *Sciurus carolinensis*, to its nearest relatives. Southeast. Asn. Game Fish Comm., Ann. Conf. 13:356–363.

———. 1959b. Relationships among living squirrels of the Sciurinae. Am. Mus. Nat. Hist., Bull. 118:155–206.

———. 1961. Geographic variation in some reproductive characteristics of diurnal squirrels. Am. Mus. Nat. Hist., Bull. 122:1–32.

———. 1968. Sympatric species of tree squirrels mix in mating chase. J. Mamm. 49:531–533.

More, Gavin. 1976. Some winter food habits of lynx (*Felis lynx*) in the southern Mackenzie District, Northwest Territories. Can. Field-Nat. 90:499–500.

Morris, R., and D. Morris. 1966. Men and pandas. Hutchinson, London.

Morris, R. D. 1969. Competitive exclusion between *Microtus* and *Clethrionomys* in the aspen parkland of Saskatchewan. J. Mamm. 50:291–301.

Morton, S. R. 1978. Torpor and nest-sharing in free-living *Sminthopsis crassicaudata* (Marsupialia) and *Mus musculus* (Rodentia). J. Mamm. 59:569.

Mosby, H. S. 1969. The influence of hunting on the population dynamics of a woodlot gray squirrel population. J. Wildl. Mgt. 33:59–73.

Moyle, J. B. 1975. The uncommon ones. Minn. Dept. Nat. Resour.:1–32.

Moyle, J. B., and E. W. Moyle. 1977. Northland wild flowers: a guide for the Minnesota region. Univ. Minn. Press, Minneapolis.

Muchlinski, A. E., and E. N. Rybak. 1978. Energy consumption of resting and hibernating meadow jumping mice. J. Mamm. 59:435–437.

Muchlinski, A. E., and K. A. Shump, Jr. 1979. The sciurid tail: a possible thermoregulatory mechanism. J. Mamm. 60:652–654.

Mullen, D. A., and F. A. Pitelka. 1972. Efficiency of winter scavengers in the Arctic. Arctic 25:225–231.

Mumford, R. E. 1958. Population turnover in wintering bats in Indiana. J. Mamm. 39:253–261.

Mumford, R. E., and J. B. Cope. 1964. Distribution and status of the Chiroptera of Indiana. Am. Midl. Nat. 72:473–489.

Mumford, R. E., and J. O. Whitaker, Jr. 1974. Seasonal activity of bats at an Indiana cave. Proc. Ind. Acad. Sci. 84:500–507.

Munkel, R. E., and C. R. Fremling. 1967. A review of the bounty system as a method of controlling undesirable animal populations in Houston County, Minnesota. J. Minn. Acad. Sci. 34:117–121.

Munyer, E. A. 1967. A parturition date for the hoary bat, *Lasiurus c. cinereus*, in Illinois and notes on the newborn young. Trans. Ill. Acad. Sci. 60:95–97.

Murie, A. 1936. Following fox trails. Misc. Publ. Mus. Zool., Univ. Mich. 32:1–45.

———. 1944. The wolves of Mount McKinley. U. S. Natl. Park Serv., Fauna Ser. 5:1–238.

———. 1961. Some food habits of the marten. J. Mamm. 42:516–521.

Murie, Jan O. 1971. Behavioral relationships between two sympatric voles (*Microtus*): relevance to habitat segregation. J. Mamm. 52:181–186.

———. 1973. Population characteristics and phenology of a Franklin ground squirrel (*Spermophilus franklinii*) colony in Alberta, Canada. Am. Midl. Nat. 90:334–340.

Murie, Jan O., and Dawn Dickinson. 1973. Behavioral interactions between two species of red-backed vole (*Clethrionomys*) in captivity. Can. Field-Nat. 87:123–129.

Murie, O. J. 1951. The elk of North America. Stackpole, Harrisburg, Pa.

Mutch, G. R. P. 1977. Locations of winter dens used by striped skunks in Delta Marsh, Manitoba. Can. Field-Nat. 91:289–291.

Mutch, G. R. P., and M. Aleksiuk. 1977. Ecological aspects of winter dormancy in the striped skunk (*Mephitis mephitis*). Can. J. Zool. 55:607–615.

Muul, Illar. 1965. Day length and food caches. Nat. Hist. 74(3):22–27.

———. 1968. Behavioral and physiological influences on the distribution of the flying squirrel, *Glaucomys volans*. Misc. Publ. Mus. Zool., Univ. Mich. 134.

———. 1969a. Mating behavior, gestation period, and development of *Glaucomys sabrinus*. J. Mamm. 50:121.

———. 1969b. Photoperiod and reproduction in flying squirrels, *Glaucomys volans*. J. Mamm. 50:542–549.

———. 1970. Intra- and inter-familial behaviour of *Glaucomys volans* (Rodentia) following parturition. Anim. Behav. 18:20–25.

———. 1974. Geographic variation in the nesting habits of *Glaucomys volans*. J. Mamm. 55:840–844.

Myers, G. T., and T. A. Vaughan. 1964. Food habits of the plains pocket gopher in eastern Colorado. J. Mamm. 45:588–598.

Myers, Judith H. 1974a. The absence of *t* alleles in feral populations of house mice. Evolution 27:702–704.

———. 1974b. Genetic and social structure of feral house mouse populations on Grizzly Island, California. Ecology 55:747–759.

Myers, Judith H., and C. J. Krebs. 1971. Genetic, behavioral, and reproductive attributes of dispersing field voles *Microtus pennsylvanicus* and *Microtus ochrogaster*. Ecol. Monogr. 41:53–78.

———. 1974. Population cycles in rodents. Sci. Am. 230(6):38–46.

Myers, Philip. 1978. Sexual dimorphism in size of vespertilionid bats. Am. Nat. 112:701–711.

Myers, R. F. 1960. *Lasiurus* from Missouri caves. J. Mamm. 41:114–117.

Myrhe, Roar, and Svein Myrberget. 1975. Diet of wolverines (*Gulo gulo*) in Norway. J. Mamm. 56:752–757.

Myton, Becky. 1974. Utilization of space by *Peromyscus leucopus* and other small mammals. Ecology 55:277–290.

Nadler, C. F. 1966. Chromosomes and systematics of American ground squirrels of the subgenus *Spermophilus*. J. Mamm. 47: 579–596.

Nadler, C. F., R. S. Hoffman, and K. R. Greer. 1971. Chromosomal divergence during evolution of ground squirrel populations (Rodentia: *Spermophilus*). Syst. Zool. 20:298–305.

Nadler, C. F., et al. 1977. Chromosomal evolution in chipmunks, with special emphasis on A and B karyotypes of the subgenus *Neotamias*. Am. Midl. Nat. 98:343–353.

Nadler, C. F., and C. E. Hughes. 1966. Chromosomes and taxonomy of the ground squirrel subgenus *Ictidomys*. J. Mamm. 47:46–53.

Nadler, C. F., et al. 1974. Evolution in ground squirrels—I. Transferrins in Holarctic populations of *Spermophilus*. Comp. Biochem. Physiol. 47A:663–681.

Nadler, C. F., et al. 1978. Biochemical relationships of the Holarctic vole genera *Clethrionomys*, *Microtus*, and *Arvicola* (Rodentia: Arvicolinae). Can. J. Zool. 56:1564–1575.

Nagel, J. W. 1972. Observations of the second record of the least weasel in Tennessee. Am. Midl. Nat. 87:553.

Neal, E. 1948. The badger. Collins, London.

Neal, T. J. 1968. A comparison of two muskrat populations. Iowa State J. Sci. 43:193–210.

———. 1977. A closed trapping season and subsequent muskrat harvests. Wildl. Soc. Bull. 5:194–196.

Nellis, C. H. 1969. Productivity of Richardson's ground squirrels near Rochester, Alberta. Can. Field-Nat. 83:246–250.

Nellis, C. H., and L. B. Keith. 1968. Hunting activities and success of lynx in Alberta. J. Wildl. Mgt. 32:718–722.

———. 1976. Population dynamics of coyotes in central Alberta, 1964–1968. J. Wildl. Mgt. 40:389–399.

Nellis, C. H., S. P. Wetmore, and L. B. Keith. 1972. Lynx-prey interactions in central Alberta. J. Wildl. Mgt. 36:320–329.

Nelson, B. A. 1945. The spring molt of the northern red squirrel. J. Mamm. 26:397–400.

Nelson, M. E. 1979. Home range location of white-tailed deer. USDA For. Serv., Res. Pap. NC-173:1–10.

Nelson, M. M., ed. 1971. The white-tailed deer in Minnesota. (Proc. of a symposium). Minn. Dept. Nat. Resour.

Nelson, U. C. 1947. Woodland caribou in Minnesota. J. Wildl. Mgt. 11:283–284.

Nero, R. W., and R. E. Wrigley. 1977. Status and habits of the cougar in Manitoba. Can. Field-Nat. 91:28–40.

Nesbit, W. H. 1975. Ecology of a feral dog pack on a wildlife refuge. Pp. 391–396 *in* M. W. Fox, ed. The wild canids. Van Nostrand Reinhold, New York.

Neumann, R. L. 1967. Metabolism in the eastern chipmunk (*Tamias striatus*) and the southern flying squirrel (*Glaucomys volans*) during the winter and summer. Pp. 64–74 *in* K. C. Fisher et al., eds. Mammalian hibernation III. American Elsevier, New York.

Nevo, E., et al. 1974. Genetic variation, selection, and speciation in *Thomomys talpoides* pocket gophers. Evolution 28:1–23.

Newby, F. E., and J. J. McDougal. 1964. Range extension of the wolverine in Montana, J. Mamm. 45:485–487.

Newby, F. E., and P. L. Wright. 1955. Distribution and status of the wolverine in Montana. J. Mamm. 36:248–253.

Newson, R., and A. de Vos. 1964. Population structure and body weight of snowshoe hares on Manitoulin Island, Ontario. Can. J. Zool. 42:975–986.

Nicholson, A. J. 1941. The homes and social habits of the wood-mouse (*Peromyscus leucopus noveboracensis*) in southern Michigan. Am. Midl. Nat. 25:196–223.

Nixon, C. M., R. W. Donohoe, and T. Nash. 1974. Overharvest of fox squirrels from two wood-lots in western Ohio. J. Wildl. Mgt. 38:67–80.

Nixon, C. M., and M. W. McClain. 1969. Squirrel population decline following a late spring frost. J. Wildl. Mgt. 33:353–357.

———. 1975. Breeding seasons and fecundity of female gray squirrels in Ohio. J. Wildl. Mgt. 39:426–438.

Nixon, C. M., M. W. McClain, and R. W. Donohoe. 1975. Effects of hunting and mast crops on a squirrel population. J. Wildl. Mgt. 39:1–25.

Nixon, C. M., D. M. Worley, and M. W. McClain. 1968. Food habits of squirrels in southeast Ohio. J. Wildl. Mgt. 32:294–305.

Norberg, Ulla M. 1969. An arrangement giving a stiff leading edge to the hand wing in bats. J. Mamm. 50:766–770.

———. 1972. Bat wing structures important for aerodynamics and rigidity. Z. Morph. Tiere. 73:45–61.

Norris-Elye, L. T. S. 1951. The cougar in Manitoba. Can. Field-Nat. 65:119.

Northcott, T. H. 1974. Dispersal of mink in insular Newfoundland. J. Mamm. 55:243–248.

———. 1977. Marten. Can. Wildl. Serv., Hinterland Who's Who.:4 pp.

Northcott, T. H., and Fawn E. Elsey. 1971. Fluctuations in black bear populations and their relationships to climate. Can. Field-Nat. 85:123–128.

Norum, D. A. 1966. A study of edaphic factors and distribution of pocket gophers of Kittson County. Masters thesis, Bemidji State Univ.

Novak, Melinda A., and L. L. Getz. 1969. Aggressive behavior of meadow voles and pine voles. J. Mamm. 50:637–639.

Novak, Milan. 1977. Determining the average size and composition of beaver families. J. Wildl. Mgt. 41:751–754.

Novakowski, N. S. 1967. The winter bioenergetics of a beaver population in northern latitudes. Can. J. Zool. 45:1107–1118.

Nowak, R. M. 1973. Return of the wolverine. Natl. Parks Cons. Mag. 47(2):20–23.

Nudds, T. D. 1978. Convergence of group size strategies by mammalian social carnivores. Am. Nat. 112:957–960.

Nugent, R. F., and J. R. Choate. 1970. Eastward dispersal of the badger, *Taxidea taxus*, into the northeastern United States. J. Mamm. 51:626–627.

Nusetti, O. A., and M. Aleksiuk. 1975. Regulation of mammalian growth in cold environments: studies of DNA synthesis in *Rattus norvegicus*. J. Mamm. 56:770–780.

Nussbaum, R. A., and Chris Maser. 1975. Food habits of the bobcat, *Lynx rufus*, in the Coast and Cascade ranges of western Oregon in relation to present management policies. Northwest Sci. 49:261–266.

O'Boyle, Michael. 1974. Rats and mice together: the predatory nature of the rat's mouse-killing response. Psychol. Bull. 81:261–269.

Odum, E. P. 1971. Fundamentals of ecology, 3d ed. Saunders, Philadelphia.

O'Farrell, T. P. 1975. Unusual fertilization of a grasshopper mouse, *Onychomys leucogaster*. Am. Midl. Nat. 93:255–256.

O'Farrell, T. P., and G. E. Cosgrove. 1975. Longevity and age-related lesions in a laboratory colony of grasshopper mice, *Onychomys leucogaster*. Am. Midl. Nat. 94:241–247.

O'Gara, B. W. 1969a. Unique aspects of reproduction in the female pronghorn (*Antilocapra americana* Ord). Am. J. Anat. 125:217–231.

———. 1969b. Horn casting by female pronghorns. J. Mamm. 50:373–375.

———. 1978. *Antilocapra americana*. Mamm. Species 90:1–7.

O'Gara, B. W., and G. Matson. 1975. Growth and casting of horns by pronghorns and exfoliation of horns by bovids. J. Mamm. 56:829–846.

O'Gara, B. W., R. Moy, and G. D. Bear. 1971. The annual testicular cycle and horn casting in the pronghorn (*Antilocapra americana*). J. Mamm. 52:537–544.

Ogilvie, R. T., and T. Furman. 1959. Effect of vegetational cover of fence rows on small mammal populations. Ecology 40:140–141.

Olsen, Arnold, and P. N. Lehner. 1978. Conditioned avoidance of prey in coyotes. J. Wildl. Mgt. 42:676–679.

Olsen, P. F. 1959. Muskrat breeding biology at Delta, Manitoba. J. Wildl. Mgt. 23:40–53.

Olsen, S. J., and J. W. Olsen. 1977. The Chinese wolf, ancestor of New World dogs. Science 197:533–535.

Olson, Sigurd F. 1938a. A study in predatory relationship with particular reference to the wolf. Sci. Monthly 46:323–336.

———. 1938b. Organization and range of the pack. Ecology 19:168–170.

Orians, G. 1969. On the evolution of mating systems in birds and mammals. Am. Nat. 103:589–603.

Orr, H. D. 1966. Behavior of translocated white-footed mice. J. Mamm. 47:500–506.

Orr, R. T. 1950. Unusual behavior and occurrence of a hoary bat. J. Mamm. 31:456–457.

Osborn, D. J. 1953. Age classes, reproduction, and sex ratios of Wyoming beaver. J. Mamm. 34:27–44.

Osgood, W. H. 1909. Revision of the mice of the American genus *Peromyscus*. N. Am. Fauna 28:1–285.

Owen-Smith, Norman. 1977. On territoriality in ungulates and an evolutionary model. Quart. Rev. Biol. 52:1–38.

Oxberry, B. A. 1975. An anatomical, histochemical, and autoradiographic study of the evergrowing molar dentition of *Microtus* with comments on the role of structure in growth and eruption. J. Morph. 147:337–353.

Ozoga, J. J. 1968. Variations in microclimate in a conifer swamp deeryard in northern Michigan. J. Wildl. Mgt. 32:574–585.

Ozoga, J. J., and L. W. Gysel. 1972. Response of white-tailed deer to winter weather. J. Wildl. Mgt. 36:892–896.

Ozoga, J. J., and E. M. Harger. 1966. Winter activities and feeding habits of northern Michigan coyotes. J. Wildl. Mgt. 30:809–818.

Pack, J. C., H. S. Mosby, and P. B. Siegel. 1967. Influence of social hierarchy on gray squirrel behavior. J. Wildl. Mgt. 31:720–728.

Packard, R. L. 1956. The tree squirrels of Kansas: ecology and economic importance. Misc. Publ., Mus. Nat. Hist., Univ. Kan. 11:1–67.

Panuska, J. A. 1959. Weight patterns and hibernation in *Tamias striatus*. J. Mamm. 40:554–566.

Paradiso, J. L. 1968. Notes on recently collected specimens of east Texas canids, with comments on the speciation and taxonomy of the red wolf. Am. Midl. Nat. 80:529–534.

———. 1969. Mammals of Maryland. N. Am. Fauna 66.

Paradiso, J. L., and A. M. Greenhall. 1967. Longevity records for American bats. Am. Midl. Nat. 78:251–252.

Parker, P. 1977. An evolutionary comparison of placental and marsupial patterns of reproduction. Pp. 273–285 *in* B. Stonehouse and D. Gilmore, eds. Biology of marsupials. Macmillan, New York.

Parmalee, P. W. 1971. Fisher and porcupine remains from cave deposits in Missouri. Trans. Ill. State Acad. Sci 64:225–229.

Patric, E. F., and W. L. Webb. 1960. An evaluation of three age determination criteria in live beavers. J. Wildl. Mgt. 24:37–44.

Patton, J. L., R. K. Selander, and M. H. Smith. 1972. Genic variation in hybridizing populations of gophers (genus *Thomomys*). Syst. Zool. 21:263–270.

Paul, J. R. 1970. Observations on the ecology, populations, and reproductive biology of the pine vole, *Microtus pinetorum*, in North Carolina. Ill. State Mus. Rep. Invest. 20:1–28.

Pauls, R. W. 1978. Behavioural strategies relevant to the energy economy of the red squirrel (*Tamiasciurus hudsonicus*). Can. J. Zool. 56:1519–1525.

Pearson, E. W., and T. R. B. Barr. 1962. Absence of rabies in some bats and shrews from southern Illinois. Trans. Ill. State Acad. Sci. 55:35–37.

Pearson, O. P. 1942. On the cause and nature of a poisonous action produced by the bite of a shrew (*Blarina brevicauda*). J. Mamm. 23:159–166.

———. 1944. Reproduction in the shrew (*Blarina brevicauda* Say). Am. J. Anat. 75:39–93.

———. 1946. Scent glands of the short-tailed shrew. Anat. Rec. 94:615–629.

———. 1964. Carnivore-mouse predation: an example of its intensity and bioenergetics. J. Mamm. 45:177–188.

Pearson, P. G. 1959. Small mammals and old field succession on the Piedmont of New Jersey. Ecology 40:249–255.

Pease, J. L., R. H. Vowles, and L. B. Keith. 1979. Interaction of snowshoe hares and woody vegetation. J. Wildl. Mgt. 43:43–60.

Peek, J. M. 1974. Initial response of moose to a forest fire in northern Minnesota. Am. Midl. Nat. 91:435–438.

Peek, J. M., R. E. LeResche, and D. R. Stevens. 1974. Dynamics of moose aggregations in Alaska, Minnesota, and Montana. J. Mamm. 55:126–137.

Peek, J. M., D. L. Urich, and R. J. Mackie. 1976. Moose habitat selection and relationships to forest management in northeastern Minnesota. Wild. Monogr. 48:1–65.

Pelton, M. R., and G. M. Burghardt. 1976. Black bears of the Smokies. Nat. Hist. 85(1):54–63.

Pelton, M. R., and E. E. Provost. 1972. Onset of breeding and breeding synchrony by Georgia cottontails. J. Wildl. Mgt. 36:544–549.

Pembleton, E. F., and R. J. Baker. 1978. Studies of a contact zone between chromosomally characterized populations of *Geomys bursarius*. J. Mamm. 59:233–242.

Penny, D. F., and E. G. Zimmerman. 1976. Genic divergence and local population differentiation by random drift in the pocket gopher genus *Geomys*. Evolution 30:473–483.

Petajan, J. H., and P. R. Morrison. 1962. Physical and physiological factors modifying the development of temperature regulation in the opossum. J. Exper. Zool. 149:45–57.

Peters, R. P., and L. D. Mech. 1975. Scent-marking in wolves. Am. Sci. 63:628–637.

Petersen, K. E., and T. L. Yates. 1980. *Condylura cristata*. Mamm. Spp. 129.

Peterson, R. L. 1955. North American moose. Univ. Toronto Press, Toronto.

———. 1962. Notes on the distribution of *Microtus chrotorrhinus*. J. Mamm. 43:420.

———. 1965. A well-preserved grizzly bear skull recovered from a late glacial deposit near Lake Simcoe, Ontario. Nature 208:1233–1234.

———. 1966. The mammals of eastern Canada. Oxford Univ. Press, Toronto.

Peterson, R. O. 1977. Wolf ecology and prey relationships on Isle Royale. Natl. Park Serv., Sci. Monogr. Ser. 11.

Peterson, R. O., and D. L. Allen. 1974. Snow conditions

as a parameter in moose-wolf relationships. Naturaliste Can. 101:481–492.

Peterson, W. J. 1981. Coming of the caribou. Minn. Volun. 44(259):17–23.

Petraborg, W. H., and D. W. Burcalow. 1965. The white-tailed deer in Minnesota. Pp. 11–48 *in* J. B. Moyle, ed. Big game in Minnesota. Minn. Dept. Cons. Tech. Bull. 9.

Petraborg, W. H., and V. E. Gunvalson. 1962. Observations on bobcat mortality and bobcat predation on deer. J. Mamm. 43:430–431.

Phillips, G. L. 1966. Ecology of the big brown bat (Chiroptera: Vespertilionidae) in northeastern Kansas. Am. Midl. Nat. 75:168–198.

Phillips, R. L., et al. 1972. Dispersal and mortality of red foxes. J. Wildl. Mgt. 36:237–248.

Phillips, R. L., W. E. Berg, and D. B. Siniff. 1973. Moose movement patterns and range use in northwestern Minnesota. J. Wildl. Mgt. 37:266–278.

Pianka, E. R. 1978. Evolutionary ecology, 2d ed. Harper and Row, New York.

Pickford, G. D., and A. H. Reid. 1943. Competition of elk and domestic livestock for summer range forage. J. Wildl. Mgt. 7:328–332.

Pidduck, E. R., and J. B. Falls. 1973. Reproduction and emergence of juveniles in *Tamias striatus* (Rodentia: Sciuridae) at two localities in Ontario, Canada. J. Mamm. 54:693–707.

Pietsch, Manfred. 1970. Vergleichende Untersuchen an Schaedeln nordamerikanischer und europaeischer Bisamratten (*Ondatra zibethicus* L. 1766). Zeitschr. Saugetierk. 35:257–288.

Pimlott, D. H. 1967. Wolf predation and ungulate populations. Am. Zool. 7:267–278.

———. 1975. The ecology of the wolf in North America. Pp. 280–285 *in* M. W. Fox, ed. The wild canids. Van Nostrand Reinhold, New York.

Pimlott, D. H., J. A. Shannon, and G. B. Kolenosky. 1969. The ecology of the timber wolf in Algonquin Provincial Park. Ont. Dept. Lands For., Res. Br. Res. Rep. (Wildl.) 87:1–92.

Pinter, A. J. 1970. Reproduction and growth for two species of grasshopper mice (*Onychomys*) in the laboratory. J. Mamm. 51:236–243.

Po-Chedley, D. S., and A. R. Shadle. 1955. Pelage of the porcupine, *Erethizon dorsatum dorsatum*. J. Mamm. 36:84–95.

Poelker, R. J., and H. D. Hartwell. 1973. The black bear of Washington. Washington State Game Dept., Biol. Bull. 14.

Polder, E. B. (E. B. Polderboer prior to 1953). 1968. Spotted skunk and weasel populations den and cover usage by northeast Iowa. Proc. Iowa Acad. Sci. 75:142–146.

Polderboer, E. B. 1937. The pocket mouse (*Perognathus flavescens*) a new species in Iowa? Proc. Iowa Acad. Sci. 44:199–200.

———. 1942. Habits of the least weasel (*Mustela rixosa*) in northeastern Iowa. J. Mamm. 23:145–147.

———. 1948a. Predation on the domestic pig by the long-tailed weasel. J. Mamm. 29:295–296.

———. 1948b. Late fall sexual activity in an Iowa least weasel. J. Mamm. 29:296.

Polderboer, E. B., L. W. Kuhn, and G. O. Hendrickson. 1941. Winter and spring habits of weasels in central Iowa. J. Wildl. Mgt. 5:115–119.

Pollack, E. M. 1951. Food habits of the bobcat in the New England states. J. Wildl. Mgt. 15:209–213.

Pond, C. M. 1977. The significance of lactation in the evolution of mammals. Evolution 31:177–199.

Poole, T., and N. Dunstone. 1976. Underwater predatory behavior of the American mink (*Mustela vison*). J. Zool., London 178:395–412.

Porter, L. S. W. 1978. Pleistocene pluvial climates as indicated by present day climatic parameters of *Cryptotis parva* and *Microtus mexicanus*. J. Mamm. 59:330–338.

Possardt, E. E., and W. E. Dodge. 1978. Stream channelization impacts on songbirds and small mammals in Vermont. Wildl. Soc. Bull. 6:18–24.

Powell, R. A. 1972. A comparison of populations of boreal red-backed vole (*Clethrionomys gapperi*) in tornado blowdown and standing forest. Can. Field-Nat. 86:377–379.

———. 1973. A model for raptor predation on weasels. J. Mamm. 54:259–263.

———. 1977. The return of the fisher. Field Mus. Nat. Hist. Bull. 48(2):8–13.

———. 1978. A comparison of fisher and weasel hunting behavior. Carnivore 1:28–34.

———. 1979a. Ecological energetics and foraging strategies of the fisher (*Martes pennanti*). J. Anim. Ecol. 48:195–212.

———. 1979b. Fishers, population models, and trapping. Wildl. Soc. Bull. 7:149–154.

Preble, N. A. 1956. Notes on the life history of *Napaeozapus*. J. Mamm. 37:196–200.

Preston, E. M. 1975. Home range defense in the red fox, *Vulpes vulpes* L. J. Mamm. 56:645–652.

Preston, F. W. 1948. Porcupines gnaw bottles. J. Mamm. 29:72–73.

Price, E. O., and S. Loomis. 1973. Maternal influence on the response of wild and domestic Norway rats to a novel environment. Dev. Psychobiol. 6:203–208.

Priewert, F. W. 1961. Record of an extensive movement by a raccoon. J. Mamm. 42:113.

Prince, L. A. 1941. Water traps capture the pigmy shrew (*Microsorex hoyi*) in abundance. Can. Field-Nat. 55:72.

Progulske, D. R., and T. S. Baskett. 1958. Mobility of Missouri deer and their harassment by dogs. J. Wildl. Mgt. 22:184–192.

Provost, E. E. 1962. Morphological characteristics of the

beaver ovary. J. Wildl. Mgt. 26:272–278.

Provost, E. E., and C. M. Kirkpatrick. 1952. Observations on the hoary bat in Indiana and Illinois. J. Mamm. 33:110–113.

Pruitt, W. O., Jr. 1954a. Aging in the masked shrew, *Sorex cinereus cinereus* Kerr. J. Mamm. 35:35–39.

———. 1954b. Notes on the shorttail shrew (*Blarina brevicauda kirtlandi*) in northern Lower Michigan. Am. Midl. Nat. 52:236–241.

———. 1959. Microclimates and local distribution of small mammals on the George Reserve, Michigan. Misc. Publ. Mus. Zool., Univ. Mich. 109:1–27.

———. 1978. Boreal ecology. University Park Press, Baltimore.

Pryor, L. B. 1956. Sarcoptic mange in wild foxes in Pennsylvania. J. Mamm. 37:90–93.

Quanstrom, W. R. 1971. Behaviour of Richardson's ground squirrel *Spermophilus richardsonii richardsonii*. Anim. Behav. 19:646–652.

Quay, W. B., and J. S. Miller. 1955. Occurrence of the red bat, *Lasiurus borealis*, in caves. J. Mamm. 36:454–455.

Quay, W. B., and D. Müller-Schwarze. 1971. Relations of age and sex to integumentary glandular regions in Rocky Mountain mule deer (*Odocoileus hemionus hemionus*). J. Mamm. 52:670–685.

Quick, H. F. 1951. Notes on the ecology of weasels in Gunnison County, Colorado. J. Mamm. 32:281–290.

———. 1952. Some characteristics of wolverine fur. J. Mamm. 33:492–493.

———. 1953. Occurrence of porcupine quills in carnivorous mammals. J. Mamm. 34:256–259.

———. 1955. Food habits of marten (*Martes americana*) in northern British Columbia. Can. Field-Nat. 69:144–147.

Quimby, Don. 1942a. *Thomomys* in Minnesota. J. Mamm. 23:216–217.

———. 1942b. Notes on a northern red bat and her young. J. Mamm. 23:448–449.

———. 1944. A comparison of overwintering populations of small mammals in a northern coniferous forest for two consecutive years. J. Mamm. 25:86–87.

———. 1951. The life history and ecology of the jumping mouse *Zapus hudsonius*. Ecol. Monogr. 21:61–95.

Rabb, G. B. 1959. Toxic salivary glands in the primitive insectivore *Solenodon*. Nat. Hist. Misc., Chicago Acad. Sci. 170:1–3.

Rabb, G. B., J. H. Woolpy, and B. E. Ginsburg. 1967. Social relationships in a group of captive wolves. Am. Zool. 7:305–311.

Radinsky, L. B. 1969. Outlines of canid and felid brain evolution. Ann. New York Acad. Sci. 167:277–288.

———. 1973. Evolution of the canid brain. Brain Behav. Evol. 7:169–202.

———. 1978. Evolution of brain size in carnivores and ungulates. Am. Nat. 112:815–831.

Ralls, Katherine. 1977. Sexual dimorphism in mammals: avian models and unanswered questions. Am. Nat. 111:917–938.

Rand, A. L. 1944. The status of the fisher, *Martes pennanti* (Erxleben), in Canada. Can. Field-Nat. 58:77–81.

Ransom, A. B. 1967. Reproductive biology of white-tailed deer in Manitoba. J. Wildl. Mgt. 31:114–123.

Rasmussen, D. I. 1964. Blood group polymorphism and inbreeding in natural populations of the deer mouse *Peromyscus maniculatus*. Evolution 18:219–229.

Rasmussen, J. L. 1969. A recent record of the lynx in Iowa. J. Mamm. 50:370–371.

Rathbun, A. P., M. C. Wells, and M. Bekoff. 1980. Cooperative predation by coyotes on badgers. J. Mamm. 61:375–376.

Rausch, R. A., and A. M. Pearson. 1972. Notes on the wolverine in Alaska and the Yukon Territory. J. Wildl. Mgt. 36:249–268.

Rausch, Robert. 1953. On the status of some Arctic mammals. Arctic 6:91–148.

Rees, J. W. 1971. Discriminatory analysis of divergence in mandibular morphology of *Odocoileus*. J. Mamm. 52:724–731.

Reeves, Brian. 1973. The concept of an altithermal cultural hiatus in northern plains prehistory. Am. Anthropol. 75:1221–1253.

Reichard, T. A. 1976. Spring food habits and feeding behavior of fox and red squirrels. Am. Midl. Nat. 96:443–450.

Reig, O. A. 1977. A proposed unified nomenclature for the enamelled components of the molar teeth of the Cricetidae (Rodentia). J. Zool., London 181:227–241.

Renouf, R. N. 1972. Waterfowl utilization of beaver ponds in New Brunswick. J. Wildl. Mgt. 36:740–744.

Reynolds, H. W., R. M. Hansen, and D. G. Peden. 1978. Diets of the Slave River lowland bison herd, Northwest Territories, Canada. J. Wildl. Mgt. 42:581–590.

Richards, S. H., and R. L. Hine. 1953. Wisconsin fox populations. Wisc. Cons. Dept., Tech. Bull. 6.

Richardson, J. H. 1973. Locomotory and feeding activity of the shrews, *Blarina brevicauda* and *Suncus murinus*. Am. Midl. Nat. 90:224–227.

Richens, V. B., and R. D. Hugie. 1974. Distribution, taxonomic status, and characteristics of coyotes in Maine. J. Wildl. Mgt. 38:447–454.

Richens, V. B., and G. R. Lavigne. 1978. Response of white-tailed deer to snowmobile trails in Maine. Can. Field-Nat. 92:334–344.

Richmond, M., and R. Stehn. 1976. Olfaction and reproductive behavior in microtine rodents. Pp. 197–217 *in* R. L. Doty, ed. Mammalian olfaction, reproductive processes, and behavior. Academic Press, New York.

Richmond, N. D. 1952. Fluctuations in gray fox popula-

tions in Pennsylvania and their relationship to precipitation. J. Wildl. Mgt. 16:198–206.
Ricklefs, R. E. 1977. Prehistoric range extension of the elk, *Cervus canadensis*. Am. Midl. Nat. 97:230–235.
de Ricqlès, A. J. 1974. Evolution of endothermy: histological evidence. Evol. Theory 1:51–80.
Riemann, H. P., et al. 1978. Toxoplasmosis and Q fever antibodies among wild carnivores in California. J. Wildl. Mgt. 42:198–202.
Riley, G. A., and R. T. McBride. 1975. A survey of the red wolf (*Canis rufus*). Pp. 263–277 *in* M. W. Fox, ed. The wild canids. Van Nostrand Reinhold, New York.
Roberts, T. S. 1935. The introduction of the beaver into Itasca State Park. Minn. Conservationist 21:10, 23–24.
———. 1941. Itasca Park's pioneering beavers. Cons. Volun. 1(6):37–40.
Robinette, W. L., J. S. Gashwiler, and O. W. Morris. 1959. Food habits of the cougar in Utah and Nevada. J. Wildl. Mgt. 23:261–273.
———. 1961. Notes on cougar productivity and life history. J. Mamm. 42:204–217.
Robins, J. D. 1971. Movement of Franklin's ground squirrel into northeastern Minnesota. J. Minn. Acad. Sci. 37:30–31.
Robinson, R. A. 1980. Rabies and pet skunk production in Minnesota. Proc. Symp. Mammal. Ecol. Hab. Mgt. Minn., Bemidji State Univ. 52–53.
Robinson, W. B. 1961. Population changes of carnivores in some coyote control areas. J. Mamm. 42:510–515.
Robinson, W. B., and E. F. Grand. 1958. Comparative movements of bobcats and coyotes as disclosed by tagging. J. Wildl. Mgt. 22:117–122.
Robinson, W. L., and G. J. Smith. 1977. Observations on recently killed wolves in Upper Michigan. Wildl. Soc. Bull. 5:25–26.
Roe, F. G. 1970. The North American buffalo. A critical study of the species in its wild state. 2d ed. Univ. Toronto Press, Toronto.
Rogers, J. G. Jr., and G. K. Beauchamp. 1976. Influence of stimuli from populations of *Peromyscus leucopus* on maturation of young. J. Mamm. 57:320–330.
Rogers, L. L. 1972. Movement patterns and social organization of black bears in northeastern Minnesota. Bull. Ecol. Soc. Am. 53(2):21.
———. 1974. Shedding of foot pads by black bears during denning. J. Mamm. 55:672–674.
———. 1975. Parasites of black bears of the Lake Superior region. J. Wildl. Dis. 11:189–192.
———. 1976. Effects of mast and berry crop failures on survival, growth, and reproductive success of black bears. Trans. N. Am. Wildl. Nat. Resour. Conf. 41:431–438.
———. 1980. Inheritance of coat color and changes in pelage coloration in black bears in northeastern Minnesota. J. Mamm. 61:324–327.
Rogers, L. L., et al. 1980. Deer distribution in relation to wolf pack territory edges. J. Wildl. Mgt. 44:253–258.
Rohn, K. H. 1978. Dance of the snowshoe hares. Minn. Volun. 41(236):52–55.
Rohwer, S. A., and D. L. Kilgore, Jr. 1973. Interbreeding in the arid-land foxes, *Vulpes velox* and *V. macrotis*. Syst. Zool. 22:157–165.
Rollings, C. T. 1945. Habits, foods, and parasites of the bobcat in Minnesota. J. Wildl. Mgt. 9:131–151.
Romer, A. S. 1966. Vertebrate paleontology, 3d ed. Univ. Chicago Press, Chicago.
———. 1969. Cynodont reptile with incipient mammalian jaw articulation. Science 166:881–882.
Rongstad, O. J. 1965. A life history of the thirteen-lined ground squirrels in southern Wisconsin. J. Mamm. 46:76–87.
———. 1966. A cottontail rabbit lens-growth curve from southern Wisconsin. J. Wildl. Mgt. 30:114–121.
———. 1968. A bull snake encounter with young ground squirrels. J. Minn. Acad. Sci. 35:95–97.
———. 1969. Gross prenatal development of cottontail rabbits. J. Wildl. Mgt. 33:164–168.
Rongstad, O. J., and J. R. Tester. 1969. Movements and habitat use of white-tailed deer in Minnesota. J. Wildl. Mgt. 33:366–379.
———. 1971. Behavior and maternal relations of young snowshoe hares. J. Wildl. Mgt. 35:338–346.
Rorabacher, J. A. 1970. The American buffalo in transition. North Star Press, St. Cloud, MN.
Roscoe, Britt, and Chris Majka. 1976. First records of the rock vole (*Microtus chrotorrhinus*) and the Gaspé shrew (*Sorex gaspensis*) from Nova Scotia and a second report of the Thompson's pygmy shrew (*Microsorex thompsoni*) from Cape Breton Island. Can. Field-Nat. 90:497–498.
Rose, G. B. 1973. Energy metabolism of adult cottontail rabbits, *Sylvilagus floridanus*, in simulated field conditions. Am. Midl. Nat. 89:473–478.
———. 1977. Mortality rates of tagged adult cottontail rabbits. J. Wildl. Mgt. 41:511–514.
Rose, R. K. 1979. Levels of wounding in the meadow vole, *Microtus pennsylvanicus*. J. Mamm. 60:37–45.
Rose, R. K., and M. S. Gaines. 1976. Levels of aggression in fluctuating populations of the prairie vole, *Microtus ochrogaster*, in eastern Kansas. J. Mamm. 57:43–57.
Rosenzweig, M. L. 1966. Community structure in sympatric Carnivora. J. Mamm. 47:602–612.
———. 1968. The strategy of body size in mammalian carnivores. Am. Midl. Nat. 80:299–315.
Rosenzweig, M. L., and R. H. MacArthur. 1963. Graphical representation and stability conditions of predator prey interactions. Am. Nat. 97:209–223.
Ross, A. J. 1961. Notes on the food habits of bats. J.

Mamm. 42:66–71.

Ross, B. A., J. R. Tester, and W. J. Breckenridge. 1968. Ecology of mima-type mounds in northwestern Minnesota. Ecology 49:172–177.

Roth, E. L. 1972. Late Pleistocene mammals from Klein Cave, Kerr County, Texas. Texas J. Sci. 24:75–84.

Rothman, R. J., and L. D. Mech. 1979. Scent-marking in lone wolves and newly formed pairs. Anim. Behav. 27:750–760.

Rounds, R. C. 1977. Population fluctuations of wapiti (*Cervus elaphus*) and moose (*Alces alces*) in Riding Mountain National Park, Manitoba, 1950–1976. Can. Field-Nat. 91:130–133.

———. 1978. Grouping characteristics of moose (*Alces alces*) in Riding Mountain National Park, Manitoba. Can. Field-Nat. 92:223–227.

Rowan, W., and L. B. Keith. 1956. Reproductive potential and sex ratios of snowshoe hares in northern Alberta. Can. J. Zool. 34:273–281.

Rowe, F. P., E. J. Taylor, and A. H. J. Chudley. 1964. The effect of crowding on the reproduction of the house-mouse (*Mus musculus* L.) living in corn-ricks. J. Anim. Ecol. 33:477–483.

Rowlands, I. W., ed. 1966. Comparative biology of reproduction in mammals. Symp. Zool. Soc. London 15.

Rudd, R. L. 1955. Age, sex, and weight comparisons in three species of shrews. J. Mamm. 36:323–339.

Ruffer, D. G. 1965a. Burrows and burrowing behavior of *Onychomys leucogaster*. J. Mamm. 46:241–247.

———. 1965b. Juvenile molt of *Onychomys leucogaster*. J. Mamm. 46:338–339.

———. 1965c. Sexual behaviour of the northern grasshopper mouse *Onychomys leucogaster*. Anim. Behav. 13:447–452.

———. 1968. Agonistic behavior of the northern grasshopper mouse (*Onychomys leucogaster breviauritus*). J. Mamm. 49:481–487.

Rusch, D. H., M. M. Gillespie, and D. I. McKay. 1978. Decline of a ruffed grouse population in Manitoba. Can. Field-Nat. 92:123–127.

Rusch, D. H., et al. 1972. Response of great horned owl populations to changing prey densities. J. Wildl. Mgt. 36:282–296.

Russell, L. S. 1928. Didelphiidae from the Lance Beds of Wyoming. J. Mamm. 9:229–232.

Russell, R. J. 1968. Evolution and classification of pocket gophers of the subfamily Geomyinae. Univ. Kan. Publ., Mus. Nat. Hist. 16:473–479.

Rust, C. C. 1962. Temperature as a modifying factor in the spring pelage change of short-tailed weasels. J. Mamm. 43:323–328.

Rust, C. C., R. M. Shackelford, and R. K. Meyer. 1965. Hormonal control of pelage cycles in the mink. J. Mamm. 46:549–565.

Ryder, R. A. 1955. Fish predation by the otter in Michigan. J. Wildl. Mgt. 19:497–498.

Rysgaard, G. N. 1941. Bats killed by severe storm. J. Mamm. 22:452–453.

———. 1942. A study of the cave bats of Minnesota with especial reference to the large brown bat, *Eptesicus fuscus fuscus* (Beauvois). Am. Midl. Nat. 28:245–267.

Sadleir, R. M. F. S. 1969. The ecology of reproduction in wild and domestic mammals. Methuen, London.

Samson, F. B., and B. D. Hill. 1979. Observations on coyote biology in north-central Missouri. Prairie Nat. 11:53–59.

Sanderson, G. C. 1956. Mule deer record for Iowa. J. Mamm. 37:457–458.

———. 1961. Estimating opossum populations by marking young. J. Wildl. Mgt. 25:20–27.

Sanderson, G. C., and A. V. Nalbandov. 1973. The reproductive cycle of the raccoon in Illinois. Ill. Nat. Hist. Surv. Bull. 31:29–85.

Sandidge, L. L. 1953. Food and dens of the opossum (*Didelphis virginiana*) in northeastern Kansas. Trans. Kan. Acad. Sci. 56:97–106.

Sargeant, A. B. 1972. Red fox spatial characteristics in relation to waterfowl predation. J. Wildl. Mgt. 36:225–236.

———. 1978. Red fox prey demands and implications to prairie duck production. J. Wildl. Mgt. 42:520–527.

———. 1980. Aspects of the ecology and population dynamics of the red fox. Proc. Symp. Mammal. Ecol. Hab. Mgt. Minn., Bemidji State Univ. 62–63.

Sargeant, A. B., and W. H. Marshall. 1959. Mammals of Itasca State Park, Minnesota. Flicker 31:116–128.

Sargeant, A. B., G. A. Swanson, and H. A. Doty. 1973. Selective predation by mink, *Mustela vison*, on waterfowl. Am. Midl. Nat. 89:208–214.

Sargeant, A. B., and D. W. Warner. 1972. Movements and denning habits of a badger. J. Mamm. 53:207–210.

Sather, J. H. 1958. Biology of the Great Plains muskrat in Nebraska. Wildl. Monogr. 2:1–35.

Satinoff, E. 1978. Neural organization and evolution of thermal regulation in mammals. Science 201:16–22.

Sauer, Peggy R. 1975. Relationship of growth characteristics to sex and age for black bears from the Adirondack region of New York. N.Y. Fish Game J. 22:81–113.

Saunders, B. P. 1973. Meningeal worm in white-tailed deer in northwestern Ontario and moose population densities. J. Wildl. Mgt. 37:327–330.

Saunders, J. K., Jr. 1963a. Food habits of the lynx in Newfoundland. J. Wildl. Mgt. 27:384–390.

———. 1963b. Movements and activities of the lynx in Newfoundland. J. Wildl. Mgt. 27:390–400.

———. 1964. Physical characteristics of the Newfoundland lynx. J. Mamm. 45:36–47.

Savage, R. J. G. 1977. Evolution in carnivorous mammals. Paleontology 20:237–271.

Savidge, I. R. 1974a. Social factors in dispersal of deer mice (*Peromyscus maniculatus*) from their natal site. Am. Midl. Nat. 91:395–405.

———. 1974b. Maternal aggressiveness and litter survival in deer mice (*Peromyscus maniculatus bairdii*). Am. Midl. Nat. 91:449–451.

Schadler, M. H., and G. M. Butterstein. 1979. Reproduction in the pine vole, *Microtus pinetorum*. J. Mamm. 60:841–844.

Schantz, Viola S. 1953. Additional information on distribution and variation of eastern badgers. J. Mamm. 34:338–389.

Scheffer, T. H. 1910. The common mole. Kan. State Agr. Coll. Exp. Sta. Bull. 168:1–36.

Scheffer, V. B. 1958. Do fossorial rodents originate Mima-type microrelief? Am. Midl. Nat. 59:505–510.

Schenkel, R. 1967. Submission: its features and functions in the wolf and dog. Am. Zool. 7:319–329.

Schindler, A-M., R. J. Low, and K. Benirscke. 1973. The chromosomes of the New World flying squirrels (*Glaucomys volans* and *Glaucomys sabrinus*), with special reference to autosomal heterochromatin. Cytologia (Tokyo) 38:137–146 (Biol. Abstr. 57:24430).

Schladweiler, J. L., and G. L. Storm. 1969. Den-use by mink. J. Wildl. Mgt. 33:1025–1026.

Schladweiler, J. L., and J. R. Tester. 1972. Survival and behavior of hand-reared mallards released in the wild. J. Wildl. Mgt. 36:1118–1126.

Schlesinger, W. H. 1976. Toxic foods and vole cycles: additional data. Am. Nat. 110:315–317.

Schloyer, C. R. 1977. Food habits of *Clethrionomys gapperi* on clearcuts in West Virginia. J. Mamm. 58:677–679.

Schmidt-Nielsen, K., W. L. Bretz, and C. R. Taylor. 1970. Panting in dogs: unidirectional air flow over evaporative surfaces. Science 169:1102–1104.

Schneider, D. G., L. D. Mech, and J. R. Tester. 1971. Movements of female raccoons and their young as determined by radio-tracking. Anim. Behav. Monogr. 4:3–43.

Schnell, J. H. 1969. Rest site selection by radio-tagged raccoons. J. Minn. Acad. Sci. 36:83–88.

Schnurrenberger, P. R., R. J. Martin, and J. M. Koch. 1970. Rabies in Illinois skunks. J. Am. Vet. Med. Asn. 157:1336–1342.

Schnurrenberger, P. R., R. J. Martin, and G. W. Lantis. 1970. Rabies in Illinois foxes. J. Am. Vet. Med. Asn. 157:1331–1335.

Schofield, R. D. 1958. Litter size and age ratios of Michigan red foxes. J. Wildl. Mgt. 22:313–315.

Schoonmaker, W. J. 1930. Porcupine eats water lily pods. J. Mamm. 11:84.

———. 1938a. Notes on the home range of the porcupine. J. Mamm. 19:378.

———. 1938b. The fisher as a foe of the porcupine in New York State. J. Mamm. 19:373–374.

Schoonover, L. J., and W. H. Marshall. 1951. Food habits of the raccoon (*Procyon lotor hirtus*) in north-central Minnesota. J. Mamm. 32:422–428.

Schorger, A. W. 1946a. Influx of bears into St. Louis County, Minnesota. J. Mamm. 27:177.

———. 1946b. Records of wolverine for Wisconsin and Minnesota. J. Mamm. 27:90.

———. 1951. *Zapus* with a white tail tip. J. Mamm. 32:362.

Schowalter, D. B., W. J. Dorward, and J. R. Gunson. 1978. Seasonal occurrence of silver-haired bats (*Lasionycteris noctivagans*) in Alberta and British Columbia. Can. Field-Nat. 92:288–291.

Schowalter, D. B., and J. R. Gunson. 1979. Reproductive biology of the big brown bat (*Eptesicus fuscus*) in Alberta. Can. Field-Nat. 93:48–54.

Schowalter, D. B., J. R. Gunson, and L. D. Harder. 1979. Life history characteristics of little brown bats (*Myotis lucifugus*) in Alberta. Can. Field-Nat. 93:243–251.

Schultz, R. D., and J. A. Bailey. 1978. Response of national park elk to human activity. J. Wildl. Mgt. 42:91–100.

Schwartz, C. C., and J. G. Nagy. 1976. Pronghorn diets relative to forage availability in northeastern Colorado. J. Wildl. Mgt. 40:469–478.

Schwartz, C. W. 1942. Breeding season of the cottontail in central Missouri. J. Mamm. 23:1–16.

Schwartz, C. W., and E. R. Schwartz. 1959. The wild mammals of Missouri. Univ. Mo. Press, Columbia.

Schwartz, G. M., and G. A. Thiel. 1963. Minnesota's rocks and waters. Univ. Minn. Press, Minneapolis.

Schwartz, J. H. 1978. Dental development, homologies, and primate phylogeny. Evol. Theory 4:1–32.

Scott, G. W., and K. C. Fisher. 1972. Hibernation of eastern chipmunks (*Tamias striatus*) maintained under controlled conditions. Can. J. Zool. 50:95–105.

Scott, J. P. 1967. The evolution of social behavior in dogs and wolves. Am. Zool. 7:373–381.

Scott, M. D., and K. Causey. 1973. Ecology of feral dogs in Alabama. J. Wildl. Mgt. 35:707–716.

Scott, T. G. 1943. Some food coactions of the northern plains red fox. Ecol. Monogr. 13:427–479.

Scott, T. G., and W. D. Klimstra. 1955. Red foxes and a declining prey population. Monogr. S. Ill. Univ. 1.

Scott, T. G., and L. F. Selko. 1939. A census of red foxes and striped skunks in Clay and Boone Counties, Iowa. J. Wildl. Mgt. 3:92–98.

Scott, W. E. 1941. Kettle eaten by porcupines. J. Mamm. 22:325–326.

Seabloom, R. W. 1978. Checklist of North Dakota mammals: comments on status and distribution of species. Prairie Nat. 10:123–125.

Seabloom, R. W., S. L. Iverson, and B. N. Turner. 1978. Adrenal response in a wild *Microtus* population: seasonal aspects. Can. J. Zool. 56:1433–1440.

Seal, U. S. 1975. Molecular approaches to taxonomic problems in the Canidae. Pp. 27–39 *in* M. W. Fox, ed. The wild canids. Van Nostrand Reinhold, New York.

Seal, U. S., and R. L. Hoskinson. 1978. Metabolic indicators of habitat condition and capture stress in pronghorns. J. Wildl. Mgt. 42:755–763.

Seal, U. S., L. D. Mech, and V. Van Ballenberghe. 1975. Blood analyses of wolf pups and their ecological and metabolic interpretation. J. Mamm. 56:64–75.

Seal, U. S., et al. 1978. Metabolic indicators of habitat differences in four Minnesota deer populations. J. Wildl. Mgt. 42:746–754.

Searing, G. F. 1977. The function of the bark call of the red squirrel. Can. Field-Nat. 91:187–189.

Seidensticker, J. C., IV, et al. 1973. Mountain lion social organization in the Idaho Primitive Area. Wildl. Monogr. 35:1–60.

Selander, R. K., W. G. Hunt, and S. Y. Yang. 1969. Protein polymorphism and genic heterozygosity in two European subspecies of the house mouse. Evolution 23:379–390.

Selander, R. K., et al. 1974. Genic and chromosomal differentiation in pocket gophers of the *Geomys bursarius* group. Evolution 28:557–564.

Selko, L. F. 1937. Food habits of Iowa skunks in the fall of 1936. J. Wild. Mgt. 1:70–76.

———. 1938a. Hibernation of the striped skunk in Iowa. J. Mamm. 19:320–324.

———. 1938b. Notes on the den ecology of the striped skunk in Iowa. Am. Midl. Nat. 20:455–463.

Seton, E. T. 1909. Life histories of northern animals—an account of the mammals of Manitoba. 2 vols. Scribners, New York.

———. 1929. Lives of game animals. 4 vols. Doubleday, Doran, Garden City, New York.

Severinghaus, C. W. 1974a. The coyote moves east. Conservationist (N.Y.) 29(2):8,36.

———. 1974b. Notes on the history of wild canids in New York. N.Y. Fish Game J. 21:117–125.

———. 1975. Occurrence of the opossum in the central Adirondacks. N.Y. Fish Game J. 22:80.

Severinghaus, C. W., and E. L. Cheatum. 1956. Life and times of the white-tailed deer. Pp. 57–186 *in* W. P. Taylor, ed. The deer of North America. Stackpole, Harrisburg, Pa.

Severinghaus, C. W., and R. W. Darrow. 1976. Failure of elk to survive in the Adirondacks. N.Y. Fish Game J. 23:98–99.

Severinghaus, C. W., and J. E. Tanck. 1948. Speed and gait of an otter. J. Mamm. 29:71.

Severinghaus, C. W., and B. F. Tullar. 1975. Wintering deer versus snowmobiles. Conservationist (N.Y.) 29:31.

Shackelford, N. 1966. Eastern chipmunk feeding on a starling. J. Mamm. 47:528.

Shadle, A. R. 1946. Copulation in the porcupine. J. Wildl. Mgt. 10:159–162.

———. 1947. Porcupine spine penetration. J. Mamm. 28:180–181.

———. 1948. Gestation period in the porcupine *Erethizon dorsatum dorsatum*. J. Mamm. 29:162–164.

———. 1950. Feeding, care, and handling of captive porcupines (*Erethizon*). J. Mamm. 31:411–416.

———. 1951. Laboratory copulations and gestations of porcupine, *Erethizon dorsatum*. J. Mamm. 32:219–221.

———. 1952. Sexual maturity and first recorded copulation of a 16-month male porcupine, *Erethizon dorsatum dorsatum*. J. Mamm. 33:239–241.

———. 1955a. Removal of foreign quills by porcupine. J. Mamm. 36:463–465.

———. 1955b. Effects of porcupine quills in humans. Am. Nat. 89:47–49.

———. 1956. Parturition in a skunk, *Mephitis mephitis hudsonica*. J. Mamm. 37:112–113.

Shadle, A. R., and W. R. Ploss. 1943. An unusual porcupine parturition and development of the young. J. Mamm. 24:492–496.

Shadle, A. R., and D. Po-Chedley. 1949. Rate of penetration of porcupine spine. J. Mamm. 30:172–173.

Shadle, A. R., M. Smelzer, and M. Metz. 1946. The sex reactions of porcupines (*Erethizon d. dorsatum*) before and after copulation. J. Mamm. 27:116–121.

Shakleton, D. M., L. V. Hills, and D. A. Hutton. 1975. Aspects of variation in cranial characters of plains bison (*Bison bison bison* Linnaeus) from Elk Island National Park, Alberta. J. Mamm. 56:871–887.

Shapiro, J. 1949. Ecological and life history notes on the porcupine in the Adirondacks. J. Mamm. 30:247–257.

Sharman, G. B., and P. J. Berger. 1970. Reproductive physiology of marsupials. Science 167:1221–1228.

Sharp, R. P. 1953. Glacial features of Cook County, Minnesota. Am. J. Sci. 251:855–883.

Sharp, W. M. 1958. Aging gray squirrels by use of tail-pelage characteristics. J. Wildl. Mgt. 22:29–34.

Sharps, Jon. 1978. Swift fox. S.D. Conserv. Dig. 45(1):20–21.

Shelden, R. M. 1972. The fate of short-tailed weasel, *Mustela erminea*, blastocysts following ovariectomy during diapause. J. Reprod. Fertil. 31:347–352.

Sheldon, Carolyn. 1934. Studies on the life histories of *Zapus* and *Napaeozapus* in Nova Scotia. J. Mamm. 15:290–300.

———. 1938. Vermont jumping mice of the genus *Napaeozapus*. J. Mamm. 19:444–453.

Sheldon, W. G., and W. G. Toll. 1964. Feeding habits of the river otter in a reservoir in central Massachusetts. J. Mamm. 45:449–455.

Sheppard, D. H. 1968. Seasonal changes in body and adrenal weights of chipmunks (*Eutamias*). J. Mamm.

49:463–474.

———. 1972a. Home ranges of chipmunks (*Eutamias*) in Alberta. J. Mamm. 53:379–380.

———. 1972b. Reproduction of Richardson's ground squirrel (*Spermophilus richardsonii*) in southern Saskatchewan. Can. J. Zool. 50:1577–1581.

Sheppard, D. H., and S. M. Swanson. 1976. Natural mortality in Richardson's ground squirrel. Can. Field-Nat. 90:170–172.

Sheppard, D. H., and S. M. Yoshida. 1971. Social behavior in captive Richardson's ground squirrels. J. Mamm. 52:793–799.

Sheppe, W. A. 1965a. Island populations and gene flow in the deer mouse, *Peromyscus leucopus*. Evolution 19:480–495.

———. 1965b. Dispersal by swimming in *Peromyscus leucopus*. J. Mamm. 46:336–337.

———. 1965c. Unseasonal breeding in artificial colonies of *Peromyscus leucopus*. J. Mamm. 46:641–646.

———. 1966. Exploration by the deer mouse, *Peromyscus leucopus*. Am. Midl. Nat. 76:257–276.

Sherburne, J. A., and J. B. Dimond. 1969. DDT persistence in wild hares and mink. J. Wild. Mgt. 33:944–948.

Shirer, H. W., and H. S. Fitch. 1970. Comparison from radiotracking of movements and denning habits of the raccoon, striped skunk, and opossum in northeastern Kansas. J. Mamm. 51:491–503.

Shoesmith, M. W. 1976. Twin fetuses in woodland caribou. Can. Field-Nat. 90:498–499.

Short, H. L. 1976. Composition and squirrel use of acorns of black and white oak groups. J. Wildl. Mgt. 40:479–483.

Shorten, Monica. 1954. Squirrels. Collins, London.

———. 1957. Squirrels in England, Wales, and Scotland. J. Anim. Ecol. 26:287–294.

Shull, A. F. 1907. Habits of the short-tailed shrew. Am. Nat. 41:495–522.

Shump, K. A., Jr. 1975. Nest construction by the western harvest mouse. Trans. Kan. Acad. Sci. 77:87–92.

Silver, Helenette, and W. T. Silver. 1969. Growth and behavior of the coyote-like canid of northern New England, with observations on canid hybrids. Wildl. Monogr. 17:1–41.

Simmons, J. A. 1979a. Echolocation and pursuit of prey by bats. Science 203:16–20.

———. 1979b. Perception of echo phase formation in bat sonar. Science 204:1336–1338.

Simms, D. A. 1978. Spring and summer food habits of an ermine (*Mustela erminea*) in the central Arctic. Can. Field-Nat. 92:192–193.

Simpson, G. G. 1945. The principles of classification and a classification of mammals. Am. Mus. Nat. Hist., Bull. 85:1–350.

Simpson, S. E. 1923. The nest and young of the star-nosed mole (*Condylura cristata*). J. Mamm. 4:167–170.

Sims, H. P., and C. H. Buckner. 1973. The effects of clear cutting and burning of *Pinus banksiana* forests on the populations of small mammals in southeastern Manitoba. Am. Midl. Nat. 90:228–231.

Sinclair, W., N. Dunstone, and T. B. Poole. 1974. Aerial and underwater visual acuity in the mink *Mustela vison* Schreber. Anim. Behav. 22:965–974.

Skryja, D. D. 1974. Reproductive biology of the least chipmunk (*Eutamias minimus operarius*) in southeastern Wyoming. J. Mamm. 55:221–224.

Slough, B. G. 1978. Beaver food cache structure and utilization. J. Wildl. Mgt. 42:644–646.

Smith, C. C. 1968. The adaptive nature of social organization in the genus of tree squirrels *Tamiasciurus*. Ecol. Monogr. 38:31–63.

———. 1970. The coevolution of pine squirrels (*Tamiasciurus*) and conifers. Ecol. Monogr. 40:349–371.

———. 1978. Structure and function of the vocalizations of tree squirrels (*Tamiasciurus*). J. Mamm. 59:793–808.

Smith, C. C., and David Follmer. 1972. Food preferences of squirrels. Ecology 53:82–91.

Smith, C. F. 1936. Notes on the habits of the long-tailed harvest mouse. J. Mamm. 17:274–278.

Smith, D. A., and L. C. Smith. 1975. Oestrus, copulation, and related aspects of reproduction in female eastern chipmunks, *Tamias striatus* (Rodentia: Sciuridae). Can. J. Zool. 53:756–767.

Smith, D. A., and S. W. Speller. 1970. The distribution and behavior of *Peromyscus maniculatus gracilis* and *Peromyscus leucopus noveboracensis* (Rodentia: Cricetidae) in a southeastern Ontario woodlot. Can. J. Zool. 48:1187–1199.

Smith, G. W. 1977. Population characteristics of the porcupine in northeastern Oregon. J. Mamm. 58:674–676.

Smith, L. C., and D. A. Smith. 1972. Reproductive biology, breeding seasons, and growth of eastern chipmunks, *Tamias striatus* (Rodentia, Sciuridae) in Canada. Can. J. Zool. 50:1069–1085.

Smith, M. C. 1968. Red squirrel response to spruce cone failure in interior Alaska. J. Wildl. Mgt. 32:305–317.

Smith, P. W. 1957. An analysis of post-Wisconsin biogeography of the Prairie Peninsula region based on distributional phenomena among terrestrial vertebrate populations. Ecology 38:205–218.

Snead, E., and G. O. Hendrickson. 1942. Food habits of the badger in Iowa. J. Mamm. 23:380–391.

Snyder, D. P. 1956. Survival rates, longevity, and population fluctuations in the white-footed mouse, *Peromyscus leucopus*, in southeastern Michigan. Misc. Publ. Mus. Zool. Univ. Mich. 95.

Snyder, L. L. 1924. Some details on the life history and behavior of *Napaeozapus insignis abietorum* (Preble). J.

Mamm. 5:233–237.

———. 1935. A badger specimen from Port Dover, Norfolk County, Ontario. Can. Field-Nat. 49:136–137.

Snyder, R. L. 1962. Reproductive performance of a population of woodchucks after a change in sex ratio. Ecology 43:506–515.

Snyder, R. L., and J. J. Christian. 1960. Reproductive cycle and litter size of the woodchuck. Ecology 41:647–656.

Snyder, R. L., D. E. Davis, and J. J. Christian. 1961. Seasonal changes in the weights of woodchucks. J. Mamm. 42:297–312.

Sokolov, I. I. 1973. Trends of evolution and the classification of the subfamily Lutrinae (Mustelidae, Fissipedia). Bull. Moscow Prir. Biol. 78(6):45–52 (cited in Jones et al. 1975).

Sollberger, D. E. 1940. Notes on the life history of the small eastern flying squirrel. J. Mamm. 21:282–293.

———. 1943. Notes on the breeding habits of the eastern flying squirrel (*Glaucomys volans volans*). J. Mamm. 24:163–173.

Soper, J. D. 1941. History, range, and home life of the northern bison. Ecol. Monogr. 11:347–412.

———. 1946. Mammals of the northern Great Plains along the international boundary in Canada. J. Mamm. 27:127–153.

Soutiere, E. C. 1979. Effects of timber harvesting on marten in Maine. J. Wildl. Mgt. 43:850–860.

Sowls, L. K. 1948. The Franklin ground squirrel, *Citellus franklinii* (Sabine), and its relationship to nesting ducks. J. Mamm. 29:113–137.

Sparks, D. R. 1968. Occurrence of milk in stomachs of young jackrabbits. J. Mamm. 49:324–325.

Speer, R. J., and T. G. Dilworth. 1978. Porcupine winter food and habitat utilization in central New Brunswick. Can. Field-Nat. 92:271–274.

Spencer, A. W., and D. Pettus. 1966. Habitat preferences of five sympatric species of long-tailed shrews. Ecology 47:677–683.

Spiess, Arthur. 1976. Labrador grizzly (*Ursus arctos* L.): first skeletal evidence. J. Mamm. 57:787–790.

Spieth, H. T. 1979. The *virilis* group of *Drosophila* and the beaver *Castor*. Am. Nat. 114:312–316.

Stains, H. J. 1956. The raccoon in Kansas, natural history, management, and economic importance. Univ. Kan. Mus. Nat. Hist., Misc. Publ. 10:1–76.

———. 1965. Female red bat carrying four young. J. Mamm. 46:333.

———. 1967. Carnivores and pinnipeds. Pp. 325–354 *in* S. Anderson and J. K. Jones, Jr., eds. Recent mammals of the world. Ronald Press, New York.

———. 1975. Distribution and taxonomy of the Canidae. Pp. 3–26 *in* M. W. Fox, ed. The wild canids. Van Nostrand Reinhold, New York.

———. 1976a. Calcanea of members of the Mustelidae: I. Mustelinae. Bull. S. Calif. Acad. Sci. 75:237–248.

———. 1976b. Calcanea of members of the Mustelidae: II. Mellivorinae, Melinae, Mephitinae, and Lutrinae. Bull. S. Calif. Acad. Sci. 75:249–257.

Stains, H. J., and R. W. Turner. 1963. Harvest mice south of the Illinois River in Illinois. J. Mamm. 44:274–275.

Stanley, W. C. 1963. Habits of the red fox in northeastern Kansas. Univ. Kan. Mus. Nat. Hist., Misc. Publ. 34:1–31.

Stapanian, M. A., and C. C. Smith. 1978. A model for seed scatter-hoarding: coevolution of fox squirrels and black walnuts. Ecology 59:884–896.

Staples, P. P., and C. R. Terman. 1977. An experimental study of movement in natural populations of *Mus musculus*, *Microtus pennsylvanicus*, and *Microtus pinetorum*. Res. Popul. Ecol. 18:267–283.

Starrett, A. 1967. Hystricoid, erethizontoid, cavioid, and chinchilloid rodents. Pp. 254–272 *in* S. Anderson and J. K. Jones, Jr., eds. Recent mammals of the world. Ronald Press, New York.

Stauffer, C. R. 1945. Some Pleistocene mammalian inhabitants of Minnesota. Proc. Minn. Acad. Sci. 13:20–43.

Stehn, R. A., E. A. Johnson, and M. E. Richmond. 1980. An antibiotic rodenticide for pine voles in orchards. J. Wildl. Mgt. 44:275–280.

Stehn, R. A., and M. E. Richmond. 1975. Male-induced pregnancy termination in the prairie vole, *Microtus ochrogaster*. Science 187:1211–1213.

Stehn, R. A., J. A. Stone, and M. E. Richmond. 1976. Feeding response of small mammal scavengers to pesticide-killed arthropod prey. Am. Midl. Nat. 95:253–256.

Stenlund, M. H. 1953. Report of Minnesota beaver die-off. J. Wildl. Mgt. 17:376–377.

———. 1955a. A recent record of the marten in Minnesota. J. Mamm. 36:133.

———. 1955b. A field study of the timber wolf (*Canis lupus*) on the Superior National Forest, Minnesota. Minn. Dept. Cons. Tech. Bull. 4:1–55.

———. 1958. A history of Itasca County deer. Cons. Volun. 21(125):19–24.

———. 1963. Hunting Minnesota's northeast. Cons. Volun. 26(152):8–10.

———. 1965. Our timber wolves. Cons. Volun. 28(164): 2–8.

———. 1974. Trials of the timber wolf. Minn. Volun. 37(213):50–61.

Stenlund, M. H., and V. E. Gunvalson. 1957. Grass, brush, trees—deer range. Cons. Volun. 20(117):41–44.

Stephenson, A. B. 1969. Temperatures within a beaver lodge in winter. J. Mamm. 50:134–136.

———. 1977. Age determination and morphological variation of Ontario otters. Can. J. Zool. 55:1577–1583.

Stewart, O. C. 1951. Burning and natural vegetation in the United States. Geog. Rev. 41:317–320.

Stewart, R. E. 1975. Breeding birds of North Dakota. Tri-College Center Envir. Studies, Fargo.

Stewart, R. W., and J. R. Bider. 1974. Reproduction and survival of ditch-dwelling muskrats in southern Quebec. Can. Field-Nat. 88:429–436.

———. 1977. Summer activity of muskrats in relation to weather. J. Wildl. Mgt. 41:487–499.

Stickel, L. F. 1968. Home range and travels. Pp. 373–411 *in* J. A. King, ed. Biology of *Peromyscus* (Rodentia). Am. Soc. Mamm., Spec. Publ. 2.

Stirton, R. A. 1938. Notes on some late Tertiary and Pleistocene antilocaprids. J. Mamm. 19:366–370.

Stoddard, H. L. 1920. The flying squirrel as a bird killer. J. Mamm. 1:95–96.

Stonehouse, B., and D. Gilmore, eds. 1977. The biology of marsupials. Univ. Park Press, Baltimore.

Stones, R. C., and L. P. Branick. 1969. Use of hearing in homing by two species of *Myotis* bats. J. Mamm. 50:157–160.

Stones, R. C., and J. E. Wiebers. 1965a. A review of temperature regulation in bats. Am. Midl. Nat. 74:155–167.

———. 1965b. Seasonal changes in food consumption of little brown bats held in captivity at a "neutral" temperature of 92° F. J. Mamm. 46:18–22.

———. 1966. Body weight and temperature regulation of *Myotis lucifugus* at a low temperature of 10° C. J. Mamm. 47:520–521.

Storer, T. I. 1937. The muskrat as native and alien. J. Mamm. 18:443–460.

Storm, G. L. 1965. Movements and activities of foxes as determined by radio-tracking. J. Wildl. Mgt. 29:1–13.

———. 1972. Daytime retreats and movements of skunks on farmlands in Illinois. J. Wildl. Mgt. 36:31–45.

Storm, G. L., et al. 1976. Morphology, reproduction, dispersal, and mortality of midwestern red fox populations. Wildl. Monogr. 49:1–82.

Storm, G. L., and G. G. Montgomery. 1975. Dispersal and social contact among red foxes: results from telemetry and computer simulation. Pp. 237–246 *in* M. W. Fox, ed. The wild canids. Van Nostrand Reinhold, New York.

Storm, G. L., and B. J. Verts. 1966. Movements of a striped skunk infected with rabies. J. Mamm. 47:705–708.

Strecker, R. L. 1954. Regulatory mechanisms in house-mouse populations: the effect of limited food supply on an unconfined population. Ecology 35:249–253.

Streubel, D. P., and J. P. Fitzgerald. 1978. *Spermophilus tridecemlineatus*. Mamm. Species 103:1–5.

Strickler, T. L. 1978a. Functional osteology and myology of the shoulder of the Chiroptera. Contrib. Vert. Evol., Kargar, Basel 4.

———. 1978b. Allometric relationships among the shoulder muscles in the Chiroptera. J. Mamm. 59:36–44.

Stroganov, S. U. 1945. Morphological characters of the auditory ossicles of Recent Talpidae. J. Mamm. 26:412–420.

Stuewer, F. W. 1943. Raccoons: their habits and management in Michigan. Ecol. Monogr. 13:203–257.

Sunquist, M. E. 1974. Winter activity of striped skunks in east-central Minnesota. Am. Midl. Nat. 92:434–446.

Surber, T. 1932. The mammals of Minnesota. Minn. Dept. Cons., St. Paul:1–84.

Sutton, D. A., and C. F. Nadler. 1969. Chromosomes of the North American chipmunk genus *Eutamias*. J. Mamm. 50:524–535.

Svendsen, G. E. 1976. Vocalizations of the long-tailed weasel (*Mustela frenata*). J. Mamm. 57:398–399.

———. 1978. Castor and anal glands of the beaver (*Castor canadensis*). J. Mamm. 59:618–620.

Svendsen, G. E., and R. H. Yahner. 1979. Habitat preference and utilization by the eastern chipmunk (*Tamias striatus*). Kirklandia 31:2–14.

Swain, A. M. 1973. A history of fire and vegetation in northeastern Minnesota as recorded in lake sediments. Quat. Res. 3:383–396.

Swanson, G. A. 1934. The little spotted skunk in northern Minnesota. J. Mamm. 15:318–319.

———. 1940. The American elk in Minnesota. Cons. Volun. 1(2):5–7.

———. 1943. Wildlife of Itasca Park—the mammals. Flicker 15:41–49.

Swanson, G. A., and C. Evans. 1936. The hibernation of certain bats in southern Minnesota. J. Mamm. 17:39–43.

Swanson, G. A., and P. O. Fryklund. 1935. The least weasel in Minnesota and its fluctuations in numbers. Am. Midl. Nat. 16:120–126.

Swanson, G. A., T. Surber, and T. S. Roberts. 1945. The mammals of Minnesota. Minn. Dept. Cons., Tech. Bull. 2:1–108.

Sweeney, J. R., R. L. Marchinton, and J. M. Sweeney. 1971. Responses of radio-monitored white-tailed deer chased by hunting dogs. J. Wildl. Mgt. 35:707–716.

Swenk, M. H. 1926. Notes on *Mustela campestris* Jackson, and on the American forms of least weasels. J. Mamm. 7:313–330.

Switzenberg, D. F. 1950. Breeding productivity in Michigan red foxes. J. Mamm. 31:194–195.

Szalay, F. S. 1977. Phylogenetic relationships and a classification of the eutherian Mammalia. Pp. 315–374 *in* M. K. Hecht et al., eds. NATO Advanced Study Inst. Ser., 14. Major patterns in vertebrate evolution. Plenum Press, New York.

Tabor, J. E., and H. M. Wight. 1977. Population status of river otter in western Oregon. J. Wildl. Mgt. 41:692–699.

Tadlock, C. C., and H. G. Klein. 1979. Nesting and food storage behavior of *Peromyscus maniculatus gracilis* and *P. leucopus noveboracensis*. Can. Field-Nat. 93:239–242.

Tamarin, R. H. 1977. Reproduction in the island beach vole, *Microtus breweri*, and the mainland meadow vole, *Microtus pennsylvanicus*, in southeastern Massachusetts. J. Mamm. 58:536–548.

———. 1978. Dispersal, population regulation, and K-selection in field mice. Am. Nat. 112:545–555.

Tamarin, R. H., and C. J. Krebs. 1969. *Microtus* population biology. II. Genetic changes at the transferrin locus in fluctuating populations of two vole species. Evolution 23:183–211.

Tanner, J. T. 1975. The stability and the intrinsic growth rates of prey and predator populations. Ecology 56:855–867.

Tanner, Ward. 1976. Ottawa Bluffs Preserve. Whispering Land 2(2):1–4.

Tarasoff, F. J., et al. 1972. Locomotory patterns and external morphology of the river otter, sea otter, and harp seal (Mammalia). Can. J. Zool. 50:915–929.

Tarasoff, F. J., and G. L. Kooyman. 1973a. Observations on the anatomy of the respiratory system of the river otter, sea otter, and harp seal: I. The topography, weight, and measurements of the lungs. Can. J. Zool. 51:163–170.

———. 1973b. Observations on the anatomy of the respiratory system of the river otter, sea otter, and harp seal: II. The trachea and bronchial tree. Can. J. Zool. 51:171–177.

Tardif, R. R., and Lincoln Gray. 1978. Feeding diversity of resident and immigrant *Peromyscus leucopus*. J. Mamm. 59:559–562.

Taube, C. M. 1947. Food habits of Michigan opossums. J. Wildl. Mgt. 11:97–103.

Taylor, Jan C. 1966. Home range and agonistic behaviour in the grey squirrel. Symp. Zool. Soc. London 18:229–235.

———. 1977. The frequency of grey squirrel (*Sciurus carolinensis*) communication by use of scent marking points. J. Zool., London 183:543–545.

Taylor, K. D., L. E. Hammon, and R. J. Quy. 1974. The reactions of common rats to four types of live-capture trap. J. Appl. Ecol. 11:453–459.

Taylor, K. D., and R. J. Quy. 1978. Long distance movements of a common rat (*Rattus norvegicus*) revealed by radio-tracking. Mammalia 42:63–71.

Telfer, E. S. 1972a. Browse selection by deer and hares. J. Wildl. Mgt. 36:1344–1349.

———. 1972b. Forage yield and browse utilization on logged areas in New Brunswick. Can. J. For. Res. 2:346–350.

———. 1974. Vertical distribution of cervid and snowshoe hare browsing. J. Wildl. Mgt. 38:944–946.

———. 1978. Cervid distribution, browse, and snow cover in Alberta. J. Wildl. Mgt. 42:352–361.

Telfer, E. S., and Anna Cairns. 1978. Stem breakage by moose. J. Wildl. Mgt. 42:639–642.

Terman, C. R. 1961. Some dynamics of spatial distribution within seminatural populations of prairie deermice. Ecology 42:288–302.

———. 1962. Spatial and homing consequences of the introduction of aliens into seminatural populations of prairie deermice. Ecology 43:216–223.

———. 1968. Population dynamics. Pp. 412–450 *in* J. A. King, ed. Biology of *Peromyscus* (Rodentia). Am. Soc. Mamm., Spec. Publ. 2.

———. 1973. Recovery of reproductive function by prairie deermice (*Peromyscus maniculatus bairdii*) from asymptotic populations. Anim. Behav. 21:443–448.

Terman, C. R., and J. F. Sassaman. 1967. Sex ratio in deer mouse populations. J. Mamm. 48:589–597.

Terres, J. K. 1939. Tree climbing techniques of the gray fox. J. Mamm. 20:256.

———. 1956. Migration records of the red bat, *Lasiurus borealis*. J. Mamm. 37:442.

Tester, J. R. 1953. Fall food habits of the raccoon in the South Platte valley of northeastern Colorado. J. Mamm. 34:500–502.

———. 1965. Effects of a controlled burn on small mammals in a Minnesota oak-savanna. Am. Midl. Nat. 74:240–243.

Tester, J. R., and W. H. Marshall. 1961. A study of certain plant and animal relations on a native prairie in northwestern Minnesota. Univ. Minn. Mus. Nat. Hist., Occ. Pap. 8.

Tester, J. R., D. W. Warner, and W. W. Cochran. 1964. A radio-tracking system for studying movements of deer. J. Wildl. Mgt. 28:42–45.

Tevis, L., Jr. 1950. Summer behavior of a family of beavers in New York State. J. Mamm. 31:40–65.

Thaeler, C. S., Jr. 1968. An analysis of three hybrid populations of pocket gophers (genus *Thomomys*). Evolution 22:543–555.

———. 1974a. Four contacts between ranges of different chromosome forms of the *Thomomys talpoides* complex (Rodentia: Geomyidae). Syst. Zool. 23:343–354.

———. 1974b. Karyotypes of the *Thomomys talpoides* complex (Rodentia: Geomyidae) from New Mexico. J. Mamm. 55:855–859.

———. 1976. Chromosome polymorphism in *Thomomys talpoides agrestis* Merriam (Rodentia—Geomyidae). Southwest. Nat. 21:105–116.

———. 1980. Chromosome numbers and systematic relations in the genus *Thomomys* (Rodentia: Geomyidae). J. Mamm. 61:414–422.

Theberge, J. B., and J. B. Falls. 1967. Howling as a means of communication in timber wolves. Am. Zool. 7:331–338.

Thibault, P. 1969. Activité éstivale de petits mammifères du Québec. Can. J. Zool. 47:817–828.

Thoma, B. L., and W. H. Marshall. 1960. Squirrel weights and populations in a Minnesota woodlot. J. Mamm.

41:272–273.

Thomas, Donna, and C. R. Terman. 1975. The effects of differential prenatal and postnatal social environments on sexual maturation of young prairie deermice (*Peromyscus maniculatus bairdii*). Anim. Behav. 23:241–248.

Thompson, D. C. 1977. Diurnal and seasonal activity of the gray squirrel (*Sciurus carolinensis*). Can. J. Zool. 55:1176–1184.

———. 1978. The social system of the grey squirrel. Behaviour 64:305–328.

Thompson, D. Q. 1952. Travel, range, and food habits of timber wolves in Wisconsin. J. Mamm. 33:429–442.

———. 1965. Food preferences of the meadow vole (*Microtus pennsylvanicus*) in relation to habitat affinities. Am. Midl. Nat. 74:76–86.

Thomson, S. C. 1974. Sight record of a cougar in northern Ontario. Can. Field-Nat. 88:87.

Thornton, W. A., and G. C. Creel. 1975. The taxonomic status of kit foxes. Tex. J. Sci. 26:127–136.

Thornton, W. A., G. C. Creel. and R. E. Trimble. 1971. Hybridization in the fox genus *Vulpes* in west Texas. Southwest. Nat. 15:473–484.

Timm, R. M. 1974. Rediscovery of the rock vole (*Microtus chrotorrhinus*) in Minnesota. Can. Field-Nat. 88:82.

———. 1975. Distribution, natural history, and parasites of mammals of Cook County, Minnesota. Bell Mus. Nat. Hist., Occ. Pap. 14:1–56.

———. 1980. Habitat requirements of microtine rodents in Minnesota. Proc. Symp. Mammal. Ecol. Hab. Mgt. Minn., Bemidji State Univ. 32–34.

Timm, R. M., L. R. Heaney, and D. D. Baird. 1977. Natural history of rock voles (*Microtus chrotorrhinus*) in Minnesota. Can. Field-Nat. 91:177–181.

Todd, Arlene. 1978. Alberta. Pp. 37–38 *in* M. K. Brown, ed. The pine marten. N.Y. State Dept. Env. Cons., Div. Fish. Wildl.

Todd, N. B. 1975. Chromosomal mechanisms in the evolution of artiodactyls. Paleobiology 1:175–188.

———. 1977. Cats and commerce. Sci. Am. 237(5):100–107.

———. 1978. An ecological, behavioral genetic model for the domestication of the cat. Carnivore 1:52–60.

Tomasi, T. E. 1978. Function of venom in the short-tailed shrew, *Blarina brevicauda*. J. Mamm. 59:852–854.

———. 1979. Echolocation by the short-tailed shrew *Blarina brevicauda*. J. Mamm. 60:751–759.

Toweill, D. E. 1974. Winter food habits of river otters in western Oregon. J. Wildl. Mgt. 38:107–111.

Toweill, D. E., and E. C. Meslow. 1977. Food habits of cougars in Oregon. J. Wildl. Mgt. 41:576–578.

Townsend, J. E. 1953. Beaver ecology in western Montana with special reference to movements. J. Mamm. 34:459–479.

Townsend, T. W., and E. D. Bailey. 1975. Parturitional, early maternal, and neonatal behavior in penned white-tailed deer. J. Mamm. 56:347–362.

Trapp, G. R., and D. R. Hallberg. 1975. Ecology of the gray fox (*Urocyon cinereoargenteus*): a review. Pp. 164–178 *in* M. W. Fox, ed. The wild canids. Van Nostrand Reinhold, New York.

Trent, T. T., and O. J. Rongstad. 1974. Home range and survival of cottontail rabbits in southwestern Wisconsin. J. Wildl. Mgt. 38:459–472.

Tryon, C. A., Jr. 1947. Behavior and post-natal development of a porcupine. J. Wildl. Mgt. 11:282–283.

Tryon, C. A., Jr., and D. P. Snyder. 1973. Biology of the eastern chipmunk, *Tamias striatus*: life tables, age distributions, and trends in population numbers. J. Mamm. 54:145–168.

Turkowski, F. J., and L. D. Mech. 1968. Radio-tracking a young male raccoon. J. Minn. Acad. Sci. 35:33–38.

Turner, B. N., and S. L. Iverson. 1973. The annual cycle of aggression in male *Microtus pennsylvanicus*, and its relation to population parameters. Ecology 54:967–981.

Turner, B. N., S. L. Iverson, and K. L. Severson. 1976. Postnatal growth and development of captive Franklin's ground squirrels (*Spermophilus franklinii*). Am. Midl. Nat. 95:93–102.

Turner, B. N., M. R. Perrin, and S. L. Iverson. 1975. Winter coexistence of voles in spruce forest: relevance of seasonal changes in aggression. Can. J. Zool. 53:1004–1011.

Turner, R. W. 1974. Mammals of the Black Hills of South Dakota and Wyoming. Misc. Publ. Mus. Nat. Hist., Univ. Kan. 60:1–178.

Twente, J. W. 1955. Aspects of a population study of cavern-dwelling bats. J. Mamm. 36:379–390.

Twitchell, A. R., and H. H. Dill. 1949. One hundred raccoons from one hundred and two acres. J. Mamm. 30:130–133.

Tyndale-Biscoe, H. 1973. Life of marsupials. American Elsevier, New York.

Ullrey, D. E., et al. 1971. Limitations of winter aspen browse for the white-tailed deer. J. Wildl. Mgt. 35:732–743.

Upham, W. 1920. Minnesota geographic names: their origin and historic significance. Coll. Minn. Hist. Soc. 17.

Urban, D. 1970. Raccoon populations, movement patterns, and predation on a managed waterfowl marsh. J. Wildl. Mgt. 34:372–382.

Valentine, G. L., and R. L. Kirkpatrick. 1970. Seasonal changes in reproductive and related organs in the pine vole, *Microtus pinetorum*, in southwestern Virginia. J. Mamm. 51:553–560.

Van Ballenberghe, V. 1972. Ecology, movements, and popu-

lation characteristics of timber wolves in northeastern Minnesota. Ph.D. thesis, Univ. Minn.
———. 1975a. Recent records of the swift fox (*Vulpes velox*) in South Dakota. J. Mamm. 56:525.
———. 1975b. The swift fox: South Dakota's rarest mammal? S.D. Cons. Dig., Jan.-Feb.:28–29.
Van Ballenberghe, V., and A. W. Erickson. 1973. A wolf pack kills another wolf. Am. Midl. Nat. 90:490–493.
Van Ballenberghe, V., A. W. Erickson, and D. Byman. 1975. Ecology of the timber wolf in northeastern Minnesota. Wildl. Monogr. 43:1–43.
Van Ballenberghe, V., and L. D. Mech. 1975. Weights, growth, and survival of timber wolf pups in Minnesota. J. Mamm. 56:44–63.
Van Ballenberghe, V., and J. M. Peek. 1971. Radiotelemetry studies of moose in northeastern Minnesota. J. Wildl. Mgt. 35:63–71.
Van Cura, N. J., and D. F. Hoffmeister. 1966. A taxonomic review of the grasshopper mice, *Onychomys*, in Arizona. J. Mamm. 47:613–630.
van den Brink, F. H. 1968. A field guide to the mammals of Britain and Europe. Houghton Mifflin, Boston.
van der Meulen, A. J. 1978. *Microtus* and *Pitymys* (Arvicolidae) from Cumberland Cave, Maryland, with a comparison of some New and Old World species. Ann. Carnegie Mus. 47:101–145.
Van Deusen, H. M., and J. K. Jones, Jr. 1967. Marsupials. Pp. 61–86 *in* S. Anderson and J. K. Jones, Jr., eds. Recent mammals of the world. Ronald Press, New York.
Van Gelder, R. G. 1953. The egg-opening technique of a spotted skunk. J. Mamm. 34:255–256.
———. 1959. A taxonomic revision of the spotted skunks (genus *Spilogale*). Am. Mus. Nat. Hist., Bull. 117:229–392.
———. 1978. A review of canid classification. Am. Mus. Novitates 2646:1–10.
Van Gelder, R. G., and W. W. Goodpaster. 1952. Bats and birds competing for food. J. Mamm. 33:491.
Van Lawick-Goodall, H., and J. Van Lawick-Goodall. 1971. Innocent killers. Houghton Mifflin, Boston.
Van Valen, Leigh. 1971a. Toward the origin of artiodactyls. Evolution 25:523–529.
———. 1971b. Adaptive zones and the orders of mammals. Evolution 25:420–428.
———. 1978. The beginning of the Age of Mammals. Evol. Theory 4:45–80.
Van Valen, Leigh, and R. E. Sloan. 1977. Ecology and the extinction of the dinosaurs. Evol. Theory 2:37–64.
Van Vleck, D. B. 1968. Movements of *Microtus pennsylvanicus* in relation to depopulated areas. J. Mamm. 49:92–103.
van Zyll de Jong, C. G. 1966. Food habits of the lynx in Alberta and the Mackenzie District, N.W.T. Can. Field-Nat. 80:18–23.
———. 1971. The status and management of the Canada lynx in Canada. Pp. 16–22 *in* S. E. Jorgensen and L. D. Mech, eds. Proceedings of a symposium on the native cats of North America, their status and management. USFWS, Bur. Sport Fish. Wildl., Region 3.
———. 1972. A systematic review of the Nearctic and Neotropical river otters (genus *Lutra*, Mustelidae, Carnivora). Life Sci. Contrib. Royal Ontario Mus. 80:1–104.
———. 1975a. Differentiation of the Canada lynx, *Felis (Lynx) canadensis subsolana*, in Newfoundland. Can. J. Zool. 53:699–705.
———. 1975b. The distribution and abundance of wolverine (*Gulo gulo*) in Canada. Can. Field-Nat. 89:431–437.
———. 1976. Are there two species of pygmy shrews (*Microsorex*)? Can. Field-Nat. 90:485–487.
———. 1980. Systematic relationships of woodland and prairie forms of the common shrew, *Sorex cinereus cinereus* Kerr and *S. c. haydeni* Baird, in the Canadian Prairie Provinces. J. Mamm. 61:66–75.
Vaughan, T. A. 1953. Unusual concentration of hoary bats. J. Mamm. 34:256.
———. 1961. Vertebrates inhabiting pocket gopher burrows in Colorado. J. Mamm. 42:171–174.
———. 1962. Reproduction in the plains pocket gopher in Colorado. J. Mamm. 43:1–13.
———. 1963. Movements made by two species of pocket gophers. Am. Midl. Nat. 69:367–372.
———. 1966. Food-handling and grooming behaviors in the plains pocket gopher. J. Mamm. 47:132–133.
———. 1967a. Two parapatric species of pocket gophers. Evolution 21:148–158.
———. 1967b. Food habits of the northern pocket gopher on shortgrass prairie. Am. Midl. Nat. 77:176–189.
———. 1969. Reproduction and population densities in a montane small mammal fauna. Misc. Publ. Univ. Kan., Mus. Nat. Hist. 51:51–74.
———. 1974. Resource allocation in some sympatric, subalpine rodents. J. Mamm. 55:764–795.
———. 1978. Mammalogy. 2d ed. Saunders, Philadelphia.
Vaughan, T. A., and R. M. Hansen. 1964. Experiments on interspecific competition between two species of pocket gophers. Am. Midl. Nat. 72:444–452.
Verme, L. J. 1968. An index of winter weather severity for northern deer. J. Wildl. Mgt. 32:566–574.
———. 1969. Reproductive patterns of white-tailed deer related to nutritional plane. J. Wildl. Mgt. 33:881–887.
———. 1977. Assessment of natal mortality in Upper Michigan deer. J. Wildl. Mgt. 41:700–708.
Verts, B. J. 1957. The population and distribution of two species of *Peromyscus* on some Illinois strip-mined land. J. Mamm. 38:53–59.
———. 1963a. Movements and populations of opossums in a cultivated area. J. Wildl. Mgt. 27:127–129.

———. 1963b. Equipment and techniques for radio-tracking striped skunks. J. Wildl. Mgt. 27:325–339.

———. 1967. The biology of the striped skunk. Univ. Ill. Press, Urbana.

Verts, B. J., and T. R. B. Barr. 1960. Apparent absence of rabies in Illinois shrews. J. Wildl. Mgt. 24:438.

Verts, B. J., and G. L. Storm. 1966. A local study of prevalence of rabies among foxes and striped skunks. J. Wildl. Mgt. 30:419–421.

Vesall, D., R. Gersch, and R. Nyman. 1947. Beaver, timber problem in Minnesota's "Big Bog." Cons. Volun. (Mar.–Apr.):45–50.

Vestal, B. M., and J. J. Hellack. 1978. Comparison of neighbor recognition in two species of deer mice (*Peromyscus*). J. Mamm. 59:339–346.

Vickery, W. L., and J. R. Bider. 1978. The effect of weather on *Sorex cinereus* activity. Can. J. Zool. 56:291–297.

Voorhies, M. R. 1974. Fossil pocket mouse burrows in Nebraska. Am. Midl. Nat. 91:492–498.

Wade, Otis. 1930. The behavior of certain spermophiles with special reference to aestivation and hibernation. J. Mamm. 11:160–188.

Walker, W. J., and C. A. Moore. 1971. Tularemia: experience in the Hamilton area. Can. Med. Asn. J. 105:390–393.

Walley, H. D., and W. L. Jarvis. 1971. Longevity record for *Pipistrellus subflavus*. Trans. Ill. State Acad. Sci. 64:305.

Walski, T. W., and W. W. Mautz. 1977. Nutritional evaluation of three winter browse species of snowshoe hares. J. Wildl. Mgt. 41:144–147.

Walters, C. J., and P. J. Bandy. 1972. Periodic harvest as a method of increasing big game yields. J. Wildl. Mgt. 36:128–134.

Walters, C. J., and J. E. Gross. 1972. Development of big game management plans through simulation modelling. J. Wildl. Mgt. 36:119–128.

Wanek, W. J. 1971. Observations on snowmobile impact. Minn. Volun. 34(199):1–9.

———. 1972–1975. A continuing study of the ecological impact of snowmobiling in northern Minnesota. Bemidji State Univ., Center Envir. Studies. 4 vols.

Ward, A. L., and J. O. Keith. 1962. Feeding habits of pocket gophers in mountain grasslands, Black Mesa, Colorado. Ecology 43:744–749.

Warren, E. R. 1927. The beaver, its works and ways. Am. Soc. Mamm. Monogr. 2.

Warren, R. J., and R. L. Kirkpatrick. 1978. Indices of nutritional status in cottontail rabbits fed controlled diets. J. Wildl. Mgt. 42:154–158.

Watkins, L. C., and R. M. Nowak. 1973. The white-tailed jack rabbit in Missouri. Southwest. Nat. 18:352–354.

Webb, R. E., and F. Horsefall, Jr. 1967. Endrin resistance in the pine mouse. Science 156:1762.

Webb, S. D. 1973. Pliocene pronghorns in Florida. J. Mamm. 54:203–221.

Webster, D. B. 1968. Relative changes in the renal cortex of the bat (*Myotis lucifugus*) under varying conditions of temperature and water balance. Anat. Rec. 160:447.

———. 1969. Comparative middle ear morphology in Heteromyidae. Anat. Rec. 163:281–282.

Webster, D. B., and Molly Webster. 1975. Auditory systems of Heteromyidae: functional morphology and evolution of the middle ear. J. Morphol. 146:343–376.

———. 1977. Auditory systems of Heteromyidae: cochlear diversity. J. Morphol. 152:153–170.

Webster, F. A., and D. R. Griffin. 1962. The role of flight membranes in insect capture by bats. Anim. Behav. 10:332–340.

Wechsler, Charles. 1974. The beaver . . . creature of controversy. Minn. Volun. 37(212):36–45.

Wecker, S. C. 1963. The role of early experience in habitat selection by the prairie deer mouse, *Peromyscus maniculatus bairdi*. Ecol. Monogr. 33:307–325.

Weckwerth, R. P., and V. D. Hawley. 1962. Marten food habits and population fluctuations in Montana. J. Wildl. Mgt. 26:55–74.

Weckwerth, R. P., and P. L. Wright. 1968. Results of transplanting fishers in Montana. J. Wildl. Mgt. 32:977–980.

Weeks, H. P., Jr., and C. M. Kirkpatrick. 1978. Salt preferences and sodium drive phenology in fox squirrels and woodchucks. J. Mamm. 59:531–542.

Weigl, P. D. 1969. The distribution of the flying squirrels, *Glaucomys volans* and *Glaucomys sabrinus*: an evaluation of the competitive exclusion idea. Ph.D. thesis, Duke Univ., Diss. Abstr. 30B:2966.

Weigl, P. D., and D. W. Osgood. 1974. Study of the northern flying squirrel, *Glaucomys sabrinus*, by temperature telemetry. Am. Midl. Nat. 92:482–486.

Weinstein, M. S. 1977. Hares, lynx, and trappers. Am. Nat. 111:806–808.

Weir, B. J., and I. W. Rowlands. 1973. Reproductive strategies of mammals. Annu. Rev. Ecol. Syst. 4:139–163.

Weir, Christine. 1974. Aspects of predatory strategy and energetics of *Taxidea taxus*. Bemidji State Coll., internship report:1–37.

Wells, M. C., and P. N. Lehner. 1978. The relative importance of the distance senses in coyote predatory behaviour. Anim. Behav. 26:251–258.

Wenberg, G. M., and J. C. Holland. 1973a. The circannual variations of some of the hormones of the woodchuck (*Marmota monax*). Comp. Biochem. Physiol. A. 46:523–535.

———. 1973b. The circannual variations in the total serum lipids and cholesterol with respect to body weight in the woodchuck (*Marmota monax*). Comp. Biochem. Physiol. A. 44:577–583.

———. 1973c. The circannual variations of thyroid activity in the woodchuck (*Marmota monax*). Comp. Biochem. Physiol. A. 44:775–780.

Wenberg, G. M., J. C. Holland, and Joanne Sewell. 1973. Some aspects of the hibernating and non-hibernating woodchuck (*Marmota monax*). Comp. Biochem. Physiol. A. 45:513–521.

Werner, R. M., and J. A. Vick. 1977. Resistance of the opossum (*Didelphis virginiana*) to envenomation by snakes of the family Crotalidae. Toxicon 15:29–33.

Wertheim, R. F., and R. H. Giles. 1971. Effects of bacterial endotoxin and crowding on pine voles and white mice. J. Mamm. 52:238–242.

Wesley, D. E., K. L. Knox, and J. G. Nagy. 1970. Energy flux and water kinetics in young pronghorn antelope. J. Wildl. Mgt. 34:908–912.

———. 1973. Energy metabolism of pronghorn antelopes. J. Wildl. Mgt. 37:563–573.

Westfall, C. Z. 1956. Foods eaten by bobcats in Maine. J. Wildl. Mgt. 20:199–200.

Wetmore, A. 1936. Hibernation of the brown bat. J. Mamm. 17:130–131.

Wetzel, J. F., J. R. Wambaugh, and J. M. Peek. 1975. Appraisal of white-tailed deer winter habitats in northeastern Minnesota. J. Wildl. Mgt. 39:59–66.

Wetzel, R. M. 1955. Speciation and dispersal of the southern bog lemming, *Synaptomys cooperi* (Baird). J. Mamm. 36:1–20.

———. 1958. Mammalian succession on midwestern floodplains. Ecology 39:262–271.

Wetzel, R. M., and H. L. Gunderson. 1949. The lemming vole, *Synaptomys borealis*, in northern Minnesota. J. Mamm. 30:437.

Wheat, J. B. 1967. A Paleo-Indian bison kill. Sci. Am. 216(1):44–52.

Whitaker, J. O., Jr. 1962. *Endogone*, *Hymenogaster*, and *Melanogaster* as small mammal foods. Am. Midl. Nat. 67:152–156.

———. 1963a. A study of the meadow jumping mouse, *Zapus hudsonius* (Zimmerman), in central New York. Ecol. Monogr. 33:215–254.

———. 1963b. Food, habitat, and parasites of the woodland jumping mouse in central New York. J. Mamm. 44:316–321.

———. 1966. Clitoris bones of *Zapus hudsonius* and *Napaeozapus insignis*. J. Mamm. 47:127.

———. 1972a. Food habits of bats from Indiana. Can. J. Zool. 50:877–883.

———. 1972b. *Zapus hudsonius*. Mamm. Species 11:1–7.

———. 1972c. Food and external parasites of *Spermophilus tridecemlineatus* in Vigo County, Indiana. J. Mamm. 53:644–648.

———. 1977. Food and external parasites of the Norway rat, *Rattus norvegicus*, in Indiana. Proc. Ind. Acad. Sci. 86:193–198.

Whitaker, J. O., Jr., and R. L. Martin. 1977. Food habits of *Microtus chrotorrhinus* from New Hampshire, New York, Labrador, and Quebec. J. Mamm. 58:99–100.

Whitaker, J. O., Jr., and R. E. Mumford. 1972a. Food and ectoparasites of Indiana shrews. J. Mamm. 53:329–335.

———. 1972b. Ecological studies on *Reithrodontomys megalotis* in Indiana. J. Mamm. 53:850–860.

Whitaker, J. O., Jr., and D. D. Pascal, Jr. 1971. External parasites of arctic shrews (*Sorex arcticus*) taken in Minnesota. J. Mamm. 52:202.

Whitaker, J. O., Jr., and L. L. Schmeltz. 1973. Food and external parasites of *Sorex palustris* and food of *Sorex cinereus* from St. Louis County, Minnesota. J. Mamm. 54:283–285.

———. 1974. Food and external parasites of the eastern mole, *Scalopus aquaticus*, from Indiana. Proc. Ind. Acad. Sci. 83:478–481.

Whitaker, J. O., Jr., and G. R. Sly. 1970. First record of *Reithrodontomys megalotis* in Indiana. J. Mamm. 51:381.

Whitaker, J. O., Jr., and R. E. Wrigley. 1972. *Napaeozapus insignis*. Mamm. Species 14:1–6.

White, J. A. 1953. Genera and subgenera of chipmunks. Univ. Kan. Publ., Mus. Nat. Hist. 5:543–561.

———. 1970. Late Cenozoic porcupines (Mammalia, Erethizontidae) of North America. Am. Mus. Novit. 2421:1–15.

Whitrow, G. C., ed. 1971. Comparative physiology of thermoregulation. Vol. 2, Mammals. Academic Press, New York.

Wiehe, J. M. 1978. Key to the skulls of North Dakota mammals. Prairie Nat. 10:1–16.

Wiehe, J. M., and J. F. Cassel. 1978. Checklist of North Dakota mammals (revised). Prairie Nat. 10:81–88.

Wight, H. M., and C. H. Conaway. 1961. Weather influences on the onset of breeding in Missouri cottontails. J. Wildl. Mgt. 25:87–89.

———. 1962. Determination of pregnancy rates of cottontail rabbits. J. Wildl. Mgt. 26:93–95.

Wilde, S. A., C. T. Youngberg, and J. H. Hovind. 1950. Changes in ground water, soil fertility, and forest growth produced by construction and removal of beaver dams. J. Wildl. Mgt. 14:123–128.

Williams, M. W. 1962. An albino short-tailed shrew from Vermont. J. Mamm. 43:424–425.

Williams, O., and B. A. Finney. 1964. *Endogone*—food for mice. J. Mamm. 45:265–271.

Williams, S. L., R. Laubach, and H. H. Genoways. 1977. A guide to the management of Recent mammal collections. Carnegie Mus. Nat. Hist., Spec. Publ. 4:1–105.

Willner, G. R., J. A. Chapman, and D. Pursley. 1979. Reproduction, physiological responses, food habits, and abundance of nutria on Maryland marshes. Wildl. Monogr. 65:1–43.

Wilson, J. T. 1976. Continents adrift and continents aground: a Scientific American Reader. Freeman, San Francisco.

Wimsatt, W. A. 1944a. Growth of the ovarian follicle and ovulation in *Myotis lucifugus lucifugus*. Am. J. Anat. 74:129–173.

———. 1944b. An analysis of implantation in the bat *Myotis lucifugus lucifugus*. Am. J. Anat. 74:355–411.

———. 1945. Notes on breeding behavior, pregnancy, and parturition in some vespertilionid bats of the eastern United States. J. Mamm. 26:23–33.

———. 1960. An analysis of parturition in Chiroptera, including new observations on *Myotis l. lucifugus*. J. Mamm. 41:183–200.

———, ed. 1970, 1977. Biology of bats. Vol. 1 and 2, 1970; vol. 3, 1977. Academic Press, New York.

Wimsatt, W. A., and H. F. Parks. 1966. Ultrastructure of the surviving follicle of hibernation and of the ovum-follicle cell relationship in the vespertilionid bat, *Myotis lucifugus*. Symp. Zool. Soc. London 15:419–454.

Windberg, L. A., and L. B. Keith. 1976a. Snowshoe hare population response to artificial high densities. J. Mamm. 57:523–553.

———. 1976b. Experimental analyses of dispersal in snowshoe hare populations. Can. J. Zool. 54:2061–2081.

———. 1978. Snowshoe hare populations in woodlot habitat. Can. J. Zool. 56:1071–1080.

Wistrand, Harry. 1974. Individual, social, and seasonal behavior of the thirteen-lined ground squirrel (*Spermophilus tridecemlineatus*). J. Mamm. 55:329–347.

Wobeser, G. A., and F. A. Leighton. 1979. A simple burrow entrance live trap for ground squirrels. J. Wildl. Mgt. 43:571–572.

Wolfe, J. F., Jr. 1977. Man-killing wolves: fact or folklore. Minn. Volun. 40(235):44–51.

Wolfe, J. L. 1966. Agonistic behavior and dominance relationships of the eastern chipmunk, *Tamias striatus*. Am. Midl. Nat. 76:190–200.

———. 1969. Observations on alertness and exploratory behavior in the eastern chipmunk. Am. Midl. Nat. 81:249–253.

Wolfe, M. L. 1977. Mortality patterns in the Isle Royale moose population. Am. Midl. Nat. 97:267–269.

Wolfe, M. L., and D. L. Allen. 1973. Continued studies of the status, socialization, and relationships of Isle Royale wolves, 1967 to 1970. J. Mamm. 54:611–635.

Wolff, J. O. 1978. Food habits of snowshoe hares in interior Alaska. J. Wildl. Mgt. 42:148–153.

Wondolleck, J. T. 1978. Forage-area separation and overlap in heteromyid rodents. J. Mamm. 59:510–518.

Wood, A. E. 1950. Porcupines, paleogeography, and parallelism. Evolution 4:87–98.

———. 1955. A revised classification of the rodents. J. Mamm. 36:165–187.

———. 1959. Are there rodent suborders? Syst. Zool. 7:169–173.

———. 1965. Grades and clades among rodents. Evolution 19:115–130.

———. 1972. An Eocene hystricognathous rodent from Texas: its significance in interpretations of continental drift. Science 175:1250–1251.

———. 1974. The evolution of the Old World and New World hystricomorphs. Symp. Zool. Soc. London 34:21–60.

———. 1975. The problem of the hystricognathous rodents. Univ. Mich. Pap. Paleontol. 12:75–80.

Wood, G. W., and R. H. Barrett. 1979. Status of wild pigs in the United States. Wildl. Soc. Bull. 74:237–246.

Wood, J. E. 1958. Age structure and productivity of a gray fox population. J. Mamm. 39:74–86.

———. 1959. Relative estimates of fox population levels. J. Wildl. Mgt. 23:53–63.

Wood, J. E., D. E. Davis, and E. V. Komarek. 1958. The distribution of fox populations in relation to vegetation in southern Georgia. Ecology 39:160–162.

Wood, T. J., and G. D. Tessier. 1974. First records of eastern flying squirrel (*Glaucomys volans*) from Nova Scotia. Can. Field-Nat. 88:83–84.

Woods, C. A. 1972. Comparative myology of jaw, hyoid, and pectoral appendicular regions of New and Old World hystricomorph rodents. Am. Mus. Nat. Hist., Bull. 147:117–198.

———. 1973. *Erethizon dorsatum*. Mamm. Species 29: 1–6.

Woolf, Alan, and J. D. Harder. 1979. Population dynamics of a captive white-tailed deer herd, with emphasis on reproduction and mortality. Wildl. Monogr. 67:1–53.

Woolpy, J. H. 1968. The social organization of wolves. Nat. Hist. 77(5):46–55.

Worth, C. B. 1975. Virginia opossums (*Didelphis virginiana*) as disseminators of the common persimmon (*Diospyros virginiana*). J. Mamm. 56:517.

Wrazen, J. A. 1980. Late summer activity changes in populations of eastern chipmunks (*Tamias striatus*). Can. Field-Nat. 94:305–310.

Wrazen, J. A., and G. E. Svendsen. 1978. Feeding ecology of a population of eastern chipmunks (*Tamias striatus*) in southeast Ohio. Am. Midl. Nat. 100:190–201.

Wright, B. S. 1959. The ghost of North America—the story of the eastern panther. Vantage Press, New York.

———. 1972. The eastern panther—a question of survival. Clarke, Irwin and Co., Toronto.

Wright, H. E., Jr. 1968. History of the Prairie Peninsula. Pp. 78–88 *in* R. E. Bergman, ed. The Quaternary of Illinois. Spec. Publ. Univ. Ill. Coll. Agric. 14.

———. 1972. Quaternary history of Minnesota. Pp. 515–548 *in* P. K. Sims and G. B. Morey, eds. Geology of Minnesota. Minn. Geol. Sur., Univ. Minn., St. Paul.

———. 1974. Landscape development, forest fires, and wilderness management. Science 186:487–495.

Wright, P. L. 1942. Delayed implantation in the long-tailed weasel (*Mustela frenata*), the short-tailed weasel (*Mustela cicognani*), and the marten (*Martes americana*). Anat. Rec. 83:341–353.

———. 1947. The sexual cycle of the male long-tailed weasel (*Mustela frenata*). J. Mamm. 28:343–352.

———. 1948. Breeding habits of captive long-tailed weasels (*Mustela frenata*). Am. Midl. Nat. 39:338–344.

———. 1953. Intergradation between *Martes americana* and *Martes caurina* in western Montana. J. Mamm. 34:74–86.

———. 1963. Variation in reproductive cycles in North American mustelids. Pp. 77–97 *in* A. C. Enders, ed. Delayed implantation. Univ. Chicago Press, Chicago.

———. 1966. Observations on the reproductive cycle of the American badger (*Taxidea taxus*). Symp. Zool. Soc. London 15:27–45.

———. 1969. The reproductive cycle of the male American badger, *Taxidea taxus*. J. Reprod. Fert., Suppl. 6:435–445.

Wright, P. L., and M. W. Coulter. 1967. Reproduction and growth in Maine fishers. J. Wildl. Mgt. 31:70–87.

Wright, P. L., and S. A. Dow, Jr. 1962. Minimum breeding age in pronghorn. J. Wildl. Mgt. 26:100–101.

Wright, P. L., and R. Rausch. 1955. Reproduction in the wolverine, *Gulo gulo*. J. Mamm. 36:346–355.

Wrigley, R. E. 1969. Ecological notes on the mammals of southern Quebec. Can. Field-Nat. 83:201–211.

———. 1972. Systematics and biology of the woodland jumping mouse, *Napaeozapus insignis*. Ill. Biol. Monogr. 47:1–117.

———. 1974. Mammals of the sandhills of southwestern Manitoba. Can. Field-Nat. 88:21–39.

Wrigley, R. E., H.-E. Drescher, and Sabine Drescher. 1973. First record of the fox squirrel in Canada. J. Mamm. 54:782–783.

Wrigley, R. E., and J. E. Dubois. 1973. Distribution of the pocket gophers *Geomys bursarius* and *Thomomys talpoides* in Manitoba. Can. Field-Nat. 87:167–169.

Wrigley, R. E., J. E. Dubois, and H. W. R. Copland. 1979. Habitat, abundance, and distribution of six species of shrews in Manitoba. J. Mamm. 60:505–520.

Wurster-Hill, D. H. 1973. Chromosomes of eight species from five families of Carnivora. J. Mamm. 54:753–760.

Wydeven, A. P., and P. R. Wydeven. 1976. The status of the least chipmunk (*Eutamias minimus jacksoni*) in central Wisconsin. Rep. Mamm. Mus. Nat. Hist. (Stevens Pt., WI) 11(3):3.

Yahner, R. H. 1975. The adaptive significance of scatter hoarding in the eastern chipmunk. Ohio J. Sci. 75:176–177.

———. 1977. Activity lull of *Tamias striatus* in summer in southeast Ohio. Ohio J. Sci. 77:143–145.

———. 1978a. Weight gain of post-emergent juvenile *Tamias striatus*. J. Mamm. 59:196–197.

———. 1978b. Burrow system and home range use by eastern chipmunks, *Tamias striatus*: ecological and behavioral considerations. J. Mamm. 59:324–329.

———. 1978c. Ethology of eastern chipmunks: interrelationships among behaviors. J. Mamm. 59:880–882.

———. 1978d. Seasonal rates of vocalizations in eastern chipmunks. Ohio J. Sci. 78:301–303.

Yahner, R. H., and G. E. Svendsen. 1978. Effects of climate on the circannual rhythm of the eastern chipmunk, *Tamias striatus*. J. Mamm. 59:109–117.

Yates, T. L., and D. J. Schmidly. 1975. Karyotype of the eastern mole (*Scalopus aquaticus*), with comments on the karyology of the family Talpidae. J. Mamm. 56:902–905.

———. 1978. *Scalopus aquaticus*. Mamm. Species 105: 1–4.

Yates, T. L., D. J. Schmidly, and K. L. Culbertson. 1976. Silver-haired bat in Mexico. J. Mamm. 57:205.

Yeager, L. E. 1938. Tree climbing by a gray fox. J. Mamm. 19:376.

———. 1943. Storing of muskrats and other foods by minks. J. Mamm. 24:100–101.

Yeaton, R. I. 1972. Social behavior and social organization in Richardson's ground squirrel (*Spermophilus richardsonii*) in Saskatchewan. J. Mamm. 53:139–147.

Yerger, R. W. 1955. Life history notes on the eastern chipmunk, *Tamias striatus lysteri* (Richardson), in central New York. Am. Midl. Nat. 53:312–323.

Young, J. Z. 1975. The life of mammals: their anatomy and physiology. 2d ed. (with M. J. Hobbs). Oxford Univ. Press, London.

Young, S. P. 1946. The wolf in North American history. Caxton Printers, Caldwell, Idaho.

———. 1958. The bobcat of North America. Stackpole, Harrisburg, Pa.

Young, S. P., and E. A. Goldman. 1944. The wolves of North America. Dover Publ., New York.

———. 1946. The puma, mysterious American cat. Dover Publ., New York.

Young, S. P., and H. H. T. Jackson. 1951. The clever coyote. Stackpole, Harrisburg, Pa.

Youngman, P. M. 1975. Mammals of the Yukon Territory. Natl. Mus. Can., Natl. Mus. Nat. Sci., Publ. Zool. 10:1–192.

Youngman, P. M., and D. A. Gill. 1968. First record of the southern flying squirrel, *Glaucomys volans volans*, from Quebec. Can. Field-Nat. 82:227–228.

Zagata, M. D., and A. N. Moen. 1974. Antler shedding by

white-tailed deer in the Midwest. J. Mamm. 55:656–659.

Zalom, F. G. 1977. Role of ambient temperature in emergence of woodchucks (*Marmota monax*) from hibernation. Am. Midl. Nat. 97:224–230.

Zelley, R. A. 1971. The sounds of the fox squirrel, *Sciurus niger rufiventer*. J. Mamm. 52:597–604.

Ziegler, A. C. 1971. Dental homologies and possible relationships of Recent Talpidae. J. Mamm. 52:50–68.

Zimen, Erik. 1975. Social dynamics of the wolf pack. Pp. 336–362 *in* M. W. Fox, ed. The wild canids. Van Nostrand Reinhold, New York.

Zimmerman, E. G. 1965. A comparison of habitat and food of two species of *Microtus*. J. Mamm. 46:605–612.

Zimmerman, E. G., C. W. Kilpatrick, and B. J. Hart. 1978. The genetics of speciation in the rodent genus *Peromyscus*. Evolution 32:565–579.

Zimny, Marilyn L. 1965. Thirteen-lined ground squirrels born in captivity. J. Mamm. 46:521–522.

Zirul. D. L., and W. A. Fuller. 1970. Winter fluctuations in size of home range of the red squirrel (*Tamiasciurus hudsonicus*). Trans. N. Am. Wildl. Nat. Resour. Conf. 35:115–127.

Index

Index

Boldface numbers indicate the first pages of accounts of species or higher taxa that occur wild in Minnesota. Topics that are included in all or most species accounts (e.g., reproduction, food habits) are not indexed. If topics are discussed generally outside the individual species accounts (e.g., endothermy), these instances are indexed.

Evan B. Hazard took a degree in wildlife conservation at Cornell University and received his Ph.D. in zoology at the University of Michigan. He is professor of biology at Bemidji State University, where he also serves as curator of the mammal and bird collections.

Nan Kane received her bachelor of science degree in wildlife biology and management at the University of Minnesota. She works as a preparator at the University's Bell Museum of Natural History and is studying painting and drawing in a program designed for apprentice artists.